HEAT CONDUCTION

Applied Mathematics and Engineering Science Texts

SERIES EDITOR

PROFESSOR ALAN JEFFREY
PhD, DSc, FIMA
University of Newcastle upon Tyne

1 Heat Conduction

J. M. HILL & J. N. DEWYNNE

Applied Mathematics and Engineering Science Texts

HEAT CONDUCTION

JAMES M. HILL &
JEFFREY N. DEWYNNE

Both of The University of Wollongong
Wollongong, NSW, Australia

BLACKWELL SCIENTIFIC PUBLICATIONS

OXFORD   LONDON   EDINBURGH

BOSTON   PALO ALTO   MELBOURNE

First published 1987

DISTRIBUTORS

USA and Canada
   Blackwell Scientific Publications Inc.
   PO Box 50009, Palo Alto
   California 94303

Australia
   Blackwell Scientific Publications
   (Australia) Pty Ltd
   107 Barry Street,
   Carlton, Victoria 3053

Set at the University of Wollongong
by the authors using the T_EX system
and printed by R. J. Acford Ltd,
Chichester

British Library
Cataloguing in Publication Data

Hill, James M.
   Heat conduction. —(Applied mathematics
   and engineering science texts, ISSN 0950-5903; v. 1)
   1. Heat equation
   I. Title   II. Dewynne, Jeffrey N.   III. Series
   536′.23′01515353   QC321.6

   ISBN 0-632-01716-3

Library of Congress
Cataloging-in-Publication Data

Hill, James M., 1945–
   Heat conduction.

   (Applied mathematics and engineering science texts)
   Bibliography: p.
   Includes index.
   1. Heat—Conduction.   I. Dewynne, Jeffrey N.
II. Title.   III. Series.
QC321.H624   1986   536′.23   86–26924
ISBN 0–632–01716–3

To our parents,
without whose sacrifices
this book would not have been possible

TABLE OF CONTENTS

PREFACE

One of the most important partial differential equations in applied mathematics is the heat or diffusion equation. Its importance in the modelling of heat conduction, diffusional processes and flow through a porous medium is well known. It arises from a probabilistic framework and emerges as the simplest approximation to bulk processes governed at the microscopic level by random spatial variations. This book is intended as a modern undergraduate text which reflects the importance of the heat equation in applied mathematics and mathematical modelling, and which also is intermediate to the now classical treatise of H.S. Carslaw and J.C. Jaeger (Conduction of heat in solids). Familiarity with the heat equation is essential for aspiring mathematical modellers, and this book is intended to support an undergraduate course for third year students in applied mathematics, science or engineering.

We have attempted throughout to provide a balanced account of solutions and results for the heat equation. Accordingly, we have adopted the strategy of bringing together the simplest and most useful results from many diverse areas of mathematics. More complicated results are summarized as problems at the end of each chapter. The first two chapters of the book are introductory and serve to summarize the essential elements of heat flow and to a certain extent diffusion and the mathematical formulation and simple general results. The next two chapters develop exact analytical solutions, obtained by Laplace transforms and Fourier series, for infinite and finite media problems respectively. The chapter thereafter deals with approximate analytical solutions based on the heat-balance integral method. The final two chapters of the book deal respectively with numerical methods for the heat equation and simple heat conduction moving boundary problems.

The literature concerning the applications and mathematical theory of the heat equation is extensive and widespread, and we have made no attempt to accommodate or acknowledge original contributions, which would in any case be inappropriate in an introductory text of this nature. Following in the footsteps of such a monumental treatise as that already cited is a daunting enough task, and we therefore emphasize that the aim of the present text is to assemble a coherent up to date body of material on the heat equation so as to provide a preliminary background sufficient to make the research literature and classical treatises more accessible to the student. We hope that the text constitutes a well balanced introduction and that the results prove to be useful in a practical context.

James M. Hill and Jeffrey N. Dewynne,
Department of Mathematics,
The University of Wollongong,
April, 1986.

Acknowledgement The authors gratefully acknowledge the financial support of one of us (J.N.D.) from a University of Wollongong post-doctoral position.

LIST OF SYMBOLS

The symbols given below generally have the stated meaning throughout the book.

a	fixed radius or length
$c, c(T)$	heat capacity
C_1, C_2	arbitrary constants
$D, D(C)$	diffusivity
E	surface emissivity
$f(x), f(\underline{x})$	prescribed initial temperature
$f(x,t), f_1(x,t), f_2(x,t)$	probability density functions
$f_\epsilon(x)$	delta convergent sequence
$g(t), g(\underline{x},t)$	prescribed surface temperature
$G(x,\xi,t,\tau), G(\underline{x},\underline{\xi},t,\tau)$	Green's functions
h	surface heat transfer coefficient
$h(t), h(\underline{x},t)$	prescribed surface heat flux
$H(x)$	Heaviside unit step function
$H(T), H(x,t)$	enthalpy function
$j(x,t)$	heat flux
$\underline{j}(\underline{x},t)$	heat flux vector (j_x, j_y, j_z)
$k, k(T)$	thermal conductivity
$K_n(\underline{x},t)$	n dimensional unit source solution $(n = 1,2,3)$
ℓ	fixed finite length
L	latent heat of fusion
$\underline{n}$	unit outward drawn normal vector (n_x, n_y, n_z)
p	Laplace transform variable
q	function of Laplace transform variable $\sqrt{p/\kappa}$
q_0	point source strength
$Q(x,t), Q(\underline{x},t)$	quantity of heat
S	surface of body

t	time
$\bar{t}$	non-dimensional time
t^*	stretched time $\lambda^2 t$
$\bar{t}_n(\overline{X})$	estimate of boundary motion from integral iteration
$T(x,t),\ T(\underline{x},t)$	temperature
$\overline{T}$	non-dimensional temperature
T^*	stretched temperature
T_f	fusion temperature
$\tilde{T}_{ij}$	finite difference approximation to temperature
$\overline{T}_{ij}$	exact temperature at finite difference mesh point
$\overline{T}_h$	finite element approximate temperature
$T_0,\ T_0(t),\ T_0(\underline{x},t)$	surrounding temperature
$\overline{T}_n^{\dagger}(\bar{x},\overline{X})$	estimates of temperature from integral iteration
$u(x,t),\ u(\underline{x},t)$	temperature arising from prescribed initial temperature and zero boundary conditions
u_0	constant initial temperature
$v(x,t),\ v(\underline{x},t)$	temperature arising from zero initial temperature and prescribed boundary conditions
v_0	constant prescribed surface temperature
V	volume of body
w_0	constant prescribed surface flux
(x,y,z)	cartesian position coordinates
$\underline{x}$	position vector (x,y,z)
x^*	stretched position λx
$X(t)$	moving boundary position
$\overline{X(t)}$	non-dimensional moving boundary position

Greek symbols

α	Stefan number or phase change parameter $L/c_1(T_f - T_0)$
β	inverse Biot modulus k_1/ha
δ	ratio $\Delta\bar{t}/(\Delta\bar{x})^2$ for finite difference schemes
δ_{ij}	Kronecker delta
$\delta(x)$	one dimensional Dirac delta function
$\delta(\underline{x})$	three dimensional Dirac delta function $\delta(x)\delta(y)\delta(z)$
ΔQ	small quantity of heat
Δt	small time increment
$\Delta\bar{t}$	non-dimensional time step
ΔT	small temperature change
Δx	small space increment
$\Delta\bar{x}$	non-dimensional space step
ϵ	parameter tending to zero in delta convergent sequence
ζ	similarity variable $x/\sqrt{t}$ or $r/\sqrt{t}$
$\theta(t)$	heat balance integral
κ	thermal diffusivity $k/\rho c$
λ	stretching variable
μ	h/k
ρ	density
$\rho(t)$	penetration depth
σ	Stefan-Boltzmann constant
$\phi(x),\ \phi(\underline{x})$	test function
ω	boundary fixing transformation $\bar{x}/\overline{X}(\bar{t})$

Throughout we generally adopt ξ, η and τ as integration variables and variables overlined by bars denote non-dimensional quantities. We also note here expressions for the Laplacian operator ∇^2 and the volume element dV in cylindrical and spherical polar coordinates. In cartesian coordinates (x, y, z) we have

$$\nabla^2 = \frac{\partial^2}{\partial x^2} + \frac{\partial^2}{\partial y^2} + \frac{\partial^2}{\partial z^2}, \qquad dV = dx\, dy\, dz.$$

In cylindrical polar coordinates (r, θ, z), where r and θ are defined by

$$x = r\cos\theta, \quad y = r\sin\theta, \quad \text{or} \quad r = \sqrt{x^2 + y^2}, \quad \theta = \tan^{-1}\frac{y}{x},$$

we have

$$\nabla^2 = \frac{\partial^2}{\partial r^2} + \frac{1}{r}\frac{\partial}{\partial r} + \frac{1}{r^2}\frac{\partial^2}{\partial \theta^2} + \frac{\partial^2}{\partial z^2}, \qquad dV = r\, dr\, d\theta\, dz.$$

In spherical polar coordinates (r, θ, ϕ) defined by

$$x = r\sin\theta\cos\phi, \quad y = r\sin\theta\sin\phi, \quad z = r\cos\theta,$$

or

$$r = \sqrt{x^2 + y^2 + z^2}, \qquad \theta = \tan^{-1}\frac{\sqrt{x^2 + y^2}}{z}, \qquad \phi = \tan^{-1}\frac{y}{x},$$

we have

$$\nabla^2 = \frac{\partial^2}{\partial r^2} + \frac{2}{r}\frac{\partial}{\partial r} + \frac{1}{r^2}\frac{\partial^2}{\partial \theta^2} + \frac{\cot\theta}{r^2}\frac{\partial}{\partial \theta} + \frac{1}{r^2\sin^2\theta}\frac{\partial^2}{\partial \phi^2},$$

and

$$dV = r^2\sin\theta\, dr\, d\theta\, d\phi.$$

Chapter One

Introduction

1.1 Historical introduction

The originator of the modern physical and mathematical theory of heat conduction, Joseph Fourier (1768-1830) first became interested in the subject around 1803 and originally considered the movement of heat between a finite number of discrete bodies arranged in a straight line. Soon afterwards he turned his attention to the real problem at issue, the propagation of heat in continuous bodies and his subsequent researches culminated in 1822 with the publication of his major work "Analytical Theory of Heat". Although Fourier is remembered primarily as a theoretician, he certainly appreciated the need for experimental investigations and indeed his intimate understanding of the relevant physical concepts (specific heat, thermal conductivity and heat flow) enabled him to correctly formulate the general three dimensional heat equation and the general expression for the conditions to be satisfied at the surface of a body. Thus for three dimensional heat flow he obtained the equation

$$\rho c \frac{\partial T}{\partial t} = k \, \nabla^2 T,\qquad(1.1)$$

governing the temperature T in a three dimensional body and

$$k \, \nabla T \cdot n + h(T - T_0) = 0,\qquad(1.2)$$

relating the temperature difference between the surface of a body and its surroundings to the temperature gradient in the body at the surface. Here ρ is the density of the material, c is the specific heat (heat capacity per unit mass), k the thermal conductivity, $T(x, t)$ the temperature at position x and time t, and h the heat transfer coefficient of the surface (or surface conductance) and n is a unit outward drawn normal vector to the surface of the body.

Crucial to the derivation of the equations governing the flow of heat in a body is the knowledge that the rate of heat flow across an isothermal surface (that is, a surface of constant temperature) per unit area is proportional to the temperature gradient at the surface. Mathematically we express this law as

$$j = -k \, \nabla T,$$

where the vector j denotes the heat flux vector. This relationship is now known as Fourier's law of heat conduction. The arduous process and controversies surrounding the development of these equations is described in detail in Herivel (1975), which also presents a fascinating account of Fourier's life.

Fourier's impact on mathematics arises from his procedure for solving equations such as (1.1) subject to the specification of the initial temperature in the body,

$$T(\underline{x}, 0) = f(\underline{x}), \tag{1.3}$$

where f is a prescribed function of position $\underline{x}$. In general terms Fourier's strategy is to look for a solution of (1.1), (1.2) and (1.3) of the form

$$T(\underline{x}, t) = T_0 + \sum_{m=0}^{\infty} C_m e^{-\lambda_m^2 t} \phi_m(\underline{x}),$$

where the λ_m are constant, and each of the ϕ_m is assumed to satisfy

$$\left.\begin{aligned} k \nabla^2 \phi_m + \rho c \lambda_m^2 \phi_m &= 0, \\ k \underline{\nabla} \phi_m \cdot \underline{n} + h \phi_m &= 0, \end{aligned}\right\}$$

so that the constants C_m must be found such that the initial condition (1.3) is satisfied, namely

$$f(\underline{x}) = T_0 + \sum_{m=0}^{\infty} C_m \phi_m(\underline{x}).$$

Thus the necessity arises for a theory of expanding an arbitrary function in terms of an infinite series of functions. Although such series solutions had been used before Fourier (for example Daniel Bernoulli for the problem of the vibrating string), it was Fourier who mastered their general use, particularly in the context of heat flow, and for this reason such series are generally referred to as Fourier series.

Unlike many of his contemporaries, Fourier's contribution to science consists almost entirely of the material published in 1822 in the "Analytical Theory of Heat", although this work and various minor additions to it had been in existence since around 1807. However, few works have contained so many original results or initiated so much research in both pure and applied mathematics and in physics. In the context of present day science it is perhaps difficult to appreciate that in spite of his outstanding success as a scientist, much of Fourier's scientific work was done part-time during his career as an administrator and governor during the reign of Napoleon.

1.2 Physical derivation of the one dimensional heat equation

The so called calorimetric definition of *heat* defines heat as being that which is transferred between a body and its surroundings by virtue of a temperature difference only. It was once believed that heat was a material substance, a fluid which flowed from hotter to colder. By the beginning of the last century it was, however, widely suspected that heat was a form of energy. In 1798 Benjamin Thompson (later Count Rumford) noticed the increase in the temperature of brass during the boring of cannon, and attributed this rise in temperature to the work (in the physicists' sense) done on the metal. It was left to Joule, however, to demonstrate conclusively the relationship between work and heat and their essential equivalence, in a series of experiments conducted between 1840 and 1849. Strictly, heat is energy moving between regions due to differences in temperature. Once the temperature difference disappears and heat flow ceases it becomes meaningless to talk of 'the heat in a body', in the same way as it is meaningless to speak about 'the mechanical work in a body'. The flow of heat will change the internal energy of a body (due to the random molecular motions in the body), but once this has occurred there is no way to distinguish the result from that which would have occurred as a result of an equivalent quantity of mechanical or electrical work being done on the body. Having said this, we now note that at various points in this book we shall refer to 'the heat in a body', by which we will mean 'the change in internal energy due to heat flow'. Since we are concerned only with heat flow, any change in internal energy must be due to heat flow and no confusion can result.

Generally, heat flow is accompanied by temperature change. If heat flows out of one region the temperature there will decrease and vice versa. One important exception occurs when a phase change (such as water freezing or ice melting) is present, and in these circumstances because of the latent heats involved there is isothermal liberation or absorption of heat. For the present we assume that heat flow is accompanied by temperature change, and consider the ratio $\Delta Q / \Delta T$ where ΔQ is the quantity of heat which has entered a body and ΔT is the change in that body's temperature which results. Under most circumstances as ΔQ approaches zero the ratio approaches a positive, finite limit, which we call the *heat capacity* of the body. In general the heat capacity varies in proportion to the mass of the body and it is convenient to work with the specific heat capacity or *specific heat* which is the heat capacity per unit mass. The specific heat of a material, $c(T)$, is given by

$$c(T) = \frac{\mathrm{d}Q}{\mathrm{d}T},\qquad(1.4)$$

that is, the rate of change of heat Q with temperature T per unit mass. The specific heat has dimensions of [energy]/[temperature] $\times$ [mass], and is usually given in units of joule/gram-degree. Table 1.1 gives the specific heats of a variety of substances under standard conditions. The molar heat capacity, that is the heat capacity per gram-mole of material, is also frequently used. As is implicit in equation (1.4), the specific heat c is usually a function of temperature (and other thermodynamic quantities), although if only small temperature differences are involved it is usually possible to regard c as a constant.

In order to formulate a mathematical theory of heat conduction we need to relate temperature difference to heat flow. *Fourier's law* provides this link, and is suggested by the results of the following experiment. Consider a thin plate consisting of an

Substance	c	k
	$\mathrm{Jg^{-1}K^{-1}}$	$\mathrm{Wm^{-1}K^{-1}}$
Aluminium	0.90	190
Copper	0.39	390
Gold	0.13	310
Iron	0.45	78
Lead	0.13	37
Nickel	0.44	89
Silver	0.23	418
Steel	0.49	46
Tin	0.23	63
Granite	0.88	2.5
Graphite	0.72	45
Ice	2.1	2.2
Limestone	0.92	1.7
Sandstone	0.96	2.5
Silicon	0.71	84
Sulphur	0.70	0.28
Water	4.2	0.60

Table 1.1. Specific heats (joule/gram-kelvin) and thermal conductivities (watt/metre-kelvin) for various elements and materials under standard conditions (297K, 1 atm) except for ice (273K, 1 atm).

homogeneous material with insulated sides of width Δx, constant cross sectional area A and with the two parallel faces of the plate held at different constant temperatures (see Figure 1.1). After these conditions have prevailed for sometime, the temperature in the bar settles down to a time independent state (steady temperature), which is the same on any plane parallel to the faces of the plate, and varies only along lines perpendicular to the faces of the plate. We can assume without loss of generality that the faces of the plate lie in the planes x constant and $x + \Delta x$ constant and the temperatures on these faces are respectively T and $T + \Delta T$. Once the temperature has ceased varying in time, it is observed experimentally that the quantity of heat ΔQ which flows through the plate in time Δt is given by

$$\Delta Q = -kA\frac{\Delta T}{\Delta x}\Delta t,$$

where k is a positive quantity known as the *thermal conductivity*, and the negative sign is due the fact that heat flows from hot to cold (that is 'down the temperature gradient'). In general the thermal conductivity is a function of temperature (and other

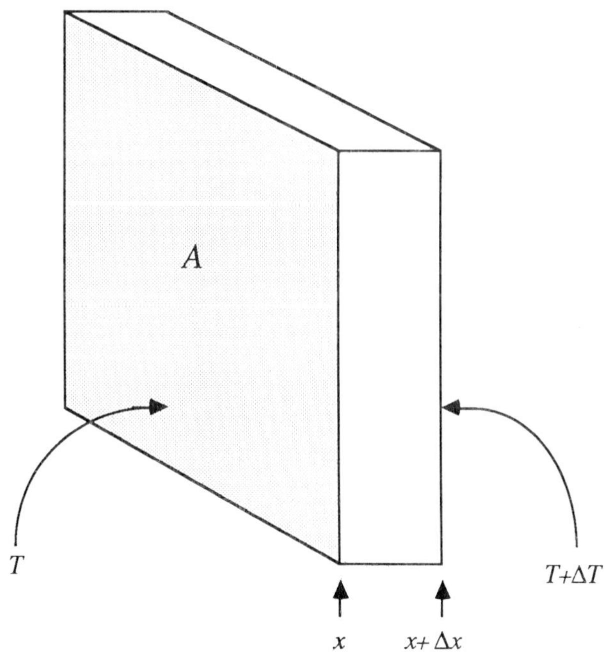

Figure 1.1. Elementary slab for one dimensional heat flow.

thermodynamic variables), but usually for small temperature differences we can regard it as a constant. It has the dimensions of [power]/[temperature] $\times$ [length] and is usually given in watt/metre-degree. If we let Δx, ΔT and Δt approach zero so that the ratio $\Delta T/\Delta x$ tends to a finite constant (that is, the temperature gradient remains finite) we find that the rate $j(x,t)$ at which heat flows through a unit area is given by

$$j(x,t) \;=\; -k(T)\frac{\partial T}{\partial x}(x,t), \tag{1.5}$$

which is the one dimensional form of Fourier's law. We call $j(x,t)$ the *heat flux* at x. We assume that this law holds instantaneously, since evidently an infinitesimally thin plate attains thermal equilibrium instantaneously.

We are now in a position to derive the one dimensional *heat equation*. Consider the one dimensional flow of heat in a rectangular box bounded by rectangular faces of unit area at x and $x + \Delta x$, and assume that the heat flows perpendicular to the faces of the box. From Fourier's law (1.5) the difference between the instantaneous rates of heat flow across the two faces is given by

$$j(x,t) - j(x + \Delta x, t) = k(T + \Delta T)\frac{\partial T}{\partial x}(x + \Delta x, t) - k(T)\frac{\partial T}{\partial x}(x,t),$$

where $T + \Delta T$ denotes the temperature at the face $x + \Delta x$. This quantity is the instantaneous net rate of heat flow into the box and in the absence of other heat sources inside the box, must equal the time rate of change of the heat content of the box by the conservation of energy. We can write

$$k(T + \Delta T)\frac{\partial T}{\partial x}(x + \Delta x, t) - k(T)\frac{\partial T}{\partial x}(x,t) = \int_{x}^{x + \Delta x} \frac{\partial Q}{\partial t}(\xi, t)\,d\xi,$$

and applying the mean value theorem to the integral, we have

$$k(T + \Delta T)\frac{\partial T}{\partial x}(x + \Delta x, t) - k(T)\frac{\partial T}{\partial x}(x,t) = \Delta x \frac{\partial Q}{\partial t}(x + \theta \Delta x, t),$$

where θ is a constant lying between zero and one. Dividing throughout by Δx and taking the limit $\Delta x \to 0$ gives

$$\frac{\partial Q}{\partial t}(x,t) \;=\; \frac{\partial}{\partial x}\Big(k(T)\frac{\partial T}{\partial x}(x,t)\Big), \tag{1.6}$$

leaving us to determine an expression for $\frac{\partial Q}{\partial t}$, the time rate of change of heat per unit volume.

If we integrate equation (1.4) with respect to temperature we find the expression

$$Q_m = \int_{T_1}^{T} c(\xi)\,d\xi,$$

for the quantity of heat Q_m necessary to change the temperature of a unit mass from T_1 to T. Multiplying c by the density of the material ρ (which we assume to independent of temperature) in order to find the heat capacity per unit volume gives

$$Q = \rho \int_{T_1}^{T} c(\xi)\,d\xi,$$

as the quantity of heat Q necessary to raise a unit volume from temperature T_1 to T. Differentiating this with respect to time, we have

$$\frac{\partial Q}{\partial t} = \rho c(T)\frac{\partial T}{\partial t},$$

which we can substitute in (1.6) to obtain the one dimensional heat equation

$$\rho c(T)\frac{\partial T}{\partial t} = \frac{\partial}{\partial x}\left(k(T)\frac{\partial T}{\partial x}\right). \tag{1.7}$$

For the special case in which the specific heat c and thermal conductivity k are constant we obtain the classical one dimensional heat equation

$$\frac{\partial T}{\partial t} = \kappa\frac{\partial^2 T}{\partial x^2}, \tag{1.8}$$

where the constant $\kappa = k/\rho c$ is known as the *thermal diffusivity*, which has the units of $[\text{length}]^2/[\text{time}]$, and is usually measured in $\text{metre}^2/\text{second}$.

The mathematical theory of diffusion is closely related to the mathematical theory of heat conduction. Heat conduction arises from the flow of heat due to a temperature gradient, whereas diffusion is the flow of matter due to a concentration gradient. It was Fick who, recognizing the similarities between the two processes, generalized the work of Fourier and established the mathematical theory of diffusion in 1855. Fick argued by analogy with Fourier's law that the rate at which a diffusing material flows per unit area is proportional to the gradient of the concentration of the diffusing material. The one dimensional form of this law, known as *Fick's law*, is

$$j(x,t) = -D\frac{\partial C}{\partial x}(x,t),$$

where $j(x,t)$ denotes the flow rate per unit area and $C(x,t)$ the concentration at position x and time t. The constant D is known as the diffusivity of the diffusing material and is generally a function of the concentration C. It has the dimensions of [length]^2/[time] and is usually measured in terms of metre^2/second. Using Fick's law and the principle of conservation of mass, namely

$$\frac{\partial C}{\partial t} + \frac{\partial j}{\partial x} = 0,$$

the one dimensional *diffusion equation*

$$\frac{\partial C}{\partial t} = \frac{\partial}{\partial x}\left(D(C)\frac{\partial C}{\partial x}\right),$$

is immediately apparent. In the case where the diffusivity D is a constant we obtain the classical one dimensional diffusion equation

$$\frac{\partial C}{\partial t} = D\frac{\partial^2 C}{\partial x^2},$$

which is of course mathematically identical to the heat equation (1.8).

1.3 One dimensional source solution

The *source solution* is of fundamental importance in heat flow. Physically we can regard the source solution as being the temperature distribution due to an instantaneous point source of heat located at the origin at time zero, that is a quantity of heat initially concentrated entirely at the origin. Clearly such a distribution is at best a mathematical abstraction, but it is nevertheless a very useful and important one. Mathematically, the one dimensional source solution arises as the solution of the following initial value problem,

$$\left.\begin{aligned}
\frac{\partial T}{\partial t} &= \kappa\frac{\partial^2 T}{\partial x^2}, \qquad -\infty < x < \infty, \\[1mm]
T(x,t) &\to 0 \text{ as } |x| \to \infty, \\[1mm]
T(x,0) &= q_0\delta(x),
\end{aligned}\right\} \tag{1.9}$$

where $\delta(x)$ is the *Dirac delta function*, and $\rho c q_0$ is the quantity of heat released at the origin. The Dirac delta function, which is in fact not a function in the normal sense at all, can be imagined to have the following properties

$$\delta(x) = 0 \text{ if } x \neq 0, \quad \text{and} \quad \int_{-\infty}^{\infty} \phi(\xi)\delta(\xi)\,d\xi = \phi(0), \tag{1.10}$$

for all test functions $\phi(x)$. By a test function, we mean any function which is infinitely differentiable on the real line and which, together with its derivatives of all orders, vanishes outside a closed bounded interval (which depends on the function). In particular, a test function and all its derivatives vanish as $|x| \to \infty$. These properties serve to define the delta function $\delta(x)$. For example, one property of the delta function is that for any constant $\lambda > 0$

$$\delta(\lambda x) = \frac{1}{\lambda}\delta(x), \tag{1.11}$$

since, with the change of variable $\eta = \lambda\xi$ we have

$$\int_{-\infty}^{\infty} \phi(\xi)\delta(\lambda\xi)\,d\xi = \int_{-\infty}^{\infty} \phi(\frac{\eta}{\lambda})\delta(\eta)\frac{d\eta}{\lambda} = \frac{1}{\lambda}\phi(0),$$

for any test function $\phi(x)$.

The property (1.11) of the delta function means that the heat conduction problem (1.9) remains invariant under the *stretching transformation* of variables

$$x^* = \lambda x, \qquad t^* = \lambda^2 t, \qquad T^* = \frac{T}{\lambda}, \tag{1.12}$$

by which we mean the problem assumes exactly the same form in terms of the starred variables as it does in the unstarred variables. This means that if our solution is $T = \Phi(x, t)$ then after the transformation (1.12) we have $T^* = \Phi(x^*, t^*)$ for the same function $\Phi(x, t)$. Accordingly, on noting the invariant quantities

$$\frac{x^*}{\sqrt{t^*}} = \frac{x}{\sqrt{t}}, \qquad T^*\sqrt{t^*} = T\sqrt{t},$$

we look for a solution of (1.9) of the form

$$T(x, t) = \frac{1}{\sqrt{t}}\psi(\frac{x}{\sqrt{t}}),$$

which is automatically invariant under the transformation (1.12).

With $\zeta = x/\sqrt{t}$ and using the partial derivatives

$$\frac{\partial T}{\partial t} = -\frac{(\zeta\psi' + \psi)}{2t^{3/2}}, \qquad \frac{\partial T}{\partial x} = \frac{\psi'}{t}, \qquad \frac{\partial^2 T}{\partial x^2} = \frac{\psi''}{t^{3/2}},$$

where primes denote differentiation with respect to ζ, we find that the partial differential equation (1.9)$_1$ reduces to the ordinary differential equation

$$2\kappa\psi'' + \zeta\psi' + \psi = 0,$$

which may be integrated immediately to give

$$2\kappa\psi' + \zeta\psi = 2\kappa C_1,$$

where C_1 is an arbitrary constant. Introducing an integrating factor, we find

$$\left(e^{\zeta^2/4\kappa}\psi\right)' = C_1 e^{\zeta^2/4\kappa},$$

and after a further integration we have

$$\psi(\zeta) = C_1 e^{-\zeta^2/4\kappa}\int_0^\zeta e^{\eta^2/4\kappa}\,\mathrm{d}\eta + C_2 e^{-\zeta^2/4\kappa},$$

where C_2 is a second integration constant. Now, from $(1.9)_2$ it is clear that $\psi(\zeta)\to 0$ as $|\zeta|\to\infty$ and therefore the constant C_1 must vanish. Altogether we have

$$T(x,t) = \frac{C_2}{\sqrt{t}}e^{-x^2/4\kappa t}.$$

We determine the constant C_2 from the condition that the quantity of heat $\rho c q_0$ initially released at the origin is conserved, that is we require that

$$\int_{-\infty}^{\infty} T(\xi,t)\,\mathrm{d}\xi = \int_{-\infty}^{\infty} T(\xi,0)\,\mathrm{d}\xi = q_0\int_{-\infty}^{\infty}\delta(\xi)\,\mathrm{d}\xi = q_0.$$

Thus

$$\int_{-\infty}^{\infty} T(\xi,t)\,\mathrm{d}\xi = \frac{C_2}{\sqrt{t}}\int_{-\infty}^{\infty} e^{-\xi^2/4\kappa t}\,\mathrm{d}\xi = 2\sqrt{\kappa}\,C_2\int_{-\infty}^{\infty} e^{-\eta^2}\,\mathrm{d}\eta,$$

where $\eta = \xi/2\sqrt{\kappa t}$. Since (see Problem 3)

$$\int_{-\infty}^{\infty} e^{-\xi^2}\,\mathrm{d}\xi = \sqrt{\pi}, \tag{1.13}$$

we can deduce that $C_2 = q_0/2\sqrt{\pi\kappa}$ and therefore that the basic one dimensional source solution of the heat equation is

$$T(x,t) = \frac{q_0}{2\sqrt{\pi\kappa t}}e^{-x^2/4\kappa t}, \tag{1.14}$$

which is shown graphically in Figure 1.2. Initially the temperature distribution has a very sharp peak at $x = 0$, but this rapidly decays. The temperature away from the origin is zero initially and yet non-zero at any time thereafter, so (1.14) predicts that heat is transmitted with infinite velocity. This is physically absurd and is due to the fact that in deriving the heat equation we did not allow for the finite rate at which heat is propagated. The quantities of heat involved are usually negligible, as evidenced by the infinitesimal temperatures at large distances from the origin. Nevertheless, this does highlight the fact that the heat equation (1.8) is only an approximation to reality albeit usually a very good approximation.

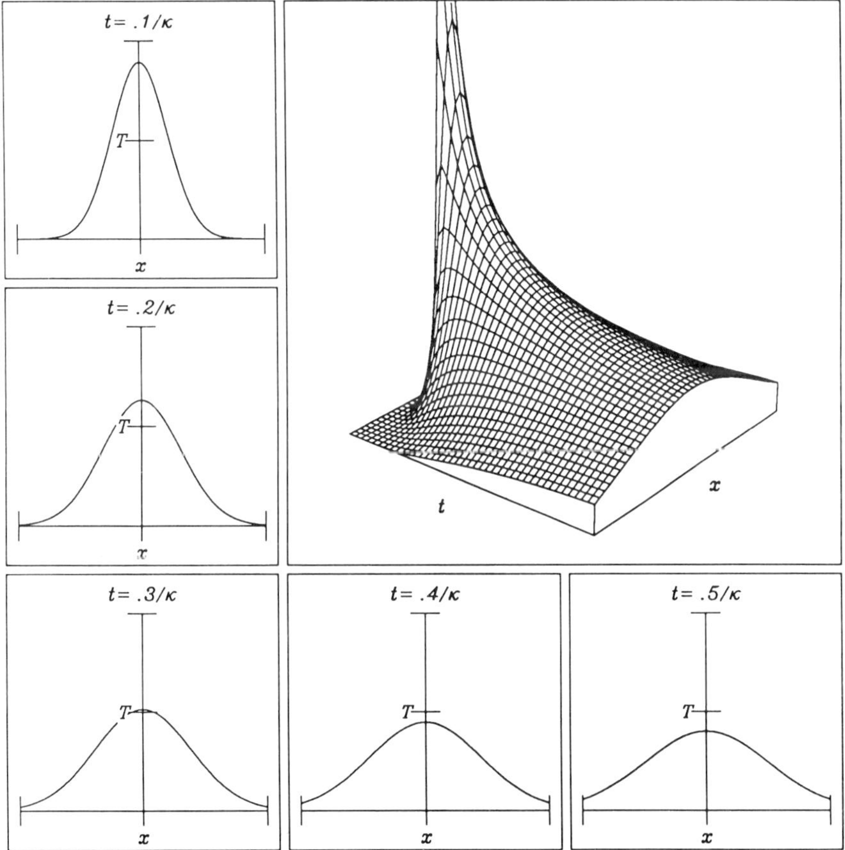

Figure 1.2. Time development of the one dimensional source solution (1.14) with q_0 unity.

As we have noted, the Dirac delta function is not a function in the normal sense of the word for no function can have both of the properties accorded to the delta function in (1.10). The Dirac delta function belongs to a class of mathematical entities known as *generalized functions* or *distributions*, which may be constructed as the 'limits' of sequences of actual functions. The basic source solution (1.14) provides an example of such a *delta convergent* sequence, as we let t approach zero the source solution 'converges' to the Dirac delta function. Various other such sequences are:

$$\textbf{(i)} \qquad f_\epsilon(x) = \frac{1}{2\sqrt{\pi\epsilon}} e^{-x^2/4\epsilon},$$

$$\textbf{(ii)} \qquad f_\epsilon(x) = \frac{1}{\pi} \frac{\epsilon}{\left(x^2 + \epsilon^2\right)},$$

$$\textbf{(iii)} \qquad f_\epsilon(x) = \begin{cases} 3(\epsilon^2 - x^2)/4\epsilon^3, & |x| < \epsilon, \\ 0, & |x| \geq \epsilon, \end{cases}$$

$$\textbf{(iv)} \qquad f_\epsilon(x) = \frac{1}{\pi} \frac{\sin(x/\epsilon)}{x},$$

which in the limit $\epsilon \to 0$ converge to the delta function. By converge to the Dirac delta function we mean that

$$\lim_{\epsilon \to 0} \int_{-\infty}^{\infty} f_\epsilon(\xi)\phi(\xi)\,\mathrm{d}\xi = \phi(0),$$

for any test function $\phi(x)$.

A (proper) function closely related to the Dirac delta function is the *Heaviside unit step function $H(x)$* defined by

$$H(x) = \begin{cases} 0, & x < 0, \\ 1, & x \geq 0, \end{cases} \tag{1.15}$$

for which we have the formal result

$$\delta(x) = \frac{\mathrm{d}H}{\mathrm{d}x}(x). \tag{1.16}$$

If we take any test function $\phi(x)$, we have on using integration by parts

$$\int_{-\infty}^{\infty} \phi(\xi)\frac{\mathrm{d}}{\mathrm{d}\xi}H(\xi)\,\mathrm{d}\xi = [\phi(\xi)H(\xi)]_{-\infty}^{\infty} - \int_{-\infty}^{\infty} H(\xi)\frac{\mathrm{d}}{\mathrm{d}\xi}\phi(\xi)\,\mathrm{d}\xi$$

$$= -\int_{0}^{\infty} \frac{\mathrm{d}\phi}{\mathrm{d}\xi}(\xi)\,\mathrm{d}\xi = -[\phi(\xi)]_{0}^{\infty} = \phi(0),$$

from which (1.16) follows immediately.

1.4 Probabilistic derivation of the one dimensional heat equation

At the molecular level the mechanisms of heat conduction and diffusion are similar. In the former case, energy is transferred by random molecular motions and in the latter case, mass is transferred by random molecular motions. These random molecular motions are governed by probabilistic laws, and it is therefore both important and instructive to derive the heat or diffusion equation and the basic source solution from probabilistic arguments. In this section we present a derivation of the diffusion equation and the basic source solution based on a discrete *random walk* model of the diffusion process. The simplest example of a symmetric random walk is a drunken student constrained to move on a one dimensional array of regularly spaced points. At regular intervals in time the student proceeds either one step to the right or one step to the left, with either outcome equally likely. After n steps the student could be at any one of $2n + 1$ sites, and we can calculate the probability of the student being at any particular site. More generally, a random walk is defined as the sum of random variables and in the former example each of the random variables takes on either of the values ± 1 with equal probability of $1/2$.

Here we first consider the following more general formulation. Consider a particle random walking a discrete lattice with a constant distance Δx between points such that at discrete time intervals of length Δt, it may move one step to the right with probability p, one step to the left with probability q or stay where it is with probability r (see Figure 1.3). For now we do not necessarily assume that $p + q + r = 1$. If $u_{m,n}$ denotes the probability of the particle being at position $m\Delta x$ at time $n\Delta t$, then for an unrestricted particle we have the discrete equations

$$u_{m,n+1} = pu_{m-1,n} + qu_{m+1,n} + ru_{m,n}, \qquad (n \geq 0). \qquad (1.17)$$

We now consider the *diffusion limit* when the particle takes vanishingly small steps at vanishingly small time intervals. Then as Δx and Δt tend to zero, $u_{m,n}$ is approximately equal to $f(x,t)\Delta x$, where $f(x,t)$ is the probability density function for the position x

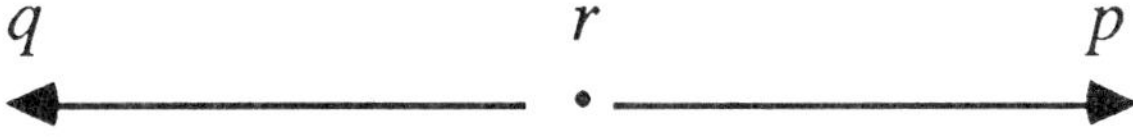

Figure 1.3. Probabilities for one dimensional random walk.

of the particle at time t, and (1.17) becomes

$$f(x, t + \Delta t) = pf(x - \Delta x, t) + qf(x + \Delta x, t) + rf(x, t).$$

Now, using Taylor's theorem we have for $f(x, t)$

$$f + \Delta t \frac{\partial f}{\partial t} = p\left(f - \Delta x \frac{\partial f}{\partial x} + \frac{(\Delta x)^2}{2} \frac{\partial^2 f}{\partial x^2}\right) + q\left(f + \Delta x \frac{\partial f}{\partial x} + \frac{(\Delta x)^2}{2} \frac{\partial^2 f}{\partial x^2}\right) + rf,$$

assuming that Δt and $(\Delta x)^2$ are of the same order of magnitude. We can write this as

$$\Delta t \frac{\partial f}{\partial t} = (p + q) \frac{(\Delta x)^2}{2} \frac{\partial^2 f}{\partial x^2} - (p - q) \Delta x \frac{\partial f}{\partial x} - (1 - (p + q + r))f, \qquad (1.18)$$

and the terms on the right hand side of this equation are respectively identified as the diffusion, convection and absorption components.

From equation (1.18) it is immediately apparent that if $p + q + r = 1$ (that is, no probability that the particle is absorbed) then the loss or absorption term disappears. If in addition we consider a symmetric random walk ($p = q$) then there is no convection term and (1.18) gives

$$\Delta t \frac{\partial f}{\partial t} = p(\Delta x)^2 \frac{\partial^2 f}{\partial x^2},$$

which clearly yields the one dimensional diffusion equation, provided that Δx and Δt tend to zero in such a way that $p(\Delta x)^2 / \Delta t$ tends to a finite constant κ. This simple formal connection, due to Einstein, can be made mathematically rigorous and is an important link in understanding diffusion processes. It is possible, using sophisticated probability theory, to solve the discrete equations (1.17) for $u_{m,n}$ and then for $p = q = \frac{1}{2}$ and the above limiting process to show that for large n, the position of the particle is normally distributed, that is, we may formally establish the one dimensional source solution. Since this procedure is beyond our scope we give the following derivation of the source solution based on the assumption that only the time variable becomes continuous. This calculation, although complicated, hinges on the rather unusual identity (1.26) and is straightforward enough for students with a fluency in the language of Bessel functions.

With only time continuous we assume that the particle makes a move to the right in time Δt with probability $\alpha \Delta t$, a move to the left with probability $\beta \Delta t$ and remains in position with probability $1 - (\alpha + \beta)\Delta t$ (that is, we assume that there is no possibility of particle absorption). If $u_m(t)$ denotes the probability of the particle being at position

m, then instead of equation (1.17) we have

$$u_m(t + \Delta t) = \alpha \Delta t\, u_{m-1}(t) + \beta \Delta t\, u_{m+1}(t) + [1 - (\alpha + \beta)\Delta t]u_m(t),$$

so that, as Δt tends to zero we obtain the classical Master equation

$$\frac{du_m}{dt}(t) = \alpha u_{m-1}(t) - (\alpha + \beta)u_m(t) + \beta u_{m+1}(t). \tag{1.19}$$

Assuming the particle is initially at the origin, so that

$$u_m(0) = \delta_{0m}, \tag{1.20}$$

where δ_{ij} is the *Kronecker delta*, defined by

$$\delta_{ij} = \begin{cases} 1, & i = j, \\ 0, & i \neq j, \end{cases} \tag{1.21}$$

we then solve (1.19), subject to (1.20) by means of the generating function

$$U(t, z) = \sum_{m=-\infty}^{\infty} u_m(t)z^m. \tag{1.22}$$

After multiplying (1.19) by z^m and summing over m we find that

$$\frac{\partial U}{\partial t} = \left[(\alpha z + \frac{\beta}{z}) - (\alpha + \beta) \right]U, \qquad U(0, z) = 1,$$

which has the solution

$$U(t, z) = e^{-(\alpha + \beta)t}\, e^{(\alpha z + \frac{\beta}{z})t}. \tag{1.23}$$

From the identity

$$e^{y(z + \frac{1}{z})} = \sum_{m=-\infty}^{\infty} I_m(2y)z^m, \tag{1.24}$$

where $I_m(x)$ denotes the usual modified Bessel function of order m, it is not difficult to establish from equations (1.22), (1.23), and (1.24)

$$u_m(t) = e^{-(\alpha + \beta)t}\left(\frac{\alpha}{\beta}\right)^{m/2} I_m\left[2\sqrt{\alpha\beta\, t}\right]. \tag{1.25}$$

For $\alpha = \beta$ we deduce the source solution (1.14) as follows. As the jump length Δx tends to zero in the diffusion limit we have the relations

$$u_m(t) \approx f(x, t)\Delta x, \qquad \alpha = \frac{\kappa}{(\Delta x)^2}, \qquad m = \frac{x}{\Delta x},$$

and we see that these, together with (1.25) yield

$$f(x,t) \approx \frac{1}{x} \lim_{m \to \infty} m e^{-\gamma m^2} I_m(\gamma m^2),$$

where γ denotes $2\kappa t / x^2$. The basic one dimensional source solution (1.23) now follows from the remarkable identity

$$\lim_{m \to \infty} m e^{-\gamma m^2} I_m(\gamma m^2) = \frac{e^{-1/(2\gamma)}}{\sqrt{2\pi\gamma}}, \tag{1.26}$$

for all real positive γ. Equation (1.26) may be proved starting with the representation

$$I_m(z) = \frac{1}{2\pi} \int_{-\pi}^{\pi} e^{z \cos\theta + im\theta} \, d\theta,$$

from which we may deduce

$$\sqrt{z}\, e^{-z} I_m(z) = \frac{\sqrt{z}}{2\pi} \int_{-\pi}^{\pi} \exp\left\{ -2z \sin^2\!\left(\frac{\theta}{2}\right) + im\theta \right\} d\theta,$$

which, with the substitution $\theta = 2\xi/m$ and $z = \gamma m^2$ gives

$$\sqrt{\gamma}\, m e^{-\gamma m^2} I_m(\gamma m^2) = \frac{\sqrt{\gamma}}{\pi} \int_{-m\pi/2}^{m\pi/2} \exp\left\{ -2\gamma\xi^2 \left[\frac{\sin^2(\xi/m)}{(\xi/m)^2} \right] + 2i\xi \right\} d\xi.$$

Now, letting m tend to infinity we may replace $\sin^2(\xi/m)/(\xi/m)^2$ by unity, while the limits of integration pass to infinity, and we have

$$\lim_{m \to \infty} m e^{-\gamma m^2} I_m(\gamma m^2) = \frac{e^{-1/(2\gamma)}}{\pi} \int_{-\infty}^{\infty} e^{-2\gamma(\xi + i/2\gamma)^2} \, d\xi,$$

and the identity (1.26) follows on making the substitution

$$\eta = \sqrt{2\gamma}\left(\xi + \frac{i}{2\gamma} \right),$$

and using the integral

$$\int_{-\infty+i\omega}^{\infty+i\omega} e^{-\eta^2} \, d\eta = \sqrt{\pi},$$

for any real ω (see Problem 3).

Although in this book we make no further use of the following derivation of the so called *telegraphers' equation*, (1.31), it is nevertheless instructive for the student to note how a modification of the simple random walk gives rise to this equation, since this equation and related wave equations are also important in applied mathematics and physics. In the above discussion there is no property that might be regarded as an analog to momentum. That is to say, the particle has the option of reversing its direction at every step, regardless of the direction of its last step. A more realistic model of the motion of a physical object or quantity needs to take into account some form of memory of where it has come from and the likelihood that it will continue to move in the same direction. The simplest such modification to the random walk is due to the British applied mathematician G.I. Taylor which he proposed in a study of turbulent diffusion. Here we deduce the telegraphers' equation (1.31) from a slightly more general framework.

Again we consider a particle moving along a straight line taking steps of length Δx every time interval Δt. However in this case we designate one direction along the straight line by the subscript 1 and the opposite direction by the subscript 2. If the particle has just moved in the 1 direction (that is, has just come from the left) then we suppose it moves at the next jump to the right with probability p_{11} and to the left with probability p_{12} where we assume that $p_{11} + p_{12} = 1$. However, if the particle has just moved in the 2 direction (that is, has just come from the right) then we suppose it moves at the next jump to the right with probability p_{21} and to the left with probability p_{22}, where again we assume that $p_{21} + p_{22} = 1$ (see Figure 1.4). Further we let $f_1(x, t)\Delta x$ be the conditional probability of the particle being at position x at time t and moving in the 1 direction (that is, arriving from the left), and we let $f_2(x, t)\Delta x$ be the conditional probability of the particle being at position x at time t and moving in

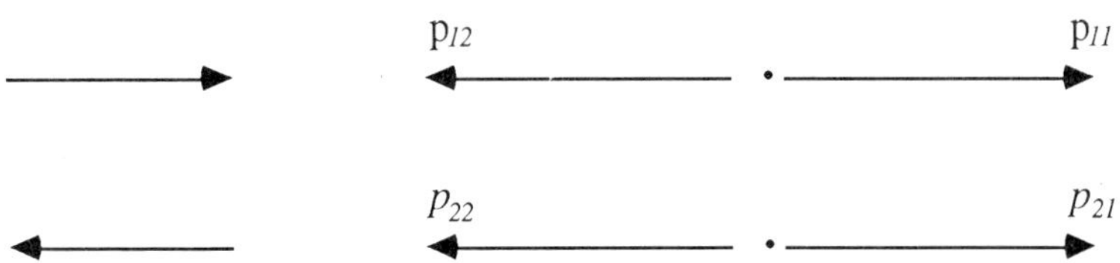

Figure 1.4. Probabilities for one dimensional correlated random walk.

the 2 direction (that is, arriving from the right). We have

$$f_1(x, t + \Delta t) = p_{11} f_1(x - \Delta x, t) + p_{21} f_2(x - \Delta x, t), \qquad (1.27)$$

$$f_2(x, t + \Delta t) = p_{12} f_1(x + \Delta x, t) + p_{22} f_2(x + \Delta x, t). \qquad (1.28)$$

Now the total probability density function $f(x, t)$ for the position of the particle at time t is given by

$$f(x, t) = f_1(x, t) + f_2(x, t), \qquad (1.29)$$

and using (1.27) and (1.28) we have

$$\begin{aligned}
f_1(x + \Delta x, t) + f_2(x - \Delta x, t) &= p_{11} f_1(x, t - \Delta t) + p_{21} f_2(x, t - \Delta t) \\
&\quad + p_{12} f_1(x, t - \Delta t) + p_{22} f_2(x, t - \Delta t) \qquad (1.30) \\
&= f(x, t - \Delta t).
\end{aligned}$$

However, if we introduce δ by

$$\delta = p_{11} - p_{21} = (1 - p_{12}) - (1 - p_{22}) = p_{22} - p_{12},$$

then δ is a measure of correlation between two consecutive steps and, moreover, equations (1.27) and (1.28) can be rewritten respectively as

$$f_1(x, t + \Delta t) = p_{11} f(x - \Delta x, t) - \delta f_2(x - \Delta x, t),$$

$$f_2(x, t + \Delta t) = p_{22} f(x + \Delta x, t) - \delta f_1(x + \Delta x, t).$$

If we add these equations and use (1.29) and (1.30) we find

$$f(x, t + \Delta t) = p_{11} f(x - \Delta x, t) + p_{22} f(x + \Delta x, t) - \delta f(x, t - \Delta t),$$

and from Taylor's theorem, by retaining terms of order up to and including $(\Delta t)^2$ and $(\Delta x)^2$ we have

$$(1 + \delta) \frac{(\Delta t)^2}{2} \frac{\partial^2 f}{\partial t^2} + (1 - \delta) \Delta t \frac{\partial f}{\partial t} = (p_{11} + p_{22}) \frac{(\Delta x)^2}{2} \frac{\partial^2 f}{\partial x^2} - (p_{11} - p_{22}) \Delta x \frac{\partial f}{\partial x}.$$

Noting the relation $p_{11} + p_{22} = 1 + \delta$ and assuming that

$$p_{11} - p_{22} = \tilde{\lambda} \Delta x, \qquad \delta - 1 - \tilde{\mu} \Delta t,$$

where $\tilde{\mu}$ and $\tilde{\lambda}$ are constants, and that Δx and Δt tend to zero in such a way that $\Delta x / \Delta t$ approaches a finite non-zero constant $\tilde{c}$ we obtain

$$\frac{\partial^2 f}{\partial t^2} + \tilde{\mu} \frac{\partial f}{\partial t} = \tilde{c}^2 \left(\frac{\partial^2 f}{\partial x^2} - \tilde{\lambda} \frac{\partial f}{\partial x} \right).$$

Evidently, the telegraphers' equation

$$\frac{\partial^2 f}{\partial t^2} + \tilde{\mu}\frac{\partial f}{\partial t} = \tilde{c}^2\frac{\partial^2 f}{\partial x^2}, \tag{1.31}$$

emerges for the isotropic case (that is, $p_{11} = p_{22}$) and this equation is the appropriate generalization of (1.8) for finite velocity heat conduction.

1.5 Initial and boundary conditions for one dimensional problems

The heat (or diffusion) equation (1.8) is the partial differential equation governing the one dimensional flow of heat in an object. However additional information, usually in the form of initial and boundary conditions, are required in order to fully pose a physically meaningful problem. The most common *initial condition*, and the only one that we will consider in this book, is the prescription of the temperature at every point in the body at some initial moment, which without loss of generality we can take to be time zero. Thus, an initial condition has the form

$$T(x,0) = f(x), \tag{1.32}$$

where $f(x)$ is a known function whose domain coincides with the region occupied by the body in question. A solution of the heat equation (1.8) is said to satisfy the initial condition (1.32) if (1.8) is satisfied for all time $t > 0$ and

$$\lim_{t \to 0} T(x,t) \to f(x).$$

This convergence need not be pointwise or uniform, a matter which we take up later.

The specification of the initial temperature (or concentration) is still in general insufficient to fully pose the problem. We generally require more information in the form of *boundary conditions* describing the behaviour of the temperature or heat flow rate, or some relationship between them, at the surface of the body. The surface in a one dimensional heat conduction problem may consist of the points (or lines, or planes depending on the actual dimension of the object involved) $x = a$ and $x = b$, for which we adopt the convention that $-\infty \le a < b \le \infty$. One might expect that it would be necessary to specify both the temperature and heat flow rate at both $x = a$ and $x = b$. This however is an over specification of the problem, that is, there is generally no solution to the heat equation under these conditions, which we discuss further in the following chapter. In fact we only need prescribe either the temperature or heat flow rate, or some relationship between the two, at the endpoints $x = a$ and $x = b$ in order to obtain a unique solution of (1.8).

First we consider the case where one or both of the boundaries lie at infinity. To obtain physically acceptable solutions of the heat or diffusion equation we require that

$$\lim_{|x| \to \infty} \frac{\partial T}{\partial x}(x, t) = 0, \quad t > 0, \tag{1.33}$$

since by the principle of conservation of energy (or mass), energy (mass) can neither appear from nor disappear into infinity. Strictly speaking, this is a physical and not a mathematical restriction and as such requires that the initial condition also has a physically sensible interpretation. It is mathematically possible, as we will show in the next chapter, to solve the heat equation (1.8) subject only to the specification of the initial temperature on the whole real line $-\infty < x < \infty$, with no boundary conditions at all. For example, the reader will have no difficulty in verifying that the function $T(x, t) = x^2 + 2\kappa t$ satisfies (1.8) for all time $t > 0$, and satisfies the initial condition $T(x, 0) = x^2$ at all points on the real line. There is heat flow at infinity, and as a consequence there is a constant increase in the heat content of any interval on the real line, which is physically unacceptable as there is no realisable source of this heat. We can regard the condition (1.33) as a constraint on the initial temperature (or concentration) in order that a meaningful solution results.

For boundary points not at infinity there are a wide variety of physically acceptable and important boundary conditions. The simplest boundary condition is a *prescribed temperature* at a boundary point, for example

$$T(a, t) = g(t), \quad t > 0, \tag{1.34}$$

where $g(t)$ is a known function of time for $t > 0$. The situation where $g(t)$ is constant is a particularly simple and important case. Prescribed boundary temperatures are frequently used to model conditions at boundaries in contact with an object of known temperature, such as a block of melting ice or a thermostatically controlled heater. This is generally a rather crude approximation and assumes that the surface and its surroundings are in perfect thermal contact, that is, that the surface attains thermal equilibrium with its environment instantaneously. It is usually very difficult to actually produce a prescribed boundary temperature in practice.

Another widely used and studied boundary condition is the prescription of the rate of heat flow at an endpoint, which in view of Fourier's law (assuming constant thermal conductivity) is equivalent to a *prescribed temperature gradient*. For example, at the boundary $x = b$ we might have the condition

$$k\frac{\partial T}{\partial x}(b,t) = h(t), \quad t > 0, \tag{1.35}$$

where $h(t)$ is some known function of time for $t > 0$. Again, it is generally difficult to produce such a condition in practice, although the case were $h(t)$ vanishes, that is,

$$\frac{\partial T}{\partial x}(b,t) = 0, \quad t > 0, \tag{1.36}$$

is frequently encountered and is particularly important as it represents an *insulated surface* at $x = b$.

Newton's law of cooling states that the rate of heat loss (or gain) at the surface of a body in a 'draught' (more generally, in a well stirred fluid) is proportional to the difference between the surface temperature and the temperature of the cooling medium, that is,

$$\left.\begin{array}{l} -k\dfrac{\partial T}{\partial x}(a,t) + h(T(a,t) - T_0(t)) = 0, \\[2em] k\dfrac{\partial T}{\partial x}(b,t) + h(T(b,t) - T_0(t)) = 0, \end{array}\right\} \quad t > 0, \tag{1.37}$$

where $T_0(t)$ denotes the temperature of the surroundings, and h is a positive constant known as the *surface heat transfer coefficient* or surface conductance, which has units of [Power]/[Temperature] $\times$ [Length]^2 and is measured in units of watt/metre^2-degree. More generally, the function $T_0(t)$ incorporates heat flux effects as well as the temperature of the surroundings. Boundary conditions of the form (1.37) are used to model a wide variety of situations where there is imperfect thermal contact between the surface of a body and its surroundings, that is, where there is some resistance to the flow of heat at the surface of the body. In general the surface heat transfer coefficient h is a function of the geometry of the surface, the temperature, the materials which comprise the body and its surroundings and many other factors. In practice it is a quantity which is well known to be difficult to measure and it is probably best to regard (1.37) as an approximation, albeit usually a superior one to either the prescribed temperature or temperature gradient conditions. As the surface conductance h approaches zero we formally recover the insulated boundary condition (1.35), whereas if h approaches infinity the prescribed temperature condition emerges. Generally the mathematical solutions of the heat equation subject to this boundary condition do not readily reveal this property.

Newton's law (1.37) is often referred to as a radiation boundary condition which is unfortunate. The rate of heat flow due to radiation between a body at absolute temperature T and an environment at absolute temperature T_0 is given by the Stefan-Boltzmann law

$$\sigma E(T^4 - T_0^4),$$

where σ is the Stefan-Boltzmann constant, $\sigma = 5.67 \times 10^{-8}$ watt/metre^2-kelvin^4, and E is a constant called the surface emissivity which depends on the body and its surroundings, and is such that $0 \le E \le 1$. Thus, a genuine *radiation boundary condition* has the form

$$-k \frac{\partial T}{\partial x}(a,t) + \sigma E\left(T^4(a,t) - T_0^4(t)\right) = 0, \quad t > 0. \tag{1.38}$$

Such conditions are non-linear and therefore extremely difficult to treat. Indeed as far as the authors are aware, there are no general solutions of the heat equation (1.8) subject to a genuine radiation boundary condition (1.38). If the difference $T - T_0$ is small compared with T and T_0, we may approximate (1.38) by

$$-k \frac{\partial T}{\partial x}(a,t) + 4\sigma E\, T_0(t)^3(T(a,t) - T_0(t)) = 0, \quad t > 0,$$

which for constant T_0 is simply Newton's law. We shall refer to the boundary conditions (1.37) as Newton heat loss or Newton cooling boundary conditions.

The boundary conditions given here, with the exception of the genuine radiation conditions (1.38), are all used in the context of diffusion as well. Thus the diffusion analogue of (1.34) is the prescribed surface concentration, while using Fick's law (1.35) represents a prescribed flow of diffusant. The special case of an insulated surface in heat conduction corresponds to the presence of an impermeable boundary. The Newton cooling boundary conditions (1.37) can be interpreted as a surface loss (or evaporation) term or as representing a resistance to the flow of diffusant at the surface. Naturally, in these situations we must replace the thermal conductivity k by the diffusivity D.

Finally we mention that it is possible to prescribe inconsistent or incompatible boundary and initial conditions, by which we mean that the boundary conditions at time zero need not necessarily agree with the initial temperature at the boundaries. A typical problem in which such a situation arises is as follows. We consider a finite bar of length ℓ which is initially at the temperature $T = 0$. At time zero both ends of the bar are raised to the temperature $T = 1$ and maintained at this temperature thereafter. The mathematical formulation of this problem is to find a function $T(x,t)$

which satisfies

$$\frac{\partial T}{\partial t} = \kappa \frac{\partial^2 T}{\partial x^2}, \quad 0 < x < \ell,$$

$$T(0,t) = 1, \quad T(\ell,t) = 1, \quad t > 0,$$

$$T(x,0) = 0, \quad 0 \le x \le \ell.$$

Now, it is clear that the initial and boundary conditions are incompatible, in the sense that

$$\lim_{t \to 0} T(0,t) \neq \lim_{x \to 0} T(x,0), \quad \lim_{t \to 0} T(\ell,t) \neq \lim_{x \to \ell} T(x,0),$$

and that as a result the temperature is discontinuous initially at $x = 0$ and $x = \ell$. Nevertheless, the solution of this problem

$$T(x,t) = 1 - \frac{4}{\pi} \sum_{n=0}^{\infty} \frac{e^{-(2n+1)^2 \pi^2 \kappa t / \ell^2}}{(2n+1)} \sin \left(\frac{(2n+1)\pi x}{\ell} \right),$$

converges for all $t > 0$ to a function which is infinitely differentiable in both x and t and which satisfies the boundary conditions at all times $t > 0$ and is such that

$$\lim_{t \to 0} T(x,t) = 0,$$

for *almost all* x between 0 and ℓ. By this we mean that the set of points where $\lim_{t \to 0} T(x,t) \neq 0$ can be enclosed in a set of open intervals of arbitrarily small total length. In general the solution of the heat equation is an infinitely smooth function for $t > 0$ even when its initial data and boundary data are incompatible or the initial temperature is itself discontinuous. Any discontinuities which occur in the initial or boundary data are smoothed out immediately after they happen. This is a consequence of the infinite velocity of heat propagation which is implicit in the heat equation (1.8).

1.6 Three dimensional heat conduction

To derive the three dimensional heat equation we need the concept of a *heat flux vector*. The heat flux at a point **P** across a surface S is the rate, per unit area, at which heat flows across S at **P** in a direction normal to S, and as such depends on both the point **P** and the surface S. The heat flux vector enables us to determine the heat flux across S at **P** provided the normal to S at **P** is known. Fourier's law is then used to relate the temperature gradient to the heat flux vector, enabling the calculation of heat flow from the temperature distribution.

Consider the tetrahedron **PXYZ** shown in Figure 1.5.   Assume that the line segments **PX**, **PY** and **PZ** are parallel to the x, y and z axes respectively, and that the normal $\underset{\sim}{n}$ to the face **XYZ** has components n_x, n_y, n_z. Let the area of **XYZ** be A, and the perpendicular distance of **P** from **XYZ** be d, so that the volume of the tetrahedron is $\frac{d}{3}A$. Projecting the surface **XYZ** onto the coordinate planes, we find that the area of the faces **PYZ**, **PXZ** and **PXZ** are respectively $n_x A$, $n_y A$ and $n_z A$. We denote the average heat fluxes across the surfaces **PYZ**, **PXZ**, **PXY** and **XYZ** by j_x, j_y, j_z and j. Then the net rate of heat flow into the tetrahedron **PXYZ** is

$$A\left(n_x j_x + n_y j_y + n_z j_z - j\right),$$

which we can equate to the rate of change of the average temperature T_{av} in the tetrahedron

$$A\left(n_x j_x + n_y j_y + n_z j_z - j\right) = \frac{d}{3}A\rho c\frac{\partial T_{av}}{\partial t},$$

so that

$$n_x j_x + n_y j_y + n_z j_z - j = \frac{d}{3}\rho c\frac{\partial T_{av}}{\partial t}. \tag{1.39}$$

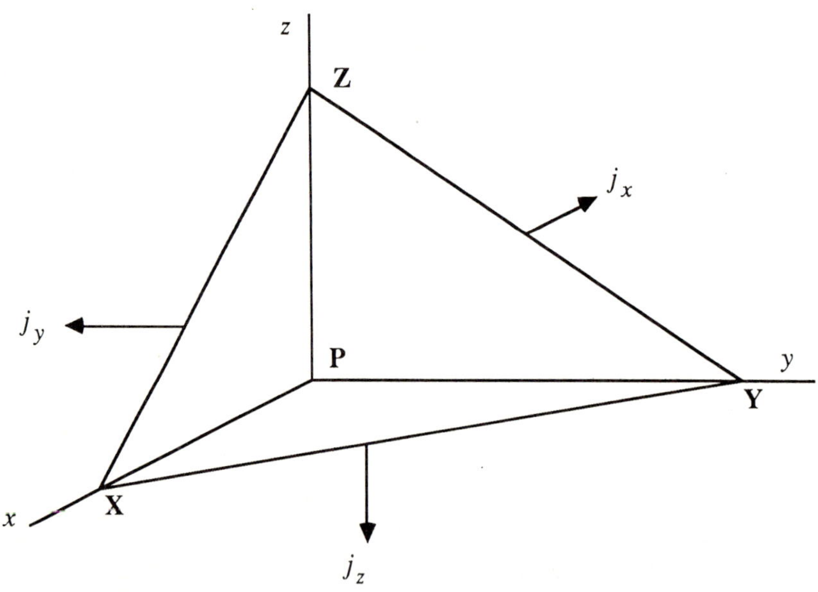

Figure 1.5. Elementary tetrahedron for heat flux vector.

If we let d approach zero, then the right hand side of (1.39) vanishes, j_x, j_y and j_z approach the heat fluxes at **P** across the three planes parallel to the coordinate planes, while j becomes the flux at **P** across the plane whose normal is (n_x, n_y, n_z). We can write this as

$$j = \underset{\sim}{j} \cdot \underset{\sim}{n}, \tag{1.40}$$

where $\underset{\sim}{j}$ is the heat flux vector, given by

$$\underset{\sim}{j} = (j_x, j_y, j_z), \tag{1.41}$$

and $\underset{\sim}{n} = (n_x, n_y, n_z)$ is the unit outward normal to the surface S at point **P**.

Equations (1.40) and (1.41) allow us to find the heat flux across a surface S at point **P** given the normal to S at **P** and the heat fluxes at **P** across the three planes parallel to the coordinate planes which pass through **P**. If we are considering a homogeneous isotropic material, a simple extension of the derivation of Fourier's law in one dimension, given in Section 1.2, shows that

$$j_x = -k\frac{\partial T}{\partial x}, \quad j_y = -k\frac{\partial T}{\partial y}, \quad j_z = -k\frac{\partial T}{\partial z},$$

where k is the thermal conductivity, which is generally a function of the temperature T. Thus we have the three dimensional form of Fourier's law (for an isotropic homogeneous material)

$$\underset{\sim}{j}(\underset{\sim}{x}, t) = -k(T)\underset{\sim}{\nabla} T(\underset{\sim}{x}, t), \tag{1.42}$$

where $\underset{\sim}{j}(\underset{\sim}{x}, t)$ denotes the heat flux vector at position $\underset{\sim}{x}$ and time t. Note that the heat flux vector is normal to the *isothermal surfaces*, that is the surfaces of constant temperature, at any time t. It follows that the heat flux at a point is greatest across a surface tangent to the isothermal surface at the point, and zero across a surface normal to the isothermal surface at the point.

To derive the three dimensional heat equation we consider an arbitrary volume V of isotropic material bounded by the surface S. The net rate at which heat is flowing into the volume is, in the absence of any internal generation of heat, the net rate at which it is flowing across the surface S

$$-\oint_S \underset{\sim}{j}(\underset{\sim}{x}, t) \cdot \underset{\sim}{n} \, \mathrm{d}S = \oint_S k(T)\underset{\sim}{\nabla} T(\underset{\sim}{x}, t) \cdot \underset{\sim}{n} \, \mathrm{d}S, \tag{1.43}$$

where $\underset{\sim}{n}$ denotes the unit outward normal to S, and the minus sign on the left hand side of (1.43) arises as we are considering the inward flow of heat. We can equate this with the rate of change of heat in the volume which is given by

$$\frac{d}{dt} \int_V Q \, dV = \int_V \rho c(T) \frac{\partial T}{\partial t} \, dV,$$

to obtain

$$\int_V \rho c(T) \frac{\partial T}{\partial t} \, dV = \oint_S k(T) \underline{\nabla} T \cdot \underline{n} \, dS. \tag{1.44}$$

Applying the divergence theorem to the surface integral in (1.44) gives

$$\int_V \left\{ \rho c(T) \frac{\partial T}{\partial t} - \underline{\nabla} \cdot (k(T) \underline{\nabla} T) \right\} dV = 0, \tag{1.45}$$

from which it follows that

$$\rho c(T) \frac{\partial T}{\partial t} = \underline{\nabla} \cdot (k(T) \underline{\nabla} T), \tag{1.46}$$

since equation (1.45) is valid for an arbitrary volume V. If we assume the specific heat $c(T)$ and thermal conductivity $k(T)$ to be constants, we obtain the classical three dimensional heat equation

$$\frac{\partial T}{\partial t} = \kappa \nabla^2 T, \tag{1.47}$$

where $\kappa = k/\rho c$ is the thermal diffusivity, and ∇^2 is *Laplace's operator*, which in cartesian coordinates (x, y, z) is given by

$$\nabla^2 = \frac{\partial^2}{\partial x^2} + \frac{\partial^2}{\partial y^2} + \frac{\partial^2}{\partial z^2}. \tag{1.48}$$

In cylindrical polar coordinates (r, θ, z) ∇^2 is given by

$$\nabla^2 = \frac{\partial^2}{\partial r^2} + \frac{1}{r} \frac{\partial}{\partial r} + \frac{1}{r^2} \frac{\partial^2}{\partial \theta^2} + \frac{\partial^2}{\partial z^2}, \tag{1.49}$$

while in spherical polar coordinates (r, θ, ϕ) we have

$$\nabla^2 = \frac{\partial^2}{\partial r^2} + \frac{2}{r} \frac{\partial}{\partial r} + \frac{1}{r^2} \frac{\partial^2}{\partial \theta^2} + \frac{\cot \theta}{r^2} \frac{\partial}{\partial \theta} + \frac{1}{r^2 \sin^2 \theta} \frac{\partial^2}{\partial \phi^2}. \tag{1.50}$$

In subsequent chapters we consider the conduction of heat in a body occupying the volume V and bounded by the surface S. An initial condition for the three dimensional heat equation consists of a prescription of the temperature T at every point $\underline{x}$ in V at some initial moment which we can take to be time zero. Thus our initial condition will have the form

$$T(\underline{x}, 0) = f(\underline{x}), \quad \underline{x} \in V, \tag{1.51}$$

where $f(\underline{x})$ is a known function for all $\underline{x} \in V$. The boundary conditions for the three dimensional heat problem consist of prescriptions of the behaviour of the temperature or temperature gradient on the surface S of the body. The boundary conditions considered in the previous section generalize to three dimensions as follows. The prescribed surface temperature condition is

$$T(\underline{x}, t) = g(\underline{x}, t), \quad \underline{x} \in S, \quad t > 0, \tag{1.52}$$

where $g(\underline{x}, t)$ is a known function defined for all $\underline{x} \in S$ and $t > 0$. The prescribed surface heat flux condition becomes

$$k \frac{\partial T}{\partial n} = h(\underline{x}, t), \quad \underline{x} \in S, \quad t > 0, \tag{1.53}$$

where $\frac{\partial T}{\partial n} = \underline{\nabla} T \cdot \underline{n}$ denotes the normal derivative of T in the outward direction and $h(\underline{x}, t)$ is a known function for all $\underline{x} \in S$ and $t > 0$, which represents the rate at which heat is entering V. In particular, for an insulated body we have

$$\frac{\partial T}{\partial n}(\underline{x}, t) = 0, \quad \underline{x} \in S, \quad t > 0. \tag{1.54}$$

A Newton cooling boundary condition at the surface S has the form

$$k \frac{\partial T}{\partial n}(\underline{x}, t) + h(T(\underline{x}, t) - T_0(\underline{x}, t)) = 0, \quad \underline{x} \in S, \quad t > 0, \tag{1.55}$$

where $T_0(\underline{x}, t)$ represents the temperature of the surroundings, and h is the surface conductance, assumed constant. The genuine radiation condition has the form

$$k \frac{\partial T}{\partial n} + \sigma E\left(T^4(\underline{x}, t) - T_0^4(\underline{x}, t)\right) = 0, \quad \underline{x} \in S, \quad t > 0, \tag{1.56}$$

where again σ is the Stefan-Boltzmann constant, E is the surface emissivity and $T_0(\underline{x}, t)$ the temperature of the surroundings. To obtain physically meaningful results, there should be no heat flux normal to any portions of S which lie at infinity. The first four boundary conditions (1.52), (1.53), (1.54) and (1.55) also occur in the mathematical theory of diffusion, as noted in the previous section.

Finally we consider the three and two dimensional source solutions which we obtain as a product of one dimensional source solutions. The three dimensional source solution arises as the solution of the problem

$$
\left.
\begin{aligned}
\frac{\partial T}{\partial t} &= \kappa\left(\frac{\partial^2 T}{\partial x^2} + \frac{\partial^2 T}{\partial y^2} + \frac{\partial^2 T}{\partial z^2}\right), \\[2mm]
T(\underline{x}, t) &\to 0 \text{ as } |\underline{x}| \to \infty, \\[2mm]
T(\underline{x}, 0) &= q_0 \delta(\underline{x}).
\end{aligned}
\right\}
\tag{1.57}
$$

As in the one dimensional case, the source solution represents the temperature distribution due to an quantity of heat $\rho c q_0$ being released at the origin at time zero. The three dimensional delta function $\delta(\underline{x})$ is defined by

$$
\delta(\underline{x}) = 0 \text{ if } \underline{x} \neq \underline{0}, \quad \text{and} \quad \int_{R^3} \delta(\underline{x})\phi(\underline{x})\,dV = \phi(\underline{0}),
\tag{1.58}
$$

where the integral is taken over all of three dimensional space and the test functions $\phi(\underline{x})$ are infinitely differentiable functions which vanish outside some closed bounded region. It follows that

$$
\delta(\underline{x}) = \delta(x)\delta(y)\delta(z),
\tag{1.59}
$$

so we may assume the solution $T(\underline{x}, t)$ of (1.57) is also a product of three functions

$$
T(\underline{x}, t) = T_1(x, t)\,T_2(y, t)\,T_3(z, t),
\tag{1.60}
$$

where each of the $T_i(x, t)$ $(i = 1, 2, 3)$ satisfies the equations (1.9) for the one dimensional source solution. Thus, the three dimensional source solution is simply a product of three one dimensional source solutions, namely

$$
T(\underline{x}, t) = \frac{q_0}{8(\pi \kappa t)^{3/2}} \exp \frac{-(x^2 + y^2 + z^2)}{4\kappa t}.
\tag{1.61}
$$

Similarly the two dimensional source solution arises as the solution of

$$
\left.
\begin{aligned}
\frac{\partial T}{\partial t} &= \kappa\left(\frac{\partial^2 T}{\partial x^2} + \frac{\partial^2 T}{\partial y^2}\right), \\[2mm]
T(x, y, t) &\to 0 \text{ as } |x|, |y| \to \infty, \\[2mm]
T(x, y, 0) &= q_0 \delta(x)\delta(y),
\end{aligned}
\right\}
\tag{1.62}
$$

and is similarly the product of two one dimensional source solutions, that is

$$
T(x, y, t) = \frac{q_0}{4\pi \kappa t} \exp \frac{-(x^2 + y^2)}{4\kappa t}.
\tag{1.63}
$$

For $q_0 = 1$ and $\kappa = 1$ this is shown graphically in Figure 1.6 at times $t = 0.050$, 0.067, 0.083 and 0.100.

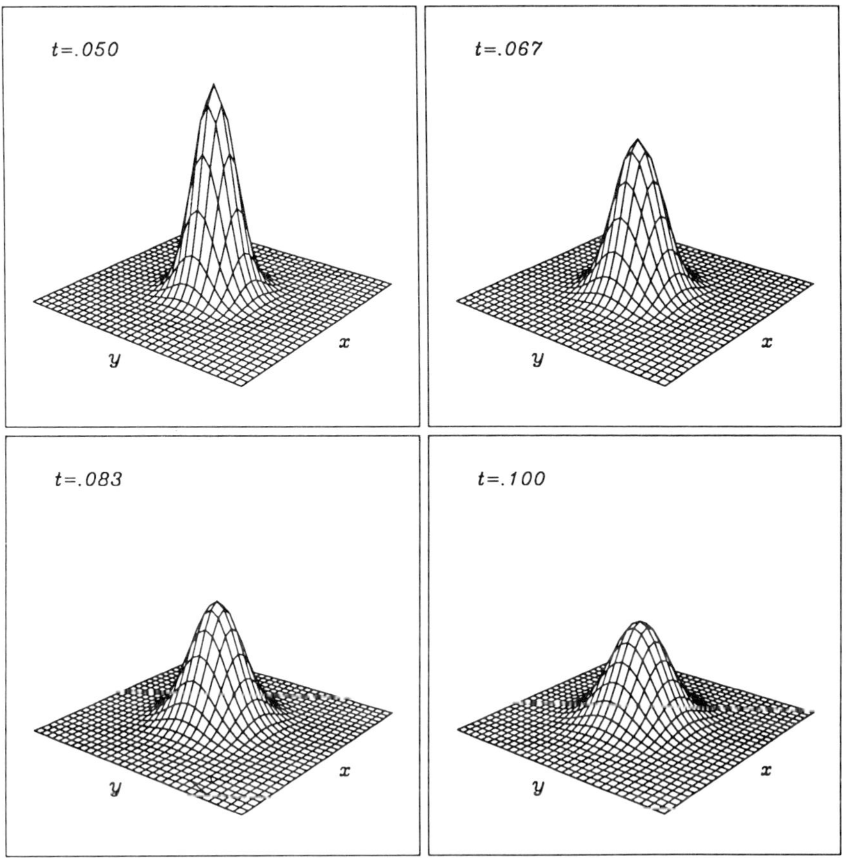

Figure 1.6. Two dimensional source solution (1.63) with q_0 and κ unity for four values of time.

PROBLEMS

1.   Show that in the limit $\epsilon \to 0$ the functions $f_\epsilon(x)$ defined by

$$\textbf{(i)} \quad f_\epsilon(x) = \begin{cases} \dfrac{1}{\epsilon}, & |x| \le \dfrac{\epsilon}{2}, \\ 0, & |x| > \dfrac{\epsilon}{2}, \end{cases}$$

$$\textbf{(ii)} \quad f_\epsilon(x) = \begin{cases} \dfrac{1}{\epsilon}(1 - \dfrac{|x|}{\epsilon}), & |x| \le \epsilon, \\ 0, & |x| > \epsilon, \end{cases}$$

each converge to the one dimensional Dirac delta function. Sketch these functions and consider their behaviour as $\epsilon \to 0$.

2.   Let $\psi(x)$ be any positive, integrable function such that

$$\int_{-\infty}^{\infty} \psi(\xi)\, d\xi = 1,$$

and define $f_\epsilon(x)$ for $\epsilon > 0$ by

$$f_\epsilon(x) = \epsilon \psi\left(\frac{x}{\epsilon}\right).$$

Show that

$$\textbf{(i)} \qquad \lim_{\epsilon \to 0} f_\epsilon(x) = 0 \text{ for any fixed } x \ne 0,$$

$$\textbf{(ii)} \qquad \lim_{\epsilon \to 0} \int_{-\infty}^{\infty} f_\epsilon(\xi)\phi(\xi)\, d\xi = \phi(0),$$

for any test function ϕ. $\Big[$Hint, recall that any test function vanishes outside some closed bounded interval.$\Big]$

3.   Show that

$$\int_{-\infty}^{\infty} e^{-\xi^2}\, d\xi = \sqrt{\pi}.$$

$\Big[$Hint, take the square of the integral, write as a double integral and evaluate this using polar coordinates.$\Big]$ Hence, by choosing an appropriate contour in the complex plane, show that for any real ω

$$\int_{-\infty+i\omega}^{\infty+i\omega} e^{-\xi^2}\, d\xi = \sqrt{\pi}.$$

4. Note that the equations (1.62) for the two dimensional source solution remain invariant under rotation about the origin, so that in polar coordinates (r, θ) these equations become

$$\frac{\partial T}{\partial t} = \kappa\left(\frac{\partial^2 T}{\partial r^2} + \frac{1}{r}\frac{\partial T}{\partial r}\right),$$

$$T(r,t) \to 0 \text{ as } r \to \infty, \quad T(r,0) = \frac{q_0}{2\pi}\frac{\delta(r)}{r}.$$

Using these equations, derive the two dimensional source solution (1.63). $\Big[$Hint, make use of the fact that this problem remains invariant under the transformation

$$r^* = \lambda r, \quad t^* = \lambda^2 t, \quad T^* = T/\lambda^2,$$

so that

$$\frac{r^*}{\sqrt{t^*}} = \frac{r}{\sqrt{t}}, \quad T^* t^* = Tt,$$

and therefore $Tt = \psi(r/\sqrt{t})$ for some function $\psi(\zeta)$.$\Big]$

5. Continuation. In spherical polar coordinates (r, θ, ϕ) deduce the three dimensional source solution $T(r, t)$ as a solution of

$$\frac{\partial T}{\partial t} = \kappa\left(\frac{\partial^2 T}{\partial r^2} + \frac{2}{r}\frac{\partial T}{\partial r}\right),$$

$$T(r,t) \to 0 \text{ as } r \to \infty, \quad T(r,0) - \frac{q_0}{4\pi}\frac{\delta(r)}{r^2},$$

$\Big[$Hint, modify the method of the previous question.$\Big]$

6. Show that the transformation $T = w/r$ reduces the spherically symmetric heat equation

$$\frac{\partial T}{\partial t} = \kappa\left(\frac{\partial^2 T}{\partial r^2} + \frac{2}{r}\frac{\partial T}{\partial r}\right),$$

to the one dimensional heat equation

$$\frac{\partial w}{\partial t} = \kappa\frac{\partial^2 w}{\partial r^2}.$$

7. If $\Psi(x,t)$ is a solution of the one dimensional heat equation

$$\frac{\partial T}{\partial t} = \kappa \frac{\partial^2 T}{\partial x^2},$$

show that another solution is given by

$$T(x,t) = \frac{1}{\sqrt{t}} \Psi\left(\frac{x}{t}, \frac{-1}{t}\right) \exp\left(\frac{-x^2}{4\kappa t}\right).$$

(The transformation $\Psi(x,t) \to T(x,t)$ is known as the one dimensional Appell transformation.)

8. Continuation. If $\Psi(x,y,z,t)$ is a solution of

$$\frac{\partial T}{\partial t} = \kappa\left(\frac{\partial^2 T}{\partial x^2} + \frac{\partial^2 T}{\partial y^2} + \frac{\partial^2 T}{\partial z^2}\right),$$

show that another solution is given by

$$T(x,y,z,t) = \frac{1}{t^{3/2}} \Psi\left(\frac{x}{t}, \frac{y}{t}, \frac{z}{t}, \frac{-1}{t}\right) \exp \frac{-(x^2 + y^2 + z^2)}{4\kappa t}.$$

(This is the three dimensional Appell transformation.)

9. A temperature which does not vary with time t is known as a steady temperature, $T(\underline{x})$, and satisfies Laplace's equation

$$\nabla^2 T = 0,$$

since the time partial derivative in the heat equation vanishes. Show that the steady one dimensional temperature distribution in the region bounded by the planes $x = 0$ and $x = \ell$, which are held at constant temperatures v_0 and v_ℓ respectively, is given by

$$T(x) = \frac{1}{\ell}[x v_\ell + (\ell - x) v_0].$$

10.   Continuation. Show that the steady state axially symmetric temperature distribution in the region between two concentric circular cylinders of radii a and b $(a < b)$ which are maintained at constant temperatures v_a and v_b respectively, is given by

$$T(r) = \frac{[v_a \log(b/r) + v_b \log(r/a)]}{\log(b/a)}.$$

11. Continuation. Show that the steady state temperature distribution in the region bounded by two concentric spheres of radii a and b $(a < b)$ which are maintained at constant temperatures v_a and v_b respectively, is given by

$$T(r) = \frac{\left[v_a\left(\frac{1}{b} - \frac{1}{r}\right) + v_b\left(\frac{1}{r} - \frac{1}{a}\right)\right]}{\left(\frac{1}{b} - \frac{1}{a}\right)}.$$

12. Suppose that in addition to heat conduction there is some mechanism producing body heating, such as an electric current passing through the body, a fluctuating magnetic field or an internal chemical reaction. Assume that this gives rise to $q(\underset{\sim}{x}, t)\,dV$ units of heat being generated per unit time in the elementary volume dV containing the point $\underset{\sim}{x}$. Assuming constant specific heat c and thermal conductivity k, modify the argument used in Section 1.6 to obtain

$$\frac{\partial T}{\partial t} = \kappa \nabla^2 T + \frac{q}{\rho c},$$

as the equation governing the flow of heat in the presence of body heat sources.

Chapter Two

Mathematical preliminaries

2.1 Introduction

In this chapter we consider the general mathematical results and methods necessary for the linear heat equation (1.47). We first note some results from the general theory of linear partial differential equations in order that the reader may see the heat equation (1.47) in this more general context. In this section we use the notation $\underline{x} = (x_0, x_1, \ldots, x_n)$ to denote a vector in $(n + 1)$-dimensional space. Where appropriate, for the heat and wave equations, we may regard x_0 as being the time variable and $x_1, \ldots, x_n$ as spatial variables. The general second order homogeneous linear partial differential equation in $(n + 1)$ variables is

$$\sum_{i=0}^{n} \sum_{j=0}^{n} A_{ij} \frac{\partial^2 w}{\partial x_i \partial x_j} + \sum_{i=0}^{n} B_i \frac{\partial w}{\partial x_i} + Cw = 0, \tag{2.1}$$

where the coefficients A_{ij}, B_i and C may be functions of $\underline{x} = (x_0, x_1, \ldots, x_n)$ but not of w nor any of its partial derivatives. This equation is homogeneous because the left hand side equals zero, and we call (2.1) *linear* because the partial derivatives of w occur linearly, and in particular if $w_1(\underline{x})$, $w_2(\underline{x})$, $\ldots$, $w_n(\underline{x})$ are solutions of (2.1), then so is any linear combination

$$w(\underline{x}) = \alpha_1 w_1(\underline{x}) + \alpha_2 w_2(\underline{x}) + \ldots + \alpha_n w_n(\underline{x}),$$

where $\alpha_1, \alpha_2, \ldots, \alpha_n$ are real constants. If we have a sequence of solutions of (2.1), say $(w_n(\underline{x}))$, then the series

$$\sum_{n=1}^{\infty} \alpha_n w_n(\underline{x}),$$

where (α_n) is a real sequence, is also a solution provided that it converges. More generally if $\{ w(\underline{x}, \xi) \}$ is a family of solutions of (2.1) indexed by the continuous variable ξ then the integral

$$\int \alpha(\xi) w(\underline{x}, \xi) \, d\xi,$$

where $\alpha(\xi)$ is a real function, is also a solution of (2.1) assuming the integral converges. In the following, we exploit all three of these features.

We assume that $A_{ij} = A_{ji}$ in (2.1), (without loss of generality assuming $\frac{\partial^2 w}{\partial x_i \partial x_j} = \frac{\partial^2 w}{\partial x_j \partial x_i}$), so that the matrix $\left(A_{ij}\right)$ of the coefficients of the second order partial derivatives in (2.1) is symmetric, and as such has $(n + 1)$ real eigenvalues (including multiple eigenvalues). In general these eigenvalues are functions of $\underline{x}$. We consider three special cases. Equation (2.1) is said to be *parabolic* at a point if one of the eigenvalues vanishes at that point. If none of the eigenvalues vanish and all have the same sign at a point, equation (2.1) is said to be *elliptic* at that point. If all the eigenvalues are non-zero but one is of different sign to all the others at a point, equation (2.1) is said to be *hyperbolic* at that point. Equation (2.1) is parabolic, elliptic or hyperbolic in a region if it is parabolic, elliptic or hyperbolic at every point in that region and is called parabolic, elliptic or hyperbolic if it is everywhere parabolic, elliptic or hyperbolic. The heat equation

$$\frac{\partial w}{\partial x_0} - \kappa \left(\frac{\partial^2 w}{\partial x_1^2} + \frac{\partial^2 w}{\partial x_2^2} + \cdots + \frac{\partial^2 w}{\partial x_n^2} \right) = 0,$$

in n spatial variables, where $\kappa > 0$, is the classical parabolic partial differential equation. The archetypal elliptic partial differential equation is Laplace's equation in $(n + 1)$ variables

$$\frac{\partial^2 w}{\partial x_0^2} + \frac{\partial^2 w}{\partial x_1^2} + \cdots + \frac{\partial^2 w}{\partial x_n^2} = 0,$$

while the wave equation in n spatial variables

$$\frac{\partial^2 w}{\partial x_0^2} - \omega \left(\frac{\partial^2 w}{\partial x_1^2} + \frac{\partial^2 w}{\partial x_2^2} + \cdots + \frac{\partial^2 w}{\partial x_n^2} \right) = 0,$$

where $\omega > 0$, is the typical hyperbolic equation.

The so called *characteristic surfaces* or characteristics associated with (2.1) are surfaces given by $\gamma(\underline{x}) = 0$, where $\gamma(\underline{x})$ satisfies the non-linear partial differential equation

$$\sum_{i=0}^{n} \sum_{j=0}^{n} A_{ij} \frac{\partial \gamma}{\partial x_i} \frac{\partial \gamma}{\partial x_j} = 0. \tag{2.2}$$

These surfaces are important in the context of the *Cauchy problem* for equation (2.1), that is, the solution of equation (2.1) subject to the prescription

$$w(\underline{x}) = f(\underline{x}), \quad \frac{\partial w}{\partial n}(\underline{x}) = g(\underline{x}), \quad \underline{x} \in S,$$

for the function $w(\underline{x})$ and its normal derivative on a given surface S. This is always possible provided the surface S is not a characteristic surface. If the surface S is characteristic, then it is sufficient to specify $w(\underline{x})$ alone on the surface S to determine $w(\underline{x})$ in the neighbourhood of S. When the surface S is characteristic, the term Cauchy problem refers to the equation (2.1) subject to the specification of $w(\underline{x})$ on S.

Finally in this section we note the celebrated *Cauchy-Kowalevski theorem* which asserts the existence and uniqueness of solutions of the Cauchy problem for equation (2.1) under quite general conditions. Unfortunately this theorem is of limited application to heat conduction problems for reasons outlined in the next section. We consider a point $\underline{x}_0$ on a surface S in $(n + 1)$ dimensional space. The surface S is taken to be analytic in the neighbourhood of $\underline{x}_0$ by which we mean that S is given by $\gamma(\underline{x}) = 0$ for some function $\gamma(\underline{x})$ which is analytic at $\underline{x}_0$, and such that $\underline{\nabla}\gamma(\underline{x}_0) \neq \underline{0}$. The coefficients A_{ij}, B_i and C occurring in (2.1) are taken to be analytic at $\underline{x}_0$. Then, given prescribed values of $w(\underline{x})$ and its normal derivative $\frac{\partial w}{\partial n}$ on S which are analytic about $\underline{x}_0$, the Cauchy-Kowalevski theorem asserts the local existence of a unique analytic solution of (2.1) in some neighbourhood of $\underline{x}_0$ which agrees with the prescribed analytic data on S provided the surface S is not characteristic at $\underline{x}_0$. The assumed analyticity of S guarantees that if S is not characteristic at $\underline{x}_0$ then it is not characteristic in a neighbourhood of $\underline{x}_0$. In the remainder of this chapter we focus on the linear heat equations (1.8) and (1.47).

2.2 The existence of solutions of the heat equation

Our common sense and physical intuition tell us that, if correctly formulated, a heat conduction or diffusion problem should have a solution. This is what we are attempting to show when we address the mathematical problem of the existence of solutions of the heat equation. A detailed discussion of the existence of solutions of general heat conduction problems is well beyond the scope of this text and our general strategy is to show the existence of solutions of particular heat conduction problems by actually constructing them explicitly, that is, by solving the problem. For problems where the Cauchy-Kowalevski theorem applies the situation is simpler, for we have only to invoke this theorem to prove the existence and uniqueness of solutions, albeit local solutions. Unfortunately this theorem requires the specification of analytic data on a non-characteristic surface, and for the heat equation, in any number of spatial variables, it is a simple matter to show that the only characteristic surfaces are given by the equation $t = $ constant (see Problem 1). In particular, the surface $t = 0$ is characteristic and therefore the Cauchy-Kowalevski theorem does not apply to the heat equation if an initial condition is specified, so other methods must be used.

As an example of a situation where the Cauchy-Kowalevski theorem does apply to the heat equation, we consider the following problem

$$\left.\begin{array}{c} \dfrac{\partial T}{\partial t} = \kappa \dfrac{\partial^2 T}{\partial x^2}, \\[2ex] T(0,t) = g(t), \quad k\dfrac{\partial T}{\partial x}(0,t) = h(t), \end{array}\right\} \tag{2.3}$$

where the functions $g(t)$ and $h(t)$ are analytic about $t = 0$. Here $h(t)$ represents the heat flow at $x = 0$ in the direction of increasing x. Since in (x,t) space the plane $x = 0$ is non-characteristic, there exists a unique analytic solution of (2.3) in some neighbourhood of the origin $x = 0$, $t = 0$. On integrating the one dimensional heat equation with respect to x we find that

$$\kappa\frac{\partial T}{\partial x}(x,t) - \kappa\frac{\partial T}{\partial x}(0,t) = \int_0^x \frac{\partial T}{\partial t}(\xi,t)\,d\xi,$$

and on using the prescribed flux $h(t)$ for $k\frac{\partial T}{\partial x}(0,t)$ and interchanging the order of integration and time partial differentiation we have

$$\frac{\partial T}{\partial x}(x,t) = \frac{1}{k}h(t) + \frac{1}{\kappa}\frac{\partial}{\partial t}\int_0^x T(\xi,t)\,d\xi.$$

A further integration and substitution of $g(t)$ for $T(0,t)$ gives

$$T(x,t) = g(t) + \frac{x}{k}h(t) + \frac{1}{\kappa}\frac{\partial}{\partial t}\int_0^x\int_0^\eta T(\xi,t)\,d\xi\,d\eta,$$

which becomes

$$T(x,t) = g(t) + \frac{x}{k}h(t) + \frac{1}{\kappa}\frac{\partial}{\partial t}\int_0^x (x-\xi)T(\xi,t)\,d\xi, \tag{2.4}$$

on changing the order of integration in the double integral. If we now substitute this expression for $T(x,t)$ into the left hand side of itself we obtain

$$T(x,t) = g(t) + \frac{x}{k}h(t) + \frac{dg}{dt}(t)\frac{x^2}{2\kappa} + \frac{1}{k}\frac{dh}{dt}(t)\frac{x^3}{6\kappa}$$

$$+ \frac{1}{\kappa^2}\frac{\partial^2}{\partial t^2}\int_0^x\int_0^\xi (x-\xi)(\xi-\eta)T(\eta,t)\,d\eta\,d\xi.$$

Repeating this process indefinitely and assuming that the remainder term vanishes we find that

$$T(x,t) = \sum_{n=0}^{\infty} g^{(n)}(t)\frac{x^{2n}}{\kappa^n(2n)!} + \frac{1}{k}\sum_{n=0}^{\infty} h^{(n)}(t)\frac{x^{2n+1}}{\kappa^n(2n+1)!}, \tag{2.5}$$

is a formal solution of (2.3), where $g^{(n)}(t)$ and $h^{(n)}(t)$ denote the nth derivatives of $g(t)$ and $h(t)$ respectively. Since the functions $g(t)$ and $h(t)$ are analytic at $t = 0$, it can be shown that (2.5) converges absolutely in a neighbourhood of the origin $x = 0$, $t = 0$, and is therefore the analytic solution of (2.3) whose existence is asserted by the Cauchy-Kowalevski theorem. While the formal solution (2.5) is of some mathematical interest, and does, for example, allow us to deduce Tychonoff's null solution (see the following section), problems of the form (2.3) are of limited practical importance since most heat conduction problems involve an initial condition.

We now consider the three dimensional heat equation with the temperature specified on the entire characteristic $t = 0$, that is, the initial value problem

$$\frac{\partial T}{\partial t} = \kappa \nabla^2 T, \quad T(\underset{\sim}{x},0) = f(\underset{\sim}{x}). \tag{2.6}$$

For brevity we introduce the following notation for the one, two and three dimensional unit source solutions respectively

$$K_1(x,t) = \frac{1}{2\sqrt{\pi \kappa t}} \exp\left(\frac{-x^2}{4\kappa t}\right), \quad t > 0,$$

$$K_2(\underset{\sim}{x},t) = \frac{1}{4\pi \kappa t} \exp\left(\frac{-|\underset{\sim}{x}|^2}{4\kappa t}\right), \quad t > 0, \tag{2.7}$$

$$K_3(\underset{\sim}{x},t) = \frac{1}{8(\pi \kappa t)^{3/2}} \exp\left(\frac{-|\underset{\sim}{x}|^2}{4\kappa t}\right), \quad t > 0.$$

Clearly the temperature at position $\underset{\sim}{x}$ and time t due to an instantaneous initial point source of strength q_0 located at the point $\underset{\sim}{\xi}$ is given by $q_0 K_3(\underset{\sim}{x} - \underset{\sim}{\xi},t)$. We can regard the initial condition in (2.6) as a continuous distribution of point sources, the strength of the source at point $\underset{\sim}{\xi}$ being $f(\underset{\sim}{\xi})\,dV_{\underset{\sim}{\xi}}$ where $dV_{\underset{\sim}{\xi}}$ denotes the elementary volume containing $\underset{\sim}{\xi}$. Thus, exploiting the linearity of the heat equation, we might expect that the solution of (2.6) could be found by integrating the contributions from each initial point source

$$f(\underset{\sim}{\xi})K_3(\underset{\sim}{x} - \underset{\sim}{\xi},t)\,dV_{\underset{\sim}{\xi}},$$

over all space, that is, that the solution of (2.6) is given by

$$T(\underline{x},t) = \int_{R^3} f(\underline{\xi})K_3(\underline{x} - \underline{\xi},t)\,dV_{\underline{\xi}}, \qquad (2.8)$$

where $dV_{\underline{\xi}}$ denotes integration with respect to $\underline{\xi}$. This is indeed the case, and (2.8) is the solution of (2.6) subject to quite mild restrictions on $f(\underline{x})$. Specifically, if $f(\underline{x})$ is a measurable function (in the sense of Lebesgue) and there are positive constants ϵ and M such that

$$|f(\underline{x})| \le Me^{\epsilon|\underline{x}|^2},$$

for all $\underline{x}$, then $T(\underline{x},t)$ given by (2.8) is infinitely differentiable in both t and $\underline{x}$ and satisfies the heat equation for all time $0 < t < 1/4\kappa\epsilon$. Moreover, $\lim_{t \to 0^+} T(\underline{x},t) = f(\underline{x})$ almost everywhere, that is, the set of $\underline{x}$ where $\lim_{t \to 0^+} T(\underline{x},t) \neq f(\underline{x})$ can be enclosed in a family of open sets of arbitrarily small total volume. In particular, $\lim_{t \to 0^+} T(\underline{x},t) = f(\underline{x})$ at any point of continuity of $f(\underline{x})$. The solution (2.8) is sometimes called the three dimensional *Poisson transform* of $f(\underline{x})$. Similar formulae apply for the one and two dimensional heat equation, with $K_1(x,t)$ and $K_2(\underline{x},t)$ respectively replacing $K_3(\underline{x},t)$.

The proof of this theorem is beyond the scope of this book, requiring some theorems concerning the Lebesgue integral. We prove the following weaker theorem for the one dimensional heat equation with bounded initial temperature. The proof generalizes in a straightforward manner to two and three dimensional problems, with $K_2(\underline{x},t)$ and $K_3(\underline{x},t)$ respectively replacing $K_1(x,t)$, and the reader familiar with the Lebesgue integral should attempt to sketch the proof of the more general theorem using the following as a guide.

<u>Theorem 2.1</u> Let $f(x)$ be a bounded measurable function defined for all real x. Then the one dimensional Poisson transform of $f(x)$

$$T(x,t) = \int_{-\infty}^{\infty} f(\xi)K_1(x - \xi,t)\,d\xi, \qquad t > 0, \qquad (2.9)$$

where $K_1(x,t)$ is the one dimensional unit source solution $(2.7)_1$, is infinitely differentiable in both x and t and satisfies the one dimensional heat equation for all $t > 0$ and moreover $\lim_{t \to 0^+} T(x,t) = f(x)$ for all points x at which $f(x)$ is continuous.

Since $f(x)$ is bounded, there is some finite positive real number M such that $|f(x)| \leq M$ for all x. Making the substitution $\eta = (x - \xi)/2\sqrt{\kappa t}$ in the integral (2.9) we have

$$\left| \int_{-\infty}^{\infty} f(\xi)K_1(x - \xi, t)\,d\xi \right| = \left| \frac{1}{\sqrt{\pi}} \int_{-\infty}^{\infty} f(x - 2\sqrt{\kappa t}\,\eta)e^{-\eta^2}\,d\eta \right|$$

$$\leq \frac{1}{\sqrt{\pi}} \int_{-\infty}^{\infty} |f(x - 2\sqrt{\kappa t}\,\eta)|e^{-\eta^2}\,d\eta$$

$$\leq \frac{M}{\sqrt{\pi}} \int_{-\infty}^{\infty} e^{-\eta^2}\,d\eta = M.$$

Thus the integral $\int_{-\infty}^{\infty} K_1(x - \xi, t)f(\xi)\,d\xi$ is absolutely convergent for all $t > 0$, allowing us to differentiate under the integral sign. Since $K_1(x - \xi, t)$ is infinitely differentiable in both x and t and satisfies the heat equation for $t > 0$, it follows that $T(x, t)$ shares these properties. Finally we note that

$$\lim_{t \to 0^+} T(x, t) = \lim_{t \to 0^+} \frac{1}{\sqrt{\pi}} \int_{-\infty}^{\infty} f(x - 2\sqrt{\kappa t}\,\eta)e^{-\eta^2}\,d\eta,$$

and since the integral is absolutely convergent, we can take the limit inside the integral, and use the continuity of $f(x)$, that is $\lim_{\epsilon \to 0} f(x + \epsilon) = f(x)$, to show that

$$\lim_{t \to 0^+} T(x, t) = \frac{1}{\sqrt{\pi}} \int_{-\infty}^{\infty} \lim_{t \to 0^+} f(x - 2\sqrt{\kappa t}\,\eta)e^{-\eta^2}\,d\eta$$

$$= \frac{1}{\sqrt{\pi}} \int_{-\infty}^{\infty} f(x)e^{-\eta^2}\,d\eta$$

$$= f(x),$$

at any point x at which $f(x)$ is continuous.

<u>Example 2.1</u> Solve the one dimensional heat conduction problem where initially the temperature is unity on the half-line $x \geq 0$ and zero on the half-line $x < 0$, that is, the problem

$$\left. \begin{array}{l} \dfrac{\partial T}{\partial t} = \kappa \dfrac{\partial^2 T}{\partial x^2}, \quad -\infty < x < \infty, \\[1.5em] T(x, 0) = \begin{cases} 0, & x < 0, \\ 1, & x \geq 0. \end{cases} \end{array} \right\} \tag{2.10}$$

We determine the solution of (2.10) by taking the Poisson transform of $f(x)$ to find

$$T(x,t) = \int_0^\infty \frac{1}{2\sqrt{\pi \kappa t}} \exp \frac{-(x-\xi)^2}{4\kappa t}\, d\xi$$

$$= \frac{1}{\sqrt{\pi}} \int_{-\infty}^{x/2\sqrt{\kappa t}} e^{-\eta^2}\, d\eta$$

$$= \frac{1}{2} + \frac{1}{\sqrt{\pi}} \int_0^{x/2\sqrt{\kappa t}} e^{-\eta^2}\, d\eta,$$

where we have made the substitution $\eta = (x-\xi)/2\sqrt{\kappa t}$ and used the fact that $\int_{-\infty}^0 e^{-\eta^2}\, d\eta = \frac{1}{2}\sqrt{\pi}$.

If we introduce the *error function* erf(x) which is defined by

$$\text{erf}(x) = \frac{2}{\sqrt{\pi}} \int_0^x e^{-\eta^2}\, d\eta, \tag{2.11}$$

then the solution of (2.10) can be written

$$T(x,t) = \frac{1}{2}\left[1 + \text{erf}\left(\frac{x}{2\sqrt{\kappa t}}\right)\right]. \tag{2.12}$$

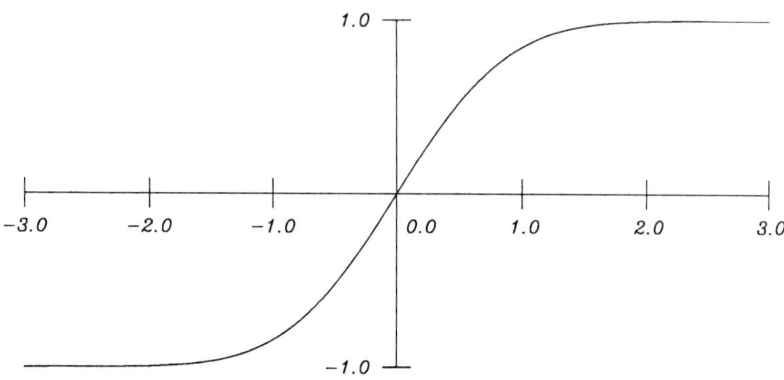

Figure 2.1. Graph of erf(x), (2.11).

The reader should verify from the definition (2.11) that $\text{erf}(x)$ is a monotonically increasing continuous function of x with the following properties (see Figure 2.1)

(i) $\text{erf}(-x) = -\text{erf}(x)$,   **(ii)** $\text{erf}(0) = 0$,   **(iii)** $\text{erf}(\infty) = 1$.

A function related to the error function is the *complementary error function*, $\text{erfc}(x)$, defined by

$$\text{erfc}(x) = \frac{2}{\sqrt{\pi}} \int_x^\infty e^{-\eta^2}\, d\eta = 1 - \text{erf}(x). \tag{2.13}$$

Thus $\text{erfc}(x)$ is a monotonically decreasing continuous function of x with the properties (see Figure 2.2)

(i) $\text{erfc}(-\infty) = 2$,   **(ii)** $\text{erfc}(0) = 1$,   **(iii)** $\text{erfc}(\infty) = 0$.

We now consider some features of the solution (2.12). In the limit $t \to 0^+$ we have

$$\lim_{t \to 0^+} T(x,t) = \frac{1}{2} + \lim_{t \to 0^+} \frac{1}{2}\,\text{erf}\!\left(\frac{x}{2\sqrt{\kappa t}}\right) = \begin{cases} 1, & x > 0, \\ \dfrac{1}{2}, & x = 0, \\ 0, & x < 0, \end{cases}$$

since $\lim_{t \to 0^+} x/2\sqrt{\kappa t}$ is infinite if $x > 0$, zero if x is zero and minus infinity if $x < 0$. So we have $\lim_{t \to 0^+} T(x,t) = f(x)$ except at the point $x = 0$ where $f(x)$ is discontinuous.

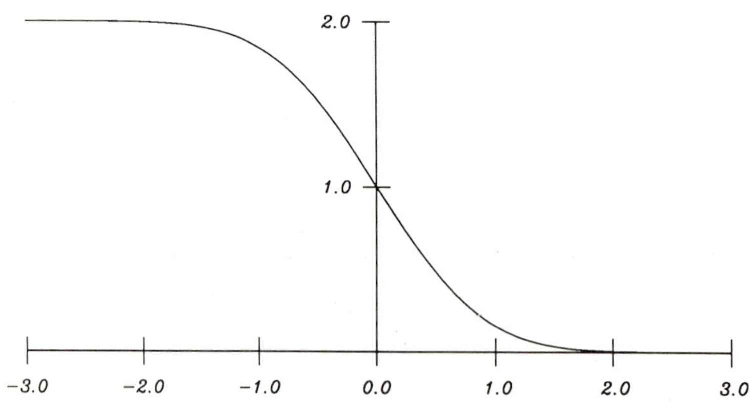

Figure 2.2. Graph of $\text{erfc}(x)$, (2.13).

At this point we have $\lim_{t \to 0^+} T(0,t) = \frac{1}{2}\left[f(0^+) + f(0^-)\right]$, which illustrates a general feature of the one dimensional Poisson transform (2.9), namely that if the initial temperature $f(x)$ has a finite jump discontinuity at x_0 then

$$\lim_{t \to 0^+} T(x_0,t) = \frac{1}{2}\left[f(x_0^+) + f(x_0^-)\right].$$

We also note that the solution (2.12) satisfies the inequality $0 \le T(x,t) \le 1$ for all x and $t > 0$, which means that $T(x,t)$ achieves its maximum and minimum values at time zero. Finally we note that as time t tends to infinity, the solution (2.12) tends to the 'average' temperature $\frac{1}{2}$.

<u>Example 2.2</u> Solve the three dimensional heat equation subject to an initial temperature which is unity in the first quadrant and zero elsewhere, namely

$$\frac{\partial T}{\partial t} = \kappa \nabla^2 T,$$

$$T(\underline{x},0) = \begin{cases} 1, & x,y,z > 0, \\ 0, & \text{otherwise.} \end{cases}$$

We solve this problem using the three dimensional Poisson transform, so that

$T(\underline{x},t)$

$$= \int_0^\infty \int_0^\infty \int_0^\infty \frac{1}{8(\kappa\pi t)^{3/2}} \exp\frac{-(x-\xi)^2}{4\kappa t} \exp\frac{-(y-\eta)^2}{4\kappa t} \exp\frac{-(z-\zeta)^2}{4\kappa t} \, d\xi \, d\eta \, d\zeta$$

$$= \frac{1}{\pi^{3/2}}\left(\int_{-\infty}^{x/2\sqrt{\kappa t}} e^{-\sigma^2}\, d\sigma\right)\left(\int_{-\infty}^{y/2\sqrt{\kappa t}} e^{-\sigma^2}\, d\sigma\right)\left(\int_{-\infty}^{z/2\sqrt{\kappa t}} e^{-\sigma^2}\, d\sigma\right)$$

$$= \frac{1}{8}\left[1 + \text{erf}\left(\frac{x}{2\sqrt{\kappa t}}\right)\right]\left[1 + \text{erf}\left(\frac{y}{2\sqrt{\kappa t}}\right)\right]\left[1 + \text{erf}\left(\frac{z}{2\sqrt{\kappa t}}\right)\right],$$

which is just the product of three functions of the form (2.12). Again we note that $\lim_{t \to 0^+} T(\underline{x},t) = f(\underline{x})$ except at points where $f(\underline{x})$ is discontinuous, and that as $t \to \infty$, $T(\underline{x},t)$ tends to the 'average' temperature $\frac{1}{8}$.

Finally in this section we draw the readers' attention to some general features of the Poisson transform (2.8). The temperature $T(\underline{x},t)$ depends on all the values of the initial temperature $f(\underline{x})$ for all time greater than zero so that the effects of the initial temperature propagate with infinite speed. This behaviour can be traced to the formulation of the heat equation where we neglected the fact that heat can only travel at finite speed. The function $T(\underline{x}, t)$ in (2.8) is also infinitely differentiable in all its variables for $t > 0$, even if it is initially discontinuous, indicating that discontinuous initial conditions cannot arise soley by heat conduction. If, however, we reverse the flow of time, that is replace t by $-t$ in the heat equation we obtain the *adjoint heat equation*

$$\frac{\partial T}{\partial t} + \kappa \nabla^2 T = 0,$$

which is unstable in the sense that it has solutions which are initially smooth but which become discontinuous after a finite time. Physically, the adjoint heat equation would represent a situation where the heat flux is proportional to the temperature gradient, but where heat flows from cold to hot. Thus heat would flow into hotter regions from colder regions, increasing the temperature in the hotter regions and increasing the temperature gradient, and hence accelerating the rate of heat flow into the hotter regions. It is therefore not surprising that many solutions of the adjoint heat equation 'blow up' after a finite time. The fact that the heat equation does not remain invariant under the transformation $t \rightarrow -t$ reflects the fact that the physical process of heat conduction is thermodynamically irreversible.

2.3 Maximum principle and uniqueness theorem

In this section we give the *maximum principle* for the heat equation, which states that the maximum temperature in a region occurs either within the region initially, or on the boundary of the region. The maximum principle is the mathematical statement of the physical tendency of heat to dissipate itself throughout a body as it flows from hot to cold. A consequence of the maximum principle is the *uniqueness theorem* for solutions of the heat equation, subject to mild restrictions on the growth of these solutions in the case of an infinite initial domain. These restrictions serve mainly to exclude the more grossly non-physical solutions, and the resulting uniqueness theorem confirms our intuition that a well posed heat conduction or diffusion problem should have only one physically meaningful solution. Finally we present Tychonoff's example of a *null solution* of the heat equation, that is, a non-zero solution of the heat equation which vanishes on the entire characteristic $t = 0$. The existence of this solution demonstrates the need for the restrictions on the growth of solutions in the uniqueness theorem.

First we will consider the maximum principle for a bounded open region V in one, two or three spatial dimensions, whose boundary is S. Let $t_0 > 0$ be some fixed time and let Ω denote the cylinder in $(\underline{x}, t)$ space defined by

$$\Omega = V \times (0, t_0) = \{(\underline{x}, t) \mid \underline{x} \in V, 0 < t < t_0\},$$

as shown (for the case of two spatial dimensions) in Figure 2.3. We divide the boundary of the cylinder Ω into two parts, the top of the cylinder V_{t_0}, corresponding to the closure $\overline{V}$ of the volume V (that is, the union of V and its boundary S, $\overline{V} = V \cup S$) at time t_0 and the base and sides of the cylinder which we denote by Γ, that is

$$V_{t_0} = \{(\underline{x}, t_0) \mid \underline{x} \in \overline{V}\}, \quad \Gamma = \{(\underline{x}, 0) \mid \underline{x} \in \overline{V}\} \bigcup \{(\underline{x}, t) \mid \underline{x} \in S, 0 \le t \le t_0\},$$

as shown in Figure 2.3. The closure of the cylinder $\overline{\Omega}$ consists of the union of the cylinder Ω and its boundaries V_{t_0} and Γ. The maximum principle states that if the function $T(\underline{x}, t)$ is continuous on the closure of the cylinder, $\overline{\Omega}$, and is sufficiently differentiable to satisfy the heat equation inside Ω, (that is for $\underline{x} \in V$ and $0 < t < t_0$), then

$$\sup_{(\underline{x}, t) \in \overline{\Omega}} T(\underline{x}, t) = \sup_{(\underline{x}, t) \in \Gamma} T(\underline{x}, t). \tag{2.14}$$

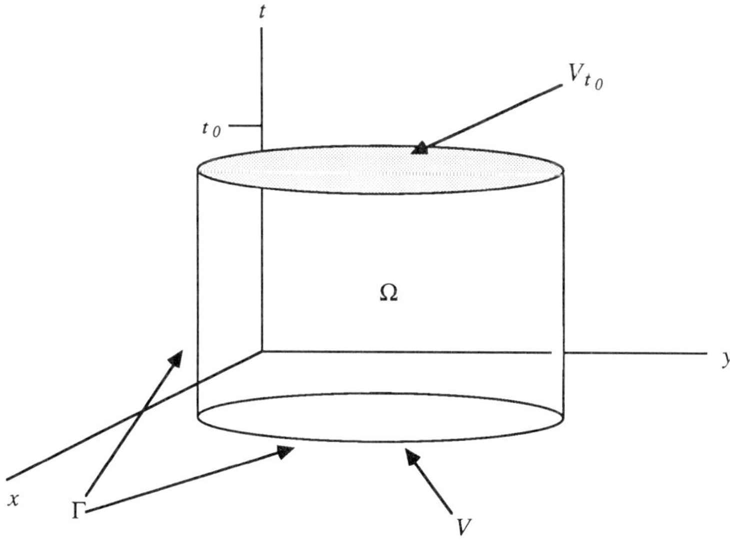

Figure 2.3. Three dimensional space-time cylinder Ω for the maximum principle.

The maximum principle therefore asserts that the temperature achieves its maximum either initially in V or on the boundary S during the period $0 \le t \le t_0$.

To prove the theorem we introduce the function $T_\epsilon(\underline{x}, t)$ defined for $\epsilon > 0$ by

$$T_\epsilon(\underline{x}, t) = T(\underline{x}, t) - \epsilon t,$$

and note that for all $\epsilon > 0$

$$\frac{\partial T_\epsilon}{\partial t} - \kappa \nabla^2 T_\epsilon = -\epsilon < 0. \tag{2.15}$$

Since $T_\epsilon(\underline{x}, t)$ is continuous in the closed region $\overline{\Omega}$, it assumes its maximum value at some point in that region, say $(\underline{x}_m, t_m)$. Assume that this point is not on Γ. Then either $(\underline{x}_m, t_m)$ is in the interior of Ω, so that it satisfies the conditions for a local maxima

$$\frac{\partial T_\epsilon}{\partial t} = 0, \qquad \nabla^2 T_\epsilon \le 0,$$

at $(\underline{x}_m, t_m)$, or it is inside V_{t_0}, the top of the cylinder, so that

$$\frac{\partial T_\epsilon}{\partial t} \ge 0, \qquad \nabla^2 T_\epsilon \le 0,$$

at $(\underline{x}_m, t_m)$. Both cases are inconsistent with (2.15) so we must have

$$\sup_{\overline{\Omega}} T_\epsilon(\underline{x}, t) = \sup_{\Gamma} T_\epsilon(\underline{x}, t),$$

for every $\epsilon > 0$. Thus

$$\sup_{\overline{\Omega}} T(\underline{x}, t) = \sup_{\overline{\Omega}} T_\epsilon(\underline{x}, t) + \epsilon t = \sup_{\Gamma} T_\epsilon(\underline{x}, t) + \epsilon t$$

$$\le \sup_{\Gamma} T_\epsilon(\underline{x}, t) + \epsilon t_0 \le \sup_{\Gamma} T(\underline{x}, t) + \epsilon t_0,$$

and in the limit $\epsilon \to 0$ we obtain (2.14).

The alert reader will have noticed that $T_\epsilon(\underline{x}, t)$ is assumed to be differentiable at time t_0 although we are only given $T(\underline{x}, t)$ as continuous at time t_0. There is no loss of generality however, for the argument given above will clearly work for any cylinder contained within $\overline{\Omega}$, as $T(\underline{x}, t)$ is differentiable in Ω. Thus the maximum principle will apply on any cylinder contained within $\overline{\Omega}$, and by the continuity of $T(\underline{x}, t)$ it must apply on all of $\overline{\Omega}$. A more general form of this theorem, which does not require $T(\underline{x}, t)$ to be continuous on Γ and therefore accommodates discontinuous initial and boundary data, states that if $T(\underline{x}, t)$ satisfies the heat equation in Ω and is such that $\liminf T(\underline{x}, t) \geq 0$ for every point $(\underline{x}, t)$ on Γ, as approached from within Ω, then $T(\underline{x}, t) \geq 0$ in Ω. We do not prove this theorem. By considering $-T(\underline{x}, t)$ we obtain from the maximum principle an analogous minimum principle for the heat equation and armed with these tools we can prove the uniqueness theorem for the heat equation posed on a bounded region

Theorem 2.2 If $T(\underline{x}, t)$ is continuous in $\overline{\Omega}$ and is sufficiently smooth to satisfy the heat equation in Ω then $T(\underline{x}, t)$ is uniquely determined by its values on Γ, that is, by its initial and boundary conditions.

To prove this theorem assume that there are two solutions of the heat equation in Ω which meet the necessary conditions and have the same initial and boundary conditions. Let their difference be $U(\underline{x}, t)$, so that $U(\underline{x}, t)$ satisfies the heat equation and meets the conditions necessary for the maximum principle to apply on $\overline{\Omega}$. Moreover, $U(\underline{x}, t)$ vanishes on Γ so that by the maximum principle $U(\underline{x}, t) \leq 0$ in $\overline{\Omega}$. Similarly we conclude that $-U(\underline{x}, t) \leq 0$ in $\overline{\Omega}$, so that $U(\underline{x}, t)$ vanishes in $\overline{\Omega}$.

We now consider the maximum principle for the heat equation with initial data on the entire characteristic $t = 0$. To obtain a maximum principle we have to impose growth restrictions on $T(\underline{x}, t)$ as $|\underline{x}| \to \infty$. Specifically we have

Theorem 2.3 If $T(\underline{x}, t)$ is continuous for all $\underline{x}$ and $0 \leq t \leq t_0$ and sufficiently smooth to satisfy the heat equation for all $\underline{x}$ and $0 < t < t_0$ then if $T(\underline{x}, 0) = f(\underline{x})$ and there exist positive constants ϵ and M such that

$$|T(\underline{x}, t)| \leq M e^{\epsilon |\underline{x}|^2},$$

for $0 \leq t \leq t_0$, then

$$T(\underline{x}, t) \leq \sup_{\underline{x}} f(\underline{x}),$$

for $0 \leq t \leq t_0$.

The proof of this theorem is beyond our scope. We do note that a corollary of this theorem is that the Poisson transform (2.8) is the unique solution of (2.6) subject to the restriction that $|T(\underline{x}, t)| \leq M e^{\epsilon |\underline{x}|^2}$ for some positive constants ϵ and M.

We now present an example due to Tychonoff of a null solution to the heat equation on the infinite line, that is a non-zero solution of the one dimensional heat equation which vanishes identically for $t \leq 0$. Such a solution can be added to any solution of the one dimensional heat equation on the infinite line (2.6) to produce a solution other than (2.9). Tychonoff's example, however, does not satisfy the growth restrictions of the maximum principle and uniqueness theorem, and is therefore excluded from that theorem. The existence of such a solution demonstrates the need for these growth restrictions in the uniqueness theorem.

Consider the formal solution of the one dimensional heat equation

$$T(x,t) = \sum_{n=0}^{\infty} g^{(n)}(t) \frac{x^{2n}}{\kappa^n (2n)!},$$ (2.16)

which is a special case of (2.5) and where $g(t)$ is an infinitely differentiable function, and $g^{(n)}(t)$ denotes the nth derivative of $g(t)$. Let us take $g(t)$ to be the function

$$g(t) = \begin{cases} e^{-1/t^2}, & t > 0, \\ 0, & t \le 0. \end{cases}$$ (2.17)

We now prove that $g(t)$ is infinitely differentiable for all real t and that with $g(t)$ so determined, (2.16) converges to a solution of the heat equation for all x and t. In particular, it is clear that for $t \le 0$ $T(x,t)$ as given by (2.16) must vanish identically, whereas for $t > 0$ it does not. Thus we can add any multiple of (2.16) to (2.9) to produce non-unique solutions of the heat equation posed on the whole real line and subject to an initial condition.

Using Cauchy's integral representation for the nth derivative $f^{(n)}(z)$ of a complex analytic function $f(z)$, namely

$$f^{(n)}(z) = \frac{n!}{2\pi i} \oint_{\gamma} \frac{f(\xi)}{(\xi - z)^{n+1}} \, d\xi,$$

where γ is a simple closed curve enclosing a domain containing z and inside which $f(z)$ is analytic, we deduce the inequality

$$|f^{(n)}(z)| \le \frac{n!}{2\pi} \oint_{\gamma} \left| \frac{f(\xi)}{(\xi - z)^{n+1}} \right| d\xi$$

$$\le \frac{n!}{2\pi} \ell(\gamma) \sup_{\xi \in \gamma} \left| \frac{f(\xi)}{(\xi - z)^{n+1}} \right|,$$

where $\ell(\gamma)$ is the length of the curve γ. Since the function

$$g(z) = e^{-1/z^2}, \quad \text{Re}(z) > 0,$$

is analytic in the half-plane $\mathrm{Re}(z) > 0$, we have for any real $t > 0$

$$|g^{(n)}(t)| \leq \frac{n!}{2\pi} \ell(\gamma) \sup_{\xi \in \gamma} \left| \frac{g(\xi)}{(\xi - t)^{n+1}} \right|,$$

where γ is some simple closed curve in the half-plane $\mathrm{Re}(z) > 0$ encircling the point t on the real axis. In particular, we can take γ to be the circle with centre t and radius ϵt where $0 < \epsilon < 1$, so that parametrically

$$\gamma = \{ t + \epsilon t e^{i\theta} \mid 0 \leq \theta < 2\pi \}.$$

Under these circumstances $\ell(\gamma) = 2\pi\epsilon t$, so we have

$$|g^{(n)}(t)| \leq \frac{n!}{(\epsilon t)^n} \sup_{0 \leq \theta < 2\pi} \left| \exp \frac{-1}{(t + \epsilon t e^{i\theta})^2} \right|.$$

Clearly we can choose some $0 < \epsilon < 1$ so that for all $0 \leq \theta < 2\pi$

$$\mathrm{Re}\left\{ \frac{1}{(t + \epsilon t e^{i\theta})^2} \right\} \geq \frac{1}{2t^2},$$

demonstrating that for all $t > 0$ there is some fixed $\epsilon > 0$, independent of t, such that

$$|g^{(n)}(t)| \leq \frac{n!}{(\epsilon t)^n} e^{-1/2t^2}.$$

Noting that $\lim_{t \to 0^+} |g^{(n)}(t)| = 0$, we see that $g(t)$ is infinitely differentiable for all t, since clearly $g^{(n)}(t) = 0$ if $t < 0$. Further, for $t > 0$ we have

$$\left| \sum_{n=0}^{\infty} g^{(n)}(t) \frac{x^{2n}}{\kappa^n (2n)!} \right| \leq \sum_{n=0}^{\infty} |g^{(n)}(t)| \frac{x^{2n}}{\kappa^n (2n)!}$$

$$\leq \sum_{n=0}^{\infty} \frac{n!}{(\epsilon t)^n} \frac{x^{2n}}{\kappa^n (2n)!} e^{-1/2t^2}$$

$$\leq e^{-1/2t^2} \sum_{n=0}^{\infty} \frac{1}{n!} \left(\frac{x^2}{\epsilon \kappa t} \right)^n$$

$$= \exp\left(\frac{x^2}{\epsilon \kappa t} - \frac{1}{2t^2} \right),$$

while for $t \leq 0$ the series vanishes identically. Thus, with $g(t)$ defined by (2.17), the series (2.16) converges uniformly on any closed bounded region in the (x, t) plane, is infinitely differentiable and satisfies the heat equation for all x and t. Obviously $T(x, t)$ is identically zero if $t \leq 0$, and non-zero if $t > 0$ as is clear if we take the temperature at $x = 0$ for $t > 0$, namely $T(0, t) = \exp(-1/t^2) > 0$. Other examples of null solutions of the heat equation are given in Widder (1975), for example (see also Problem 6).

2.4 Laplace transforms

We assume the reader has some familiarity with the Laplace transform and its application to initial value problems so that in this section we simply list the results concerning the Laplace transform which we need, and refer the reader to a more complete account such as Widder (1946) for further details. If $\psi(t)$ is a function defined for all $t > 0$ its *Laplace transform*, which we denote by $\mathscr{L}[\psi]$ or more briefly $\hat{\psi}(p)$, is given by

$$\hat{\psi}(p) = \mathscr{L}[\psi](p) = \int_0^\infty e^{-pt}\psi(t)\,\mathrm{d}t. \tag{2.18}$$

The transform variable p is generally complex, and we assume that it is such that the integral in (2.18) exists. If $\psi(t)$ is a measurable function and there exist real numbers p_0 and $M > 0$ such that

$$|\psi(t)| \le Me^{p_0 t},$$

for large enough t, then $\hat{\psi}(p)$ exists for all complex p in the half plane $\mathrm{Re}(p) > p_0$. In the following we assume that $\psi(t)$ and its derivatives meet these requirements and we refer to a function $\psi(t)$ as transformable if there is some real p_0 such that $\hat{\psi}(p)$ exists for all $\mathrm{Re}(p) \ge p_0$.

Example 2.3

(a) The Laplace transform of the constant 1 is

$$\mathscr{L}[1](p) = \int_0^\infty e^{-pt}\,\mathrm{d}t = \frac{1}{p}. \tag{2.19}$$

(b) The Laplace transform of the function $\psi(t) = H(t - t_0)$, where $H(t)$ is the Heaviside unit step function (1.15) and $t_0 > 0$, is

$$\mathscr{L}[H(t - t_0)](p) = \int_0^\infty e^{-pt}H(t - t_0)\,\mathrm{d}t$$

$$= \int_{t_0}^\infty e^{-pt}\,\mathrm{d}t \tag{2.20}$$

$$= \frac{1}{p}e^{-pt_0}.$$

(c) The Laplace transform of $\psi(t) = t^\alpha$, where $\alpha > -1$, is

$$
\begin{aligned}
\mathscr{L}[t^\alpha](p) &= \int_0^\infty e^{-pt} t^\alpha \, dt \\
&= \int_0^\infty e^{-\xi} \left(\frac{\xi}{p}\right)^\alpha \frac{d\xi}{p} \\
&= \frac{1}{p^{\alpha+1}} \int_0^\infty e^{-\xi} \xi^\alpha \, d\xi \\
&= \frac{\Gamma(\alpha+1)}{p^{\alpha+1}},
\end{aligned}
\tag{2.21}
$$

where we have made the substitution $\xi = pt$ and $\Gamma(x)$ denotes the gamma function defined by

$$
\Gamma(x) = \int_0^\infty t^{x-1} e^{-t} \, dt.
$$

(d) The Laplace transform of the Dirac delta function $\delta(t - t_0)$ which is a generalized function is

$$
\mathscr{L}[\delta(t - t_0)](p) = \int_0^\infty \delta(t - t_0) e^{-pt} \, dt = e^{-pt_0},
\tag{2.22}
$$

for $t_0 \geq 0$, even though strictly e^{-pt} is not a test function.

Some important properties of the Laplace transform are:

(i) The Laplace transform is a linear transform, that is

$$
\mathscr{L}[\alpha_1 \psi_1 + \alpha_2 \psi_2](p) = \alpha_1 \mathscr{L}[\psi_1](p) + \alpha_2 \mathscr{L}[\psi_2](p),
\tag{2.23}
$$

for any transformable functions $\psi_1(t)$ and $\psi_2(t)$ and real constants α_1 and α_2.

(ii) If $\psi(t)$ is differentiable we find

$$
\begin{aligned}
\mathscr{L}\left[\frac{d\psi}{dt}\right](p) &= \int_0^\infty e^{-pt} \frac{d\psi}{dt} \, dt \\
&= \left[e^{-pt} \psi(t)\right]_{t=0}^{t=\infty} + p \int_0^\infty e^{-pt} \psi(t) \, dt \\
&= -\psi(0) + p \, \mathscr{L}[\psi](p),
\end{aligned}
\tag{2.24}
$$

and for higher derivatives, a similar calculation yields

$$\mathscr{L}\left[\frac{\mathrm{d}^n\psi}{\mathrm{d}t^n}\right](p) = -\frac{\mathrm{d}^{n-1}\psi}{\mathrm{d}t^{n-1}}(0) - p\frac{\mathrm{d}^{n-2}\psi}{\mathrm{d}t^{n-2}}(0) - \cdots - p^{n-1}\psi(0) + p^n\,\mathscr{L}[\psi](p).$$

Note that this also applies to generalized functions as inspection of the Laplace transforms of $H(t - t_0)$ and derivative $\delta(t - t_0)$ will demonstrate.

(iii) For integrals of functions we find that

$$\mathscr{L}\left[\int_0^t \psi(\tau)\,\mathrm{d}\tau\right](p) = \frac{1}{p}\,\mathscr{L}[\psi](p). \tag{2.25}$$

(iv) If a is an arbitrary constant, then

$$\mathscr{L}\left[e^{-at}\psi(t)\right](p) = \int_0^\infty e^{-(p+a)t}\psi(t)\,\mathrm{d}t = \mathscr{L}[\psi](p + a), \tag{2.26}$$

and if $t_0 \geq 0$ is constant, we have

$$\begin{aligned}
\mathscr{L}[H(t - t_0)\psi(t - t_0)](p) &= \int_0^\infty e^{-pt}H(t - t_0)\psi(t - t_0)\,\mathrm{d}t \\[2mm]
&= \int_{t_0}^\infty e^{-pt}\psi(t - t_0)\,\mathrm{d}t \\[2mm]
&= \int_0^\infty e^{-p(\tau + t_0)}\psi(\tau)\,\mathrm{d}\tau \\[2mm]
&= e^{-pt_0}\,\mathscr{L}[\psi](p),
\end{aligned} \tag{2.27}$$

where $H(t)$ is the Heaviside unit step function (1.15) and we have used the substitution $\tau = t - t_0$. This can be verified for the case $\psi(t) = 1$ by inspection of (2.19) and (2.20).

(v) If $\psi_1(t)$ and $\psi_2(t)$ are two transformable functions whose Laplace transforms $\hat{\psi}_1(p)$ and $\hat{\psi}_2(p)$ share the same domain and satisfy $\hat{\psi}_1(p) = \hat{\psi}_2(p)$ for every p in that domain, then $\psi_1(t) = \psi_2(t)$ for almost all $t > 0$. In particular $\psi_1(t) = \psi_2(t)$ at any point t where $\psi_1(t)$ and $\psi_2(t)$ are both continuous. This implies that we can uniquely invert the Laplace transform of a continuous transformable function, and we can uniquely invert the Laplace transform of any piecewise continuous transformable function except at its points of discontinuity. We use the notation $\mathscr{L}^{-1}$ to denote the *inverse Laplace transform* so that

$$\mathscr{L}^{-1}[\hat{\psi}](t) = \psi(t), \tag{2.28}$$

The most common means of inverting the Laplace transform is to consult a table of Laplace transforms, such as the table at the end of this book. If $\hat{\psi}(p)$ is not tabulated, it can frequently be written in terms of tabulated expressions by use of partial fraction decompositions or Taylor series expansions. Since the inverse Laplace transform is linear, such a function can then be inverted term by term. We assume the reader is familiar with these methods and we employ them as the need arises.

(vi) (Convolution and the convolution theorem.) If $\psi_1(t)$ and $\psi_2(t)$ are two functions, their *convolution* $(\psi_1 * \psi_2)(t)$ is defined by the integral

$$(\psi_1 * \psi_2)(t) = \int_0^t \psi_1(\tau)\psi_2(t - \tau)\,d\tau, \tag{2.29}$$

and the change of variable $\nu = t - \tau$ in the integral shows that convolution is commutative

$$(\psi_1 * \psi_2)(t) = (\psi_2 * \psi_1)(t). \tag{2.30}$$

The convolution defined by (2.29) is sometimes referred to as the *unilateral convolution*.

Example 2.4 If $\psi_1(t) = t$ and $\psi_2(t) = e^t$ then

$$(\psi_1 * \psi_2)(t) = \int_0^t \tau e^{t-\tau}\,d\tau = e^t - t - 1,$$

and note that this is not equal to the point-wise product of $\psi_1(t)$ and $\psi_2(t)$ which is te^t.

The convolution theorem states that if $\psi_1(t)$ and $\psi_2(t)$ are transformable then

$$\mathscr{L}[(\psi_1 * \psi_2)](p) = \hat{\psi}_1(p)\hat{\psi}_2(p), \tag{2.31a}$$

or equivalently that

$$\mathscr{L}^{-1}\left[\hat{\psi}_1\hat{\psi}_2\right](t) = (\psi_1 * \psi_2)(t). \tag{2.31b}$$

We prove the result (2.31a) from first principles by changing the order of integration followed by a change of variable. We have

$$\mathscr{L}[(\psi_1 * \psi_2)](p) = \int_0^\infty \int_0^t e^{-pt}\psi_1(\tau)\psi_2(t-\tau)\,\mathrm{d}\tau\,\mathrm{d}t$$

$$= \int_0^\infty \int_\tau^\infty e^{-pt}\psi_1(\tau)\psi_2(t-\tau)\,\mathrm{d}t\,\mathrm{d}\tau$$

$$= \int_0^\infty \int_0^\infty e^{-p(\tau+\nu)}\psi_1(\tau)\psi_2(\nu)\,\mathrm{d}\nu\,\mathrm{d}\tau$$

$$= \int_0^\infty e^{-p\tau}\psi_1(\tau)\,\mathrm{d}\tau \int_0^\infty e^{-p\nu}\psi_2(\nu)\,\mathrm{d}\nu$$

$$= \hat{\psi}_1(p)\hat{\psi}_2(p),$$

where $\nu = t - \tau$.

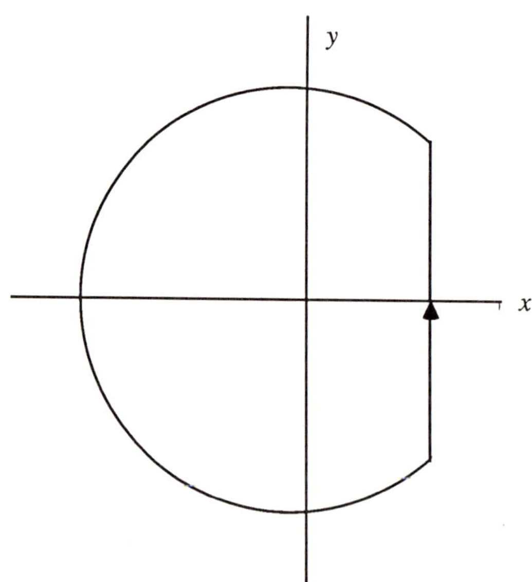

Figure 2.4. Bromwich contour for function with no branch cuts.

(vii) Most generally, however, to find the inverse Laplace transform we have the *inversion theorem* which states that

$$\psi(t) = \frac{1}{2\pi i} \int_{\omega - i\infty}^{\omega + i\infty} e^{pt}\hat{\psi}(p)\,dp, \tag{2.32}$$

where the real number ω is assumed sufficiently large to ensure that all the poles of $\hat{\psi}(p)$ lie to the left of the line of integration in the complex plane. To actually evaluate (2.32) we use contour integration to evaluate

$$\frac{1}{2\pi i} \oint_{\gamma} e^{pt}\hat{\psi}(p)\,dp,$$

where γ is the Bromwich contour shown in Figure 2.4 if $\hat{\psi}(p)$ has no branch cuts or the modified Bromwich contour (Figure 2.5) if $\hat{\psi}(p)$ has a branch cut along the negative real axis.

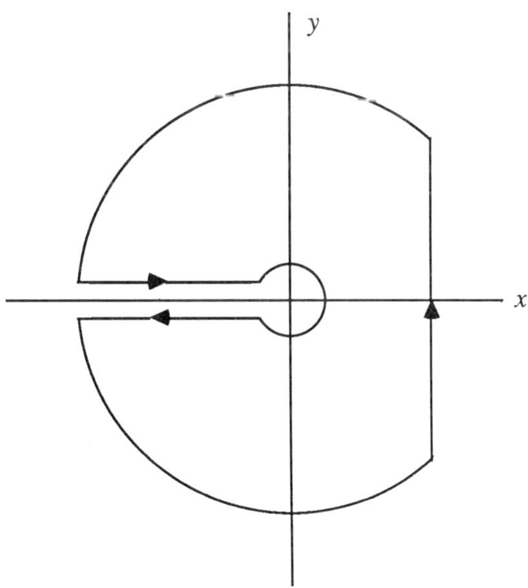

Figure 2.5. Modified Bromwich contour for functions with a branch cut.

We now consider in general terms the application of the Laplace transform to the heat equation and when we speak of the Laplace transform of a function of several variables, it is understood that the transform is with respect to the time variable t unless otherwise stated. As a consequence of (2.24), the Laplace transform reduces the one dimensional heat equation (1.8) to the ordinary differential equation

$$\frac{\partial^2 \hat{T}}{\partial x^2} - \frac{p}{\kappa}\hat{T} = -\frac{1}{\kappa}f(x), \tag{2.33}$$

where $\hat{T}(x,t)$ denotes the Laplace transform of the temperature

$$\hat{T}(x,p) = \int_0^\infty e^{-pt}\, T(x,t)\,\mathrm{d}t, \tag{2.34}$$

and $f(x)$ is the initial temperature. We assume that it is legitimate to interchange the order of differentiation and integration to obtain (2.33). In addition to transforming the heat equation, it is also necessary to transform any imposed boundary conditions, so that for example, the boundary conditions

$$T(a,t) = g(t), \quad k\frac{\partial T}{\partial x}(b,t) = h(t),$$

would be transformed to

$$\hat{T}(a,p) = \hat{g}(p), \quad k\frac{\partial \hat{T}}{\partial x}(b,p) = \hat{h}(p), \tag{2.35}$$

leaving us with the boundary value problem (2.33) and (2.35) rather that the heat equation to solve. In two or three spatial dimensions, the Laplace transform reduces the heat equation (1.47) to a lower order partial differential equation

$$\nabla^2 \hat{T} - \frac{p}{\kappa}\hat{T} = -\frac{1}{\kappa}f(\underline{x}), \tag{2.36}$$

where $f(\underline{x})$ denotes the initial temperature and $\hat{T}(\underline{x},p)$ is the Laplace transform of the temperature

$$\hat{T}(\underline{x},p) = \int_0^\infty e^{-pt}\, T(\underline{x},t)\,\mathrm{d}t. \tag{2.37}$$

It is also necessary to Laplace transform any boundary conditions imposed on $T(\underline{x},t)$ in order to obtain the boundary conditions subject to which (2.36) must be solved in order to determine the transformed temperature $\hat{T}(\underline{x},p)$.

Finally in this section we derive the Laplace transform of a number of functions which are of particular importance in the study of the heat equation. This is made less difficult by the use of the following remarkable result

$$\mathscr{L}\left[\frac{1}{\sqrt{\pi \kappa t}} \int_0^\infty \psi(\xi) \exp\left(\frac{-\xi^2}{4\kappa t}\right) d\xi\right](p) = \frac{1}{\kappa q} \hat{\psi}(q), \tag{2.38}$$

where we use the conventional notation, that is

$$q = \sqrt{\frac{p}{\kappa}}. \tag{2.39}$$

By definition, we have

$$\mathscr{L}\left[\frac{1}{\sqrt{\pi \kappa t}} \int_0^\infty \psi(\xi) \exp\left(\frac{-\xi^2}{4\kappa t}\right) d\xi\right](p) = \int_0^\infty \int_0^\infty \frac{\psi(\xi)}{\sqrt{\pi \kappa t}} \exp\left(-pt - \frac{\xi^2}{4\kappa t}\right) d\xi \, dt$$

$$= \int_0^\infty \psi(\xi) \int_0^\infty \frac{1}{\sqrt{\pi \kappa t}} \exp\left(-pt - \frac{\xi^2}{4\kappa t}\right) dt \, d\xi,$$

so that (2.38) will follow if we can show that

$$\int_0^\infty \frac{1}{\sqrt{\pi \kappa t}} \exp\left(-pt - \frac{\xi^2}{4\kappa t}\right) dt = \frac{1}{\kappa q} e^{-\xi q}, \tag{2.40}$$

where q is given by (2.39). To this end, we let $\eta = \sqrt{pt}$, so that

$$\int_0^\infty \frac{1}{\sqrt{\pi \kappa t}} \exp\left(-pt - \frac{\xi^2}{4\kappa t}\right) dt = \frac{2}{\sqrt{\pi \kappa p}} \int_0^\infty \exp\left(-\eta^2 - \frac{\xi^2 p}{4\kappa \eta^2}\right) d\eta$$

$$= \frac{2}{\kappa q \sqrt{\pi}} e^{\xi q} \int_0^\infty \exp -\left(\eta + \frac{\xi q}{2\eta}\right)^2 d\eta,$$

after completing the square. If we avail ourselves of the result that

$$\int_0^\infty \exp -\left(\eta + \frac{x}{\eta}\right)^2 d\eta = \frac{\sqrt{\pi}}{2} e^{-4x}, \tag{2.41}$$

(see Problem 7) with $x = \xi q / 2$ then (2.40) and (2.38) follow.

<u>Example 2.5</u> If we notice that

$$\frac{1}{\sqrt{\pi \kappa t}} \exp\left(\frac{-x^2}{4\kappa t}\right) = \frac{1}{\sqrt{\pi \kappa t}} \int_0^\infty \delta(\xi - x) \exp\left(\frac{-\xi^2}{4\kappa t}\right) d\xi,$$

and apply (2.38) with $\psi(\xi) = \delta(\xi - x)$, we find that the Laplace transform of the one dimensional source solution is

$$\mathscr{L}\left[\frac{1}{2\sqrt{\pi\kappa t}}\exp\left(\frac{-x^2}{4\kappa t}\right)\right](p) = \frac{1}{2\kappa q}e^{-qx}, \tag{2.42}$$

using (2.22) for the Laplace transform of the delta function $\delta(\xi - x)$.

<u>Example 2.6</u> Assuming $x \geq 0$ we can find the Laplace transform of

$$\mathrm{erfc}\left(\frac{x}{2\sqrt{\kappa t}}\right) = \frac{1}{\sqrt{\pi\kappa t}}\int_0^\infty H(\xi - x)\exp\left(\frac{-\xi^2}{4\kappa t}\right)\mathrm{d}\xi,$$

using the Laplace transform of the Heaviside unit step function (2.20) and (2.38) with $\psi(\xi) = H(\xi - x)$. We find

$$\mathscr{L}\left[\mathrm{erfc}\left(\frac{x}{2\sqrt{\kappa t}}\right)\right](p) = \frac{1}{\kappa q^2}e^{-qx} = \frac{1}{p}e^{-qx}, \tag{2.43}$$

and since $\mathrm{erf}(x) = 1 - \mathrm{erfc}(x)$, we have from (2.19) and (2.43)

$$\mathscr{L}\left[\mathrm{erf}\left(\frac{x}{2\sqrt{\kappa t}}\right)\right](p) = \mathscr{L}[1](p) - \mathscr{L}\left[\mathrm{erfc}\left(\frac{x}{2\sqrt{\kappa t}}\right)\right](p) = \frac{1}{p}\left(1 - e^{-qx}\right). \tag{2.44}$$

2.5 Fourier series

An adequate account of the historical development and complete theory of Fourier series is well beyond our scope. Here we simply state Fourier's theorem, provide some elements of a heuristic proof and state the main results needed in subsequent chapters. The interested reader might consult the appropriate chapter of Whittaker and Watson (1963) or Carslaw (1950) for additional information at an elementary level. Series of the form

$$\frac{a_0}{2} + \sum_{n=1}^\infty (a_n \cos nx + b_n \sin nx), \tag{2.45}$$

where a_n and b_n are constants are called *trigonometrical series*. If there exists a function $\psi(x)$ such that

$$a_n = \frac{1}{\pi}\int_{-\pi}^{\pi}\psi(\xi)\cos n\xi \, \mathrm{d}\xi, \quad n = 0, 1, 2, \ldots$$

$$\tag{2.46}$$

$$b_n = \frac{1}{\pi}\int_{-\pi}^{\pi}\psi(\xi)\sin n\xi \, \mathrm{d}\xi, \quad n = 1, 2, 3, \ldots$$

then the trigonometrical series (2.45) is called a *Fourier series*. In the "Analytical Theory of Heat" Fourier investigated a number of trigonometrical series and showed that in a large number of cases the Fourier series (2.45) with coefficients given by (2.46) actually converged to the function $\psi(x)$. However, it was Dirichlet who in 1829 provided the first rigorous proof of the convergence of Fourier series, for the class of functions $\psi(x)$ defined and bounded on the interval $[-\pi, \pi)$ with only a finite number of local maxima and minima and a finite number of jump discontinuities on this interval. Further, by extending $\psi(x)$ outside the interval $[-\pi, \pi)$ by

$$\psi(x + 2\pi) = \psi(x), \tag{2.47}$$

so that $\psi(x)$ is periodic with period 2π on the real line, Dirichlet showed that the sum (2.45) with coefficients (2.46) converges to the sum $\left[\psi(x^+) + \psi(x^-)\right]/2$. A theorem of this type, concerning the expansion of a function of a real variable as a trigonometrical series is known as a Fourier's theorem.

We now state a Fourier's theorem, due to Jordan, with less stringent conditions than those assumed by Dirichlet. In order to do so, however, we must explain the concept of a function of *bounded variation*.

<u>Definition 2.1</u> If $\psi(x)$ is a bounded real valued function defined on the open interval (a, b) and $\{x_0, x_1, \ldots, x_n\}$ is a finite subdivision of the interval (a, b), that is

$$a = x_0 < x_1 < \ldots < x_n = b,$$

we say that $\psi(x)$ is of bounded variation on the interval (a, b) if the sum

$$\sum_{i=1}^{n} |\psi(x_i) - \psi(x_{i-1})|,$$

is bounded above for all possible finite subdivisions of the interval (a, b).

<u>Theorem 2.4</u> (Fourier's theorem) Suppose that

(i) $\quad \psi(x)$ is defined for $-\pi \le x < \pi$ and is extended to the whole real line by (2.47), so that $\psi(x)$ is a periodic function with period 2π,

(ii) $\quad$ the integral $\int_{-\pi}^{\pi} |\psi(\xi)|\,d\xi$ exists as either a proper or improper integral,

(iii) $\quad a_n$ and b_n are as defined by (2.46),

then if x is an interior point of any open interval (a,b) on which $\psi(x)$ has bounded variation, the Fourier series (2.45) is convergent and its sum is $\frac{1}{2}\left[\psi(x^+) + \psi(x^-)\right]$. If $\psi(x)$ is continuous at x the Fourier series converges to $\psi(x)$.

Although we do not present a rigorous proof of this theorem, the essential details of the argument are as follows. We define $S_m(x)$ to be the mth partial sum of the series (2.45), that is

$$S_m(x) = \frac{a_0}{2} + \sum_{n=1}^{m} (a_n \cos nx + b_n \sin nx). \tag{2.48}$$

From (2.46) we have

$$\begin{aligned}
S_m(x) &= \frac{1}{2\pi} \int_{-\pi}^{\pi} \left\{ \psi(\xi) + 2 \sum_{n=1}^{m} \psi(\xi)\{\cos n\xi \cos nx + \sin n\xi \sin nx\} \right\} d\xi \\
&= \frac{1}{2\pi} \int_{-\pi}^{\pi} \psi(\xi) \left\{ 1 + 2 \sum_{n=1}^{m} \cos n(x - \xi) \right\} d\xi \\
&= \frac{1}{2\pi} \int_{-\pi}^{\pi} \psi(\xi) \frac{\sin(m + \frac{1}{2})(x - \xi)}{\sin \frac{1}{2}(x - \xi)} d\xi,
\end{aligned} \tag{2.49}$$

where we have used the identity (see Problem 10)

$$1 + 2 \sum_{n=1}^{m} \cos 2n\theta = \frac{\sin(2m + 1)\theta}{\sin \theta}. \tag{2.50}$$

Since the integrand in (2.49) is periodic with period 2π, the value of the integral is the same over any interval of length 2π, so

$$S_m(x) = \frac{1}{2\pi} \int_{x-\pi}^{x+\pi} \psi(\xi) \frac{\sin(m + \frac{1}{2})(x - \xi)}{\sin \frac{1}{2}(x - \xi)} d\xi.$$

We now split this integral into two integrals, one over the range $(x - \pi, x)$ and the other over the range $(x, x + \pi)$ and make the substitutions $\xi = x - 2\theta$ and $\xi = x + 2\theta$ respectively, in these two integrals to obtain

$$S_m(x) = \frac{1}{\pi} \int_{0}^{\frac{\pi}{2}} \frac{\sin(2m + 1)\theta}{\sin \theta} \psi(x + 2\theta)\, d\theta + \frac{1}{\pi} \int_{0}^{\frac{\pi}{2}} \frac{\sin(2m + 1)\theta}{\sin \theta} \psi(x - 2\theta)\, d\theta.$$

From (2.50) we have

$$\int_{0}^{\frac{\pi}{2}} \frac{\sin(2m + 1)\theta}{\sin \theta}\, d\theta = \int_{0}^{\frac{\pi}{2}} \left\{ 1 + 2 \sum_{n=1}^{m} \cos 2n\theta \right\} d\theta = \frac{\pi}{2},$$

and hence

$$S_m(x) - \tfrac{1}{2}\big[\psi(x^+) + \psi(x^-)\big] = \frac{1}{\pi}\int_0^{\frac{\pi}{2}} \frac{\sin(2m+1)\theta}{\sin\theta}\big[\psi(x+2\theta) - \psi(x^+)\big]\,d\theta$$

$$+ \frac{1}{\pi}\int_0^{\frac{\pi}{2}} \frac{\sin(2m+1)\theta}{\sin\theta}\big[\psi(x-2\theta) - \psi(x^-)\big]\,d\theta,$$

and thus Fourier's theorem reduces to proving that

$$\lim_{m\to\infty} \int_0^{\frac{\pi}{2}} \frac{\sin(2m+1)\theta}{\sin\theta}\,\Psi(\theta)\,d\theta = 0,$$

where $\Psi(\theta)$ is either of $\big[\psi(x+2\theta) - \psi(x^+)\big]$ or $\big[\psi(x-2\theta) - \psi(x^-)\big]$. The proof of this result, which relies on the properties of functions of bounded variation and the fact that x is an interior point of the interval in question, can be found in Whittaker and Watson (1963) for example. We note that if $\psi(x)$ qualifies as a test function, that is, $\psi(x)$ is infinitely differentiable and vanishes outside a closed bounded interval, then the result follows immediately as the function

$$f_m(x) = \frac{1}{2\pi}\frac{\sin(2m+1)x}{\sin x},$$

is delta convergent in the interval $[0, \pi/2]$ as $m \to \infty$.

If a function $\psi(x)$ can be expressed in the form

$$\psi(x) = \frac{a_0}{2} + \sum_{n=0}^{\infty} (a_n \cos nx + b_n \sin nx),$$

then the formulae (2.46) for the Fourier coefficients follow by noting the elementary integrals

$$\int_{-\pi}^{\pi} \sin m\xi\,\sin n\xi\,d\xi = \pi\delta_{mn},$$

$$\int_{-\pi}^{\pi} \cos m\xi\,\cos n\xi\,d\xi = \begin{cases} 2\pi, & m = n = 0, \\ \pi\delta_{mn}, & m > 0 \end{cases}$$

$$\int_{-\pi}^{\pi} \sin m\xi\,\cos n\xi\,d\xi = 0,$$

where δ_{mn} is the Kronecker delta (1.21) and m and n are non-negative integer constants.

In general a Fourier series such as (2.45) with Fourier coefficients given by (2.46) converges to the function $\psi(x)$ in the sense that

$$\lim_{m \to \infty} \int_{-\pi}^{\pi} |\psi(\xi) - S_m(\xi)|^2 \, d\xi \to 0,$$

where $S_m(x)$ denotes the mth partial sum (2.48) of (2.45). This form of convergence is known as *convergence in the mean*, and does not necessarily imply pointwise convergence, although in the case of a continuous function $\psi(x)$ of bounded variation, the convergence is pointwise on the interior of $(-\pi, \pi)$ as shown in the previous theorem. A function $\psi(x)$ defined on $(-\pi, \pi)$ is known as *square integrable* if the (Lebesgue) integral

$$\int_{-\pi}^{\pi} |\psi(\xi)|^2 \, d\xi,$$

exists, and a sequence $\{c_n\}$ is known as *square summable* if the series

$$\sum_{n=0}^{\infty} |c_n|^2,$$

converges. The Riesz-Fischer theorem, as applied to trigonometric Fourier series, states that if a function $\psi(x)$ is square integrable on $(-\pi, \pi)$, then the sequences $\{a_n\}$ and $\{b_n\}$ defined by (2.46) are square summable and that the Fourier series (2.45) converges in the mean to $\psi(x)$. Conversely, the theorem states that if $\{a_n\}$ and $\{b_n\}$ are square summable sequences, then the trigonometrical series (2.45) converges in the mean to a function $\psi(x)$ which is square integrable on $(-\pi, \pi)$ and that the terms a_n and b_n of the sequences are the Fourier coefficients of the function $\psi(x)$. As well, if $\psi(x)$ is a square integrable function on $(-\pi, \pi)$ and a_n and b_n are its Fourier coefficients defined by (2.46), then

$$\int_{-\pi}^{\pi} |\psi(\xi)|^2 \, d\xi = \frac{a_0^2}{4} + \sum_{n=1}^{\infty} (a_n^2 + b_n^2).$$

One important consequence of the Riesz-Fischer theorem is that any two square integrable functions which have exactly the same Fourier coefficients are equal almost everywhere.

Finally in this section we note the Fourier series for the general interval (a, b) and the Fourier cosine and sine series for even and odd functions, respectively, on the interval $(-\ell, \ell)$. Applying a linear transformation which maps the interval $(-\pi, \pi)$

onto (a, b), we have from our previous results that if

$$\psi(x) = \frac{a_0}{2} + \sum_{n=1}^{\infty} \left\{ a_n \cos \frac{n\pi(2x - a - b)}{(b - a)} + b_n \sin \frac{n\pi(2x - a - b)}{(b - a)} \right\}, \quad (2.51)$$

then

$$a_n = \frac{2}{(b - a)} \int_a^b \psi(\xi) \cos \frac{n\pi(2\xi - a - b)}{(b - a)} \, d\xi, \quad n = 0, 1, 2, \ldots$$

$$(2.52)$$

$$b_n = \frac{2}{(b - a)} \int_a^b \psi(\xi) \sin \frac{n\pi(2\xi - a - b)}{(b - a)} \, d\xi, \quad n = 1, 2, 3, \ldots$$

Further, if $\psi(x)$ is an even function $[\psi(-x) = \psi(x)]$ defined on $(-\ell, \ell)$, then the Fourier sine coefficients b_n are zero and we obtain the *Fourier cosine series*

$$\psi(x) = \frac{a_0}{2} + \sum_{n=1}^{\infty} a_n \cos \frac{n\pi x}{\ell}, \quad (2.53)$$

where the a_n are given by

$$a_n = \frac{2}{\ell} \int_0^\ell \psi(\xi) \cos \frac{n\pi\xi}{\ell} \, d\xi. \quad (2.54)$$

If $\psi(x)$ is an odd function $[\psi(-x) - \psi(x)]$ defined on $(-\ell, \ell)$, then the Fourier cosine coefficients a_n vanish, and we have the *Fourier sine series*

$$\psi(x) = \sum_{n=1}^{\infty} b_n \sin \frac{n\pi x}{\ell}, \quad (2.55)$$

where the coefficients b_n are given by

$$b_n = \frac{2}{\ell} \int_0^\ell \psi(\xi) \sin \frac{n\pi\xi}{\ell} \, d\xi. \quad (2.56)$$

Given an arbitrary function $\psi(x)$ defined on $[0, \ell)$, we can extend $\psi(x)$ to the interval $(-\ell, \ell)$ by the rule

$$\psi(-x) = \psi(x),$$

so that the resulting extended function is even. Thus a function defined only on $[0, \ell)$ can be represented by a Fourier cosine series (2.53) with coefficients given by (2.54). Similarly, we can extend $\psi(x)$ so that it is odd on the interval $(-\ell, \ell)$, in which case we can expand $\psi(x)$ as a Fourier sine series (2.55) with coefficients determined by (2.56).

Assuming $\psi(x)$ to be continuous and of bounded variation on $[0, \ell)$ it follows from Fourier's theorem that the series (2.55) converges to $\psi(x)$ in the open interval $(0, \ell)$ and vanishes at $x = 0$. This point has some important consequences which will manifest themselves in a subsequent chapter. Provided the function $\psi(x)$ is square integrable, the series will converge almost everywhere to $\psi(x)$, and if $\psi(x)$ is of bounded variation, the Fourier's theorem will apply.

2.6 Green's functions

The use of Green's functions is a powerful technique for the analytic solution, numerical solution and mathematical analysis of heat conduction problems. The application of Green's functions as a means of solving and analysing initial and boundary value problems is not confined to the study of heat conduction but occurs in almost all branches of mathematical physics. In this section we give the broad details of the technique and some of its consequences for the heat equation in a manner which is more intuitive that rigorous. The idea underlying the method of Green's functions is similar to that underlying the source solution, and in fact the Poisson transform given in Section 2.2 is a special case of the application of a Green's function. A *Green's function* $G(\underset{\sim}{x}, \underset{\sim}{\xi}, t, \tau)$ for the heat equation can be regarded as the temperature at position $\underset{\sim}{x}$ and time t due in part to an instantaneous point source of unit strength at position $\underset{\sim}{\xi}$ occurring at time τ. In addition, we can require that $G(\underset{\sim}{x}, \underset{\sim}{\xi}, t, \tau)$ or its normal derivative should vanish when $\underset{\sim}{\xi}$ lies on some given surface S. Specifically, we require that a Green's function should satisfy

$$\frac{\partial G}{\partial t} - \kappa \nabla^2_{\underset{\sim}{x}} G = \delta(\underset{\sim}{x} - \underset{\sim}{\xi})\delta(t - \tau), \tag{2.57}$$

and the adjoint equation

$$\frac{\partial G}{\partial \tau} + \kappa \nabla^2_{\underset{\sim}{\xi}} G = -\delta(\underset{\sim}{x} - \underset{\sim}{\xi})\delta(t - \tau), \tag{2.58}$$

where $\nabla^2_{\underset{\sim}{x}}$ denotes the Laplacian with respect to $\underset{\sim}{x}$ and $\nabla^2_{\underset{\sim}{\xi}}$ denotes the Laplacian with respect to $\underset{\sim}{\xi}$. As well the condition

$$G(\underset{\sim}{x}, \underset{\sim}{\xi}, t, \tau) = 0, \quad t \leq \tau, \tag{2.59}$$

must hold, and possibly various boundary conditions in $\underset{\sim}{\xi}$, depending on the purpose for which the Green's function is intended.

<u>Example 2.7</u> Show that the function

$$G(x, \xi, t, \tau) = H(t - \tau)K_1(x - \xi, t - \tau), \tag{2.60}$$

where $H(t)$ is the Heaviside unit step function (1.15) and where $K_1(x, t)$ is the one dimensional unit source solution $(2.7)_1$, is a Green's function for the one dimensional heat equation.

Clearly $G(x, \xi, t, \tau) = 0$ if $t \leq \tau$ because of the Heaviside step function in (2.60). We also note that both $K_1(x, t)$ and 0 are solutions of the one dimensional heat equation, and that as $t \to 0^+$, $K_1(x, t)$ and all its derivatives vanish for any x except $x = 0$, so that

$$\frac{\partial G}{\partial t} - \kappa \frac{\partial^2 G}{\partial x^2} = 0, \quad (x, t) \neq (\xi, \tau). \tag{2.61}$$

Suppose that $\phi(x, t)$ is any continuous function of x and t, then if ϵ and δ are any strictly positive constants, we can conclude from (2.60) and (2.61) that

$$\int_{-\infty}^{\infty} \int_{-\infty}^{\infty} \phi(x, t) \left(\frac{\partial G}{\partial t} - \kappa \frac{\partial^2 G}{\partial x^2} \right) dx\, dt = \int_{\tau}^{\tau + \epsilon} \int_{\xi - \delta}^{\xi + \delta} \phi(x, t) \left(\frac{\partial G}{\partial t} - \kappa \frac{\partial^2 G}{\partial x^2} \right) dx\, dt.$$

Using the mean value theorem, we can find ξ' and τ' satisfying $\xi - \delta \leq \xi' \leq \xi + \delta$, $\tau \leq \tau' \leq \tau + \epsilon$ and such that

$$\int_{\tau}^{\tau + \epsilon} \int_{\xi - \delta}^{\xi + \delta} \phi(x, t) \left(\frac{\partial G}{\partial t} - \kappa \frac{\partial^2 G}{\partial x^2} \right) dx\, dt = \phi(\xi', \tau') \int_{\tau}^{\tau + \epsilon} \int_{\xi - \delta}^{\xi + \delta} \left(\frac{\partial G}{\partial t} - \kappa \frac{\partial^2 G}{\partial x^2} \right) dx\, dt.$$

By integration we have

$$\int_{\tau}^{\tau + \epsilon} \int_{\xi - \delta}^{\xi + \delta} \left(\frac{\partial G}{\partial t} - \kappa \frac{\partial^2 G}{\partial x^2} \right) dx\, dt = \int_{\xi - \delta}^{\xi + \delta} G(x, \xi, \tau + \epsilon, \tau)\, dx - \kappa \int_{\tau}^{\tau + \epsilon} \left[\frac{\partial G}{\partial x} \right]_{x = \xi - \delta}^{x = \xi + \delta} dt,$$

and with the change of variable $\sigma = (x - \xi)/2\sqrt{\kappa \epsilon}$ we find

$$\int_{\xi - \delta}^{\xi + \delta} G(x, \xi, \tau + \epsilon, \tau)\, dx = \frac{1}{2\sqrt{\pi \kappa \epsilon}} \int_{\xi - \delta}^{\xi + \delta} \exp \frac{-(x - \xi)^2}{4\kappa \epsilon}\, dx$$

$$= \frac{1}{\sqrt{\pi}} \int_{-\delta/2\sqrt{\kappa \epsilon}}^{\delta/2\sqrt{\kappa \epsilon}} e^{-\sigma^2}\, d\sigma$$

$$= \operatorname{erf}\left(\frac{\delta}{2\sqrt{\kappa \epsilon}} \right),$$

while the change of variable $\sigma = \delta/2\sqrt{\kappa(t - \tau)}$ gives

$$\int_{\tau}^{\tau+\epsilon} \frac{\partial G}{\partial x}(\xi \pm \delta, \xi, t, \tau)\, dt = \frac{\mp \delta}{4\sqrt{\pi}} \int_{\tau}^{\tau+\epsilon} \frac{1}{[\kappa(t-\tau)]^{3/2}} \exp\left(\frac{-\delta^2}{4\kappa(t-\tau)}\right) dt$$

$$= \frac{\mp 1}{\kappa\sqrt{\pi}} \int_{\delta/2\sqrt{\kappa\epsilon}}^{\infty} e^{-\sigma^2}\, d\sigma$$

$$= \frac{\mp 1}{2\kappa}\operatorname{erfc}\left(\frac{\delta}{2\sqrt{\kappa\epsilon}}\right).$$

Thus altogether and for any $\epsilon, \delta > 0$ we have

$$\int_{\tau}^{\tau+\epsilon} \int_{\xi-\delta}^{\xi+\delta} \left(\frac{\partial G}{\partial t} - \kappa\frac{\partial^2 G}{\partial x^2}\right) dx\, dt = \operatorname{erf}\left(\frac{\delta}{2\sqrt{\kappa\epsilon}}\right) + \operatorname{erfc}\left(\frac{\delta}{2\sqrt{\kappa\epsilon}}\right) = 1.$$

In particular it follows that for any $\epsilon, \delta > 0$ we can find ξ' and τ' satisfying $\xi - \delta \leq \xi' \leq \xi + \delta$ and $\tau \leq \tau' \leq \tau + \epsilon$ and such that

$$\int_{-\infty}^{\infty} \int_{-\infty}^{\infty} \phi(x,t)\left(\frac{\partial G}{\partial t} - \kappa\frac{\partial^2 G}{\partial x^2}\right) dx\, dt = \phi(\xi', \tau'),$$

and if we let ϵ and δ tend to zero we obtain

$$\int_{-\infty}^{\infty} \int_{-\infty}^{\infty} \phi(x,t)\left(\frac{\partial G}{\partial t} - \kappa\frac{\partial^2 G}{\partial x^2}\right) dx\, dt = \phi(\xi, \tau).$$

Thus we have shown that

$$\frac{\partial G}{\partial t} - \kappa\frac{\partial^2 G}{\partial x^2} = \delta(x-\xi)\delta(t-\tau), \tag{2.62}$$

and a similar argument shows that

$$\frac{\partial G}{\partial \tau} + \kappa\frac{\partial^2 G}{\partial \xi^2} = -\delta(x-\xi)\delta(t-\tau). \tag{2.63}$$

We now consider some of the consequences of (2.63). For the purposes of illustration, we consider the problem

$$\left.\begin{aligned}
\frac{\partial T}{\partial t} &= \kappa\frac{\partial^2 T}{\partial x^2}, \quad a < x < b, \quad t > 0, \\[2mm]
T(x,0) &= f(x), \quad a \leq x \leq b, \\[2mm]
T(a,t) &= g_1(t), \quad T(b,t) = g_2(t), \quad t > 0.
\end{aligned}\right\} \tag{2.64}$$

On multiplying the heat equation in (2.64) by $G(x,\xi,t,\tau)$ and integrating, we have

$$\int_0^t \int_a^b G(x,\xi,t,\tau)\left(\frac{\partial T}{\partial \tau} - \kappa\frac{\partial^2 T}{\partial \xi^2}\right)d\xi\, d\tau = 0, \tag{2.65}$$

whereas, multiplying (2.63) by $T(\xi,\tau)$ and integrating gives

$$\int_0^t \int_a^b T(\xi,\tau)\left(\frac{\partial G}{\partial \tau} + \kappa\frac{\partial^2 G}{\partial \xi^2}\right)d\xi\, d\tau = -T(x,t). \tag{2.66}$$

If we add (2.65) and (2.66) and use the relations

$$\int_0^t \int_a^b \left(G\frac{\partial T}{\partial \tau} + T\frac{\partial G}{\partial \tau}\right)d\xi\, d\tau = \int_a^b [TG]_{\tau=0}^{\tau=t}\, d\xi = -\int_a^b G(x,\xi,t,0)T(\xi,0)\,d\xi,$$

and

$$\int_0^t \int_a^b \left(T\frac{\partial^2 G}{\partial \xi^2} - G\frac{\partial^2 T}{\partial \xi^2}\right)d\xi\, d\tau = \int_0^t \left[T\frac{\partial G}{\partial \xi} - G\frac{\partial T}{\partial \xi}\right]_{\xi=a}^{\xi=b}\, d\tau,$$

we find that on using the boundary and initial conditions in (2.64) we have

$$T(x,t) = \int_a^b G(x,\xi,t,0)f(\xi)\,d\xi$$

$$+ \kappa\int_0^t \frac{\partial G}{\partial \xi}(x,a,t,\tau)g_1(\tau)\,d\tau - \kappa\int_0^t \frac{\partial G}{\partial \xi}(x,b,t,\tau)g_2(\tau)\,d\tau \tag{2.67}$$

$$- \kappa\int_0^t G(x,a,t,\tau)\frac{\partial T}{\partial \xi}(a,\tau)\,d\tau + \kappa\int_0^t G(x,b,t,\tau)\frac{\partial T}{\partial \xi}(b,\tau)\,d\tau.$$

Since $\frac{\partial T}{\partial x}(a,t)$ and $\frac{\partial T}{\partial x}(b,t)$ are unknown, our use of (2.67) depends upon our purpose. For the mathematical or numerical analysis of the problem (2.64) we can take the Green's function $G(x,\xi,t,\tau)$ appearing in (2.67) to be some relatively simple known function such as (2.60), leaving us with the two unknowns $\frac{\partial T}{\partial x}(a,t)$ and $\frac{\partial T}{\partial x}(b,t)$. Applying the boundary conditions from (2.64) to (2.67) leads to a pair of coupled integral equations involving these unknown functions. (However, care must be taken here, for in deriving (2.67) we have tacitly assumed that $a < x < b$. If in fact $x = a$ or $x = b$, then the integrals of (2.67) actually only equal one half of the temperature, a matter which we do not pursue at this point, but return to in Section 6.5.) For the purposes of the mathematical analysis of the problem (2.64) the resulting system of integral equations is often easier to deal with than the original problem. Many proofs of existence of solutions to the heat equation posed on a finite spatial domain are based on this idea. From a numerical point of view, we can solve numerically for $\frac{\partial T}{\partial x}(a,t)$ and $\frac{\partial T}{\partial x}(b,t)$ and use these to determine $T(x,t)$ from (2.67). Numerical methods based on this procedure are known as boundary integral methods which are discussed briefly in Section 6.5.

If we could find a Green's function which vanished when $\xi = a$ and $\xi = b$, so that

$$G(x,a,t,\tau) = G(x,b,t,\tau) = 0, \quad a < x < b, \tau < t, \tag{2.68}$$

then the unknown derivatives of $T(x,t)$ occurring in (2.67) would be eliminated and we would have

$$T(x,t) = \int_a^b G(x,\xi,t,0)f(\xi)\,d\xi$$

$$+ \kappa \int_0^t g_1(\tau)\frac{\partial G}{\partial \xi}(x,a,t,\tau)\,d\tau - \kappa \int_0^t g_2(\tau)\frac{\partial G}{\partial \xi}(x,b,t,\tau)\,d\tau. \tag{2.69}$$

One of the most important features such a Green's function is that it depends only on the interval (a,b) and the class of boundary conditions applied at the endpoints, in this case prescribed temperature conditions. It does not depend on $f(x)$, $g_1(t)$ and $g_2(t)$, so that once such a Green's function $G(x,\xi,t,\tau)$ is found, (2.69) represents the general solution to (2.64), independent of any particular choice of initial or prescribed surface temperature conditions. Similarly, we can find Green's functions for various other classes of boundary conditions.

In order to apply Green's functions to the heat equation in two or more spatial dimensions, we require Green's theorem, namely

$$\int_V (\phi \nabla^2 \psi - \psi \nabla^2 \phi)\,dV = \oint_S (\phi \underline{\nabla} \psi - \psi \underline{\nabla} \phi) \cdot d\underline{S}, \tag{2.70}$$

where V is an arbitrary volume, S its surface and $\phi(\underline{x})$ and $\psi(\underline{x})$ sufficiently smooth functions. Green's theorem (2.70) is a simple consequence of the identity

$$\underline{\nabla} \cdot (\phi \underline{\nabla} \psi) = \phi \nabla^2 \psi + \underline{\nabla} \phi \cdot \underline{\nabla} \psi,$$

and the divergence theorem. If $T(\underline{x},t)$ satisfies the heat equation in V and $G(\underline{x},\underline{\xi},t,\tau)$ is a Green's function for the heat equation satisfying (2.57), (2.58) and (2.59), then

$$\int_0^t \int_V G(\underline{x},\underline{\xi},t,\tau)\left(\frac{\partial T}{\partial \tau} - \kappa \nabla_{\underline{\xi}}^2 T\right)dV_{\underline{\xi}}\,d\tau = 0, \tag{2.71}$$

and

$$\int_0^t \int_V T(\underset{\sim}{\xi},\tau)\left(\frac{\partial G}{\partial \tau} + \kappa \nabla_{\underset{\sim}{\xi}}^2 G\right) dV_{\underset{\sim}{\xi}} \, d\tau = -T(\underset{\sim}{x},t). \tag{2.72}$$

Adding (2.71) and (2.72) we have

$$-T(\underset{\sim}{\xi},t) = \int_0^t \int_V \left[\frac{\partial}{\partial \tau}(G\,T) + \kappa\left(T\,\nabla_{\underset{\sim}{\xi}}^2 G - G\,\nabla_{\underset{\sim}{\xi}}^2 T\right)\right] dV_{\underset{\sim}{\xi}} \, d\tau,$$

and on applying (2.59) and Green's theorem (2.70) we have

$$T(\underset{\sim}{x},t) = \int_V G(\underset{\sim}{x},\underset{\sim}{\xi},t,0)f(\underset{\sim}{\xi})\,dV_{\underset{\sim}{\xi}} + \kappa \int_0^t \oint_S \left(G\frac{\partial T}{\partial n_{\underset{\sim}{\xi}}} - T\frac{\partial G}{\partial n_{\underset{\sim}{\xi}}}\right) dS_{\underset{\sim}{\xi}} \, d\tau, \tag{2.73}$$

where $f(\underset{\sim}{\xi})$ denotes the temperature at time zero. It should be emphasized that the temperature $T(\underset{\sim}{x},t)$ given by (2.73) can be interpreted as consisting of two independent contributions. Firstly there is the component due to the initial temperature in the body

$$\int_V G(\underset{\sim}{x},\underset{\sim}{\xi},t,0)f(\underset{\sim}{\xi})\,dV_{\underset{\sim}{\xi}},$$

which is independent of the boundary conditions, and secondly there is the component due to the boundary conditions

$$\kappa \int_0^t \oint_S \left(G\frac{\partial T}{\partial n_{\underset{\sim}{\xi}}} - T\frac{\partial G}{\partial n_{\underset{\sim}{\xi}}}\right) dS_{\underset{\sim}{\xi}} \, d\tau,$$

which is independent of the initial temperature.

As before, we can choose either a simple known Green's function such as

$$G(\underset{\sim}{x},\underset{\sim}{\xi},t,\tau) = H(t - \tau)K_3(\underset{\sim}{x} - \underset{\sim}{\xi},t - \tau),$$

where $K_3(\underset{\sim}{x},t)$ is the three dimensional unit source solution $(2.7)_3$, and obtain an integral equation for $T(\underset{\sim}{x},t)$ in terms of the initial and boundary values of the temperature and heat flux from (2.73) or we can impose boundary conditions on $G(\underset{\sim}{x},\underset{\sim}{\xi},t,\tau)$ to eliminate unknown temperatures or heat fluxes on the surface S. Specifically, if we have a prescribed temperature surface condition

$$T(\underset{\sim}{x},t) = g(\underset{\sim}{x},t), \quad \underset{\sim}{x} \in S, \quad t > 0,$$

then by imposing the condition

$$G(\underset{\sim}{x},\underset{\sim}{\xi},t,\tau) = 0, \quad \underset{\sim}{\xi} \in S,$$

we reduce (2.73) to

$$T(\underset{\sim}{x},t) = \int_V G(\underset{\sim}{x},\underset{\sim}{\xi},t,0)f(\underset{\sim}{\xi})\,dV_{\underset{\sim}{\xi}} - \kappa \int_0^t \oint_S g(\underset{\sim}{\xi},\tau)\frac{\partial G}{\partial n_{\underset{\sim}{\xi}}}(\underset{\sim}{x},\underset{\sim}{\xi},t,\tau)\,dS_{\underset{\sim}{\xi}}\,d\tau.$$

Similarly, if we have a prescribed heat flux condition at the surface

$$k\frac{\partial T}{\partial n}(\underset{\sim}{x},t) = h(\underset{\sim}{x},t), \quad \underset{\sim}{x} \in S, \quad t > 0,$$

then by choosing G so that

$$\frac{\partial G}{\partial n_{\underset{\sim}{\xi}}}(\underset{\sim}{x},\underset{\sim}{\xi},t,\tau) = 0, \quad \underset{\sim}{\xi} \in S,$$

we reduce (2.73) to

$$T(\underset{\sim}{x},t) = \int_V G(\underset{\sim}{x},\underset{\sim}{\xi},t,0)f(\underset{\sim}{\xi})\,dV_{\underset{\sim}{\xi}} + \frac{\kappa}{k} \int_0^t \oint_S h(\underset{\sim}{\xi},\tau)G(\underset{\sim}{x},\underset{\sim}{\xi},t,\tau)\,dS_{\underset{\sim}{\xi}}\,d\tau.$$

PROBLEMS

1. Show that for the heat equation in n-spatial dimensions

$$\frac{\partial T}{\partial t} = \kappa\left(\frac{\partial^2 T}{\partial x_1^2} + \frac{\partial^2 T}{\partial x_2^2} + \cdots + \frac{\partial^2 T}{\partial x_n^2}\right),$$

the characteristics are given by $t = c$ where c is a constant. $\Big[$Hint, assume the characteristics are given by the equation $\gamma(x_1, x_2, \ldots, x_n, t) = t - \phi(x_1, x_2, \ldots, x_n) = 0$ and use equation (2.2). $\Big]$

2. Consider the initial value problem

$$\left.\begin{array}{c} \dfrac{\partial T}{\partial t} = \kappa\dfrac{\partial^2 T}{\partial x^2}, \\[2ex] T(x,0) = \dfrac{1}{1 + x^2}, \quad -\infty < x < \infty. \end{array}\right\}$$

Following Kowalevski show that the solution of this problem (as given, for example, by the one dimensional Poisson transform) is not analytic about the point $(0,0)$ in (x,t) space, even though the initial temperature $1/(1 + x^2)$ is analytic about $x = 0$. $\Big[$Hint, assume that $T(x,t)$ is analytic about $x = 0$, $t = 0$, so that the series

$$T(x,t) = \sum_{m=0}^{\infty}\sum_{n=0}^{\infty} c_{m,n}x^m t^n,$$

converges in some neighbourhood of $(x,t) = (0,0)$, show that the coefficients $c_{m,n}$ satisfy the recursion relations

$$c_{m,n+1} = \kappa\frac{(m+2)(m+1)}{(n+1)}c_{m+2,n}, \qquad m \geq 0, \quad n \geq 0,$$

and

$$c_{2m,0} = (-1)^m, \qquad c_{2m+1,0} = 0, \quad m \geq 0.$$

Hence show that

$$c_{2m,n} = \kappa^n(-1)^{m+n}\frac{(2m + 2n)!}{n!(2m)!},$$

and thus prove that the assumed series has zero radius of convergence so that $T(x,t)$ is not analytic at $(x,t) = (0,0)$. $\Big]$Why does this not contradict the Cauchy-Kowalevski theorem?

3.   Use the formal series solution (2.5) to show that the one dimensional Cauchy problem

$$\frac{\partial T}{\partial t} = \kappa \frac{\partial^2 T}{\partial x^2},$$

$$T(0,t) = e^{-t}, \qquad \frac{\partial T}{\partial x}(0,t) = 0,$$

has the solution $T(x,t) = e^{-t}\cos(x/\sqrt{\kappa})$.

4.   Using the three dimensional Poisson transform (2.8), solve

$$\left.\begin{array}{c}
\dfrac{\partial T}{\partial t} = \kappa \nabla^2 T, \quad -\infty < x < \infty, \\[2ex]
T(\underline{x},t) \to 0 \text{ as } |x| \to \infty, \\[2ex]
T(\underline{x},0) = q_0\delta(x)\delta(y).
\end{array}\right\}$$

This represents the situation where initially $\rho c q_0$ units of heat are distributed uniformly along the z axis, that is, there is a uniform line source along the z axis. Compare your answer with the two dimensional source solution (1.63).

5.   Solve the problem

$$\frac{\partial T}{\partial t} = \kappa \frac{\partial^2 T}{\partial x^2},$$

$$T(x,0) = \begin{cases} 0, & |x| > 1, \\ 1, & |x| \leq 1, \end{cases}$$

and express your answer in terms of error functions. Discuss the behaviour of the solution $T(x,t)$ as $t \to 0$, paying particular attention to the points $x = 1$ and $x = -1$. Show also that $\lim\limits_{|x| \to \infty} \frac{\partial T}{\partial x}(x,t) \to 0$ for all $t \geq 0$.

6.   If $K_1(x,t)$ denotes the one dimensional source solution of unit strength, the derived source solution $H(x,t)$ is defined by

$$H(x,t) \equiv \frac{\partial K_1}{\partial x}(x,t) = \frac{1}{4\sqrt{\pi}} \frac{-x}{(\kappa t)^{3/2}} \exp\left(\frac{-x^2}{4\kappa t}\right).$$

Show that for any fixed x

$$\lim_{t \to 0} H(x,t) = 0.$$

Show also that $H(x,t)$ satisfies the heat equation for $t > 0$. (Recall that the one dimensional source solution $K_1(x,t)$ satisfies the heat equation and is infinitely differentiable for all $t > 0$.) Thus the derived source solution $H(x,t)$ vanishes on the characteristic $t = 0$, but is non-zero for $t > 0$. Show that there are no constants $\epsilon > 0$ and $M > 0$ such that $|H(x,t)| \leq Me^{\epsilon x^2}$ for any time interval $0 < t < t_0$.

7. Let

$$\Psi(x) = \int_0^\infty \exp -\left(\eta + \frac{x}{\eta}\right)^2 d\eta.$$

(i) For $x > 0$ show that

$$\Psi(x) = \int_0^\infty \frac{x}{\xi^2} \exp -\left(\xi + \frac{x}{\xi}\right)^2 d\xi,$$

$$\left[\text{Hint, let } \xi = x/\eta.\right]$$

(ii) Using

$$\Psi(x) = \frac{1}{2} \int_0^\infty \left(1 + \frac{x}{\eta^2}\right) \exp -\left(\eta + \frac{x}{\eta}\right)^2 d\eta,$$

show that for $x > 0$ $\Psi(x)$ satisfies the ordinary differential equation

$$\frac{d\Psi}{dx} = -4\Psi,$$

and hence as $\Psi(x)$ is continuous at $x = 0$, deduce that

$$\Psi(x) = \frac{\sqrt{\pi}}{2} e^{-4x}.$$

8. Show that the Laplace transform of the function

$$\psi(t) = \delta(t) - \beta e^{-\beta t},$$

is

$$\hat{\psi}(p) = 1 - \frac{\beta}{p + \beta}.$$

Hence, using (2.27) show that the Laplace transform of

$$\psi(t) = \delta(t - x) - \beta H(t - x)e^{-\beta(t - x)},$$

is

$$\hat{\psi}(p) = \frac{p}{p + \beta} e^{-px}.$$

9.   Continuation. If

$$\psi(\xi) = \delta(\xi - x) - \beta H(\xi - x)e^{-\beta(\xi - x)},$$

show that

$$\int_0^\infty \frac{\psi(\xi)}{\sqrt{\pi\kappa t}}\exp\!\left(\frac{-\xi^2}{4\kappa t}\right)d\xi = \frac{1}{\sqrt{\pi\kappa t}}\exp\!\left(\frac{-x^2}{4\kappa t}\right) - \beta\exp\!\left(\beta x + \beta^2\kappa t\right)\mathrm{erfc}\!\left(\frac{x}{2\sqrt{\kappa t}} + \beta\sqrt{\kappa t}\right).$$

Thus from (2.38) and using the results of the previous question, show that the Laplace transform of

$$\frac{1}{\sqrt{\pi\kappa t}}\exp\!\left(\frac{-x^2}{4\kappa t}\right) - \beta\exp\!\left(\beta x + \beta^2\kappa t\right)\mathrm{erfc}\!\left(\frac{x}{2\sqrt{\kappa t}} + \beta\sqrt{\kappa t}\right),$$

is

$$\frac{e^{-qx}}{\kappa(q + \beta)}.$$

10.   Show that

$$1 + 2\sum_{n=1}^{m}\cos 2n\theta = \frac{\sin(2m + 1)\theta}{\sin\theta}.$$

$$\left[\text{Hint, evaluate } 1 + 2\sum_{n=1}^{m}e^{2n\theta i}.\right]$$

11.   Expand the function

$$f(x) = x^2, \quad 0 \le x \le \pi,$$

in a Fourier cosine series and show that

$$f(x) = \frac{\pi^2}{3} + 4\sum_{n=1}^{\infty}\frac{(-1)^n}{n^2}\cos nx.$$

Hence deduce that

$$\sum_{n=1}^{\infty}\frac{(-1)^n}{n^2} = \frac{-\pi^2}{12}, \qquad \sum_{n=1}^{\infty}\frac{1}{n^2} = \frac{\pi^2}{6}.$$

12.   Show that the solution of the heat conduction problem in a volume V with surface S and Newton cooling into a medium at temperature $T_0(\underline{x}, t)$, namely

$$\left.\begin{aligned}
\frac{\partial T}{\partial t} &= \kappa\nabla^2 T, \quad \underline{x} \in V, \\[4pt]
T(\underline{x}, 0) &= f(\underline{x}), \quad \underline{x} \in V, \\[4pt]
\frac{\partial T}{\partial n}(\underline{x}, t) + \mu T(\underline{x}, t) &= \mu T_0(\underline{x}, t), \quad \underline{x} \in S,
\end{aligned}\right\}$$

where $\mu = h/k$, can be written as

$$T(\underline{x}, t) = \int_V f(\underline{\xi}) G(\underline{x}, \underline{\xi}, t, 0) \, dV_{\underline{\xi}} - \kappa \int_0^t \oint_S T_0(\underline{\xi}, \tau) \frac{\partial G}{\partial n_{\underline{\xi}}} (\underline{x}, \underline{\xi}, t, \tau) \, dS_{\underline{\xi}} \, d\tau,$$

where $G(\underline{x}, \underline{\xi}, t, \tau)$ is a Green's function satisfying

$$\frac{\partial G}{\partial n_{\underline{\xi}}} (\underline{x}, \underline{\xi}, t, \tau) + \mu G(\underline{x}, \underline{\xi}, t, \tau) = 0, \quad \underline{\xi} \in S.$$

$\left[\text{Observe that this is equivalent to} \right.$

$$T(\underline{x}, t) = \int_V f(\underline{\xi}) G(\underline{x}, \underline{\xi}, t, 0) \, dV_{\underline{\xi}} + \mu \kappa \int_0^t \oint_S T_0(\underline{\xi}, \tau) G(\underline{x}, \underline{\xi}, t, \tau) \, dS_{\underline{\xi}} \, d\tau. \left. \right]$$

Chapter Three

Exact analytical solutions for semi-infinite media

3.1 Introduction

In this chapter we consider three general problems for the semi-infinite one dimensional region $0 \leq x < \infty$ and, by way of illustration, one special two dimensional problem for the semi-infinite wedge. In the next three sections we consider the semi-infinite one dimensional region $0 \leq x < \infty$, initially at temperature $f(x)$ with the surface $x = 0$ subject to respectively the boundary conditions of prescribed temperature, prescribed heat flux and Newton cooling into a medium of prescribed temperature. Thus we consider the following three problems for the temperature $T(x,t)$

$$
\left.
\begin{aligned}
\frac{\partial T}{\partial t} &= \kappa \frac{\partial^2 T}{\partial x^2}, \quad 0 < x < \infty, \\[2mm]
T(x,0) &= f(x), \quad 0 \leq x < \infty, \\[2mm]
\frac{\partial T}{\partial x}(x,t) &\to 0 \text{ as } x \to \infty,
\end{aligned}
\right\}
$$

and one of the following boundary conditions for $t > 0$ at $x = 0$

$$
\text{(i)} \qquad T(0,t) = g(t),
$$

$$
\text{(ii)} \qquad -k\frac{\partial T}{\partial x}(0,t) = h(t),
$$

$$
\text{(iii)} \qquad -\frac{\partial T}{\partial x}(0,t) + \mu T(0,t) = \mu T_0(t),
$$

where $g(t)$, $h(t)$ and $T_0(t)$ are assumed known functions of time and we use the notation that $\mu = h/k$ where h is the surface conductance and k is the thermal conductivity.

Formal solutions of these problems are deduced by means of the Laplace transform with respect to time, namely

$$
\hat{T}(x,p) = \int_0^\infty e^{-pt} T(x,t)\,dt,
$$

and for the inversion of the Laplace transform we make use of the table of Laplace transforms. Of particular importance are the transforms

$$e^{-qx} = \int_0^\infty e^{-pt} \frac{x}{2\sqrt{\pi \kappa t^3}} \exp\left(\frac{-x^2}{4\kappa t}\right) dt, \quad \frac{e^{-qx}}{q} = \int_0^\infty e^{-pt} \sqrt{\frac{\kappa}{\pi t}} \exp\left(\frac{-x^2}{4\kappa t}\right) dt,$$

which are used extensively and where as usual q denotes $\sqrt{p/\kappa}$. The Green's functions for these problems are noted and the general solutions are interpreted in terms of them. In this and subsequent chapters we follow the convention of letting u denote the temperature due to the initial condition and zero boundary conditions and v denote the temperature arising from a zero initial temperature and non-zero boundary conditions so that in in general $T = u + v$.

In the final two sections of the chapter we deal with the problem of the semi-infinite wedge of angle θ_0 which is initially at zero temperature and its faces $\theta = 0$ and $\theta = \theta_0$ are subjected to a prescribed constant temperature v_0. Since the problem remains unchanged by a stretching transformation, the temperature $v(x, y, t)$ may be reduced to a function of two variables only, namely $x/\sqrt{t}$ and $y/\sqrt{t}$ and the problem may then be solved by a Fourier sine series. This problem involves a number of Bessel function formulae for which we adopt the procedure of carefully stating all such essential formulae as and when they are required.

3.2 Surface $x = 0$ with prescribed temperature $g(t)$

We first consider linear flow for the semi-infinite solid $0 \le x < \infty$ with an arbitrary initial temperature $f(x)$ and arbitrary prescribed temperature on the surface $x = 0$. Thus we require the solution of the problem

$$\left. \begin{aligned} \frac{\partial T}{\partial t} &= \kappa \frac{\partial^2 T}{\partial x^2}, \quad 0 < x < \infty, \\ T(x, 0) &= f(x), \quad 0 \le x < \infty, \\ T(0, t) &= g(t), \quad t > 0, \qquad \frac{\partial T}{\partial x}(x, t) \to 0 \text{ as } x \to \infty. \end{aligned} \right\} \tag{3.1}$$

On taking the Laplace transform of (3.1) with respect to time, the problem for $\hat{T}(x, p)$ becomes (see (2.33) and (2.35))

$$\frac{\partial^2 \hat{T}}{\partial x^2} - q^2 \hat{T} = -\frac{f(x)}{\kappa}, \quad 0 < x < \infty,$$

$$\hat{T}(0,p) = \hat{g}(p), \quad \frac{\partial \hat{T}}{\partial x}(x,p) \to 0 \text{ as } x \to \infty, \tag{3.2}$$

where as usual $q = \sqrt{p/\kappa} > 0$.

We solve $(3.2)_1$ by the method of *variation of parameters*. Since the general solution of

$$\frac{\mathrm{d}^2 \tilde{T}}{\mathrm{d}x^2} - q^2 \tilde{T} = 0,$$

is

$$\tilde{T}(x) = C_1 e^{-qx} + C_2 e^{qx},$$

where C_1 and C_2 are arbitrary constants, we assume that the general solution of $(3.2)_1$ takes the form

$$\hat{T}(x,p) = A(x,p)e^{-qx} + B(x,p)e^{qx}, \tag{3.3}$$

where $A(x,p)$ and $B(x,p)$ are unknown functions of x and p. If we assume that

$$\frac{\partial A}{\partial x}e^{-qx} + \frac{\partial B}{\partial x}e^{qx} = 0, \tag{3.4}$$

and substitute (3.3) into $(3.2)_1$, we find that

$$-\frac{\partial A}{\partial x}e^{-qx} + \frac{\partial B}{\partial x}e^{qx} = \frac{-f(x)}{\kappa q}, \tag{3.5}$$

giving us a system of two differential equations in two unknowns. Subtracting (3.5) from (3.4) leads to

$$\frac{\partial A}{\partial x} = \frac{f(x)}{2\kappa q}e^{qx},$$

which can be integrated to give

$$A(x,p) = \frac{1}{2\kappa q}\int_0^x f(\xi)e^{q\xi}\,\mathrm{d}\xi + C(p),$$

where $C(p)$ is an as yet undetermined function of p. Similarly, adding (3.4) and (3.5) gives

$$\frac{\partial B}{\partial x} = \frac{-f(x)}{2\kappa q}e^{-qx},$$

which integrates to give

$$B(x,p) = \frac{-1}{2\kappa q} \int_0^x f(\xi)e^{-q\xi}\, d\xi + D(p),$$

where $D(p)$ is an undetermined function of p. Altogether we have

$$\hat{T}(x,p) = \frac{1}{2\kappa q} \int_0^x f(\xi)e^{q(\xi-x)}\, d\xi - \frac{1}{2\kappa q} \int_0^x f(\xi)e^{-q(\xi-x)}\, d\xi$$

$$+ C(p)e^{-qx} + D(p)e^{qx} \tag{3.6}$$

$$= \frac{1}{\kappa q} \int_0^x f(\xi)\sinh q(\xi-x)\, d\xi + C(p)e^{-qx} + D(p)e^{qx}.$$

We now determine the functions $C(p)$ and $D(p)$ from the boundary conditions $(3.2)_2$ and $(3.2)_3$, namely

$$C(p) + D(p) = \hat{g}(p), \quad D(p) = \frac{1}{2\kappa q} \int_0^\infty e^{-q\xi} f(\xi)\, d\xi,$$

where we have obtained the expression for $D(p)$ by differentiating (3.6) and equating the exponentially increasing terms as $x \to \infty$ in the resulting expression for $\frac{\partial \hat{T}}{\partial x}$. Altogether the solution of (3.2) becomes

$$\hat{T}(x,p) = \frac{e^{-qx}}{\kappa q} \int_0^x f(\xi)\sinh q\xi\, d\xi + \frac{\sinh qx}{\kappa q} \int_x^\infty e^{-q\xi} f(\xi)\, d\xi + \hat{g}(p)e^{-qx}. \tag{3.7}$$

On using the table of Laplace transforms for

$$e^{-qx} \text{ and } \frac{e^{-qx}}{q},$$

and the convolution theorem, we may invert this expression to obtain

$$T(x,t) = \frac{1}{2\sqrt{\pi\kappa t}} \int_0^\infty f(\xi)\left\{\exp\frac{-(x-\xi)^2}{4\kappa t} - \exp\frac{-(x+\xi)^2}{4\kappa t}\right\} d\xi$$

$$+ \frac{x}{2\sqrt{\pi\kappa}} \int_0^t \frac{g(\tau)}{(t-\tau)^{3/2}} \exp\frac{-x^2}{4\kappa(t-\tau)}\, d\tau. \tag{3.8}$$

We observe that the infinite integral arises from the first two terms in the right hand side of (3.7), since on taking the Laplace transform of this integral we have

$$\int_0^\infty \frac{e^{-pt}}{2\sqrt{\pi\kappa t}} \int_0^\infty f(\xi)\left\{\exp\frac{-(x-\xi)^2}{4\kappa t} - \exp\frac{-(x+\xi)^2}{4\kappa t}\right\}d\xi\,dt$$

$$= \frac{1}{2\kappa q}\int_0^\infty f(\xi)\left\{e^{-q|x-\xi|} - e^{-q(x+\xi)}\right\}d\xi$$

$$= \frac{1}{2\kappa q}\left\{\int_0^x f(\xi)e^{-q(x-\xi)}\,d\xi + \int_x^\infty f(\xi)e^{-q(\xi-x)}\,d\xi\right.$$

$$\left. - \int_0^x f(\xi)e^{-q(x+\xi)}\,d\xi - \int_x^\infty f(\xi)e^{-q(x+\xi)}\,d\xi\right\}$$

$$= \frac{e^{-qx}}{\kappa q}\int_0^x f(\xi)\sinh q\xi\,d\xi + \frac{\sinh qx}{\kappa q}\int_x^\infty e^{-q\xi}f(\xi)\,d\xi.$$

We note that (3.8) consists of two parts, namely the temperature due to the initial condition $f(x)$ with a prescribed temperature of zero at $x = 0$ and the temperature arising from a zero initial temperature and a prescribed surface temperature $g(t)$ at $x = 0$. Accordingly, the first integral may be deduced from (2.9) by assuming that the solid is continued along the negative x-axis such that the initial temperature at $-x$ is $-f(x)$, so that $f(x)$ is extended to an odd function. By symmetry the solution given by (2.9) vanishes at the origin for all time $t > 0$ if the initial temperature is an odd function. Thus from (2.9) we have the contribution

$$\frac{1}{2\sqrt{\pi\kappa t}}\left\{\int_0^\infty f(\xi)\exp\frac{-(x-\xi)^2}{4\kappa t}\,d\xi + \int_{-\infty}^0 -f(-\xi)\exp\frac{-(x-\xi)^2}{4\kappa t}\,d\xi\right\},$$

which reduces to the first integral in (3.8). Further, we observe that by means of the substitution

$$\eta = \frac{x}{2\sqrt{\kappa(t-\tau)}}, \tag{3.9}$$

so that

$$\tau = t - \frac{x^2}{4\kappa\eta^2}, \quad d\eta = \frac{x\,d\tau}{4\sqrt{\kappa(t-\tau)^3}}, \tag{3.10}$$

and the second integral in (3.8) admits the alternative form

$$\frac{2}{\sqrt{\pi}}\int_{x/2\sqrt{\kappa t}}^\infty e^{-\eta^2}g\left(t - \frac{x^2}{4\kappa\eta^2}\right)d\eta.$$

Thus altogether, with some simple rearrangement of the first integral (3.8) becomes

$$T(x,t) = \frac{e^{-x^2/4\kappa t}}{\sqrt{\pi \kappa t}} \int_0^\infty f(\xi) \exp\left(\frac{-\xi^2}{4\kappa t}\right) \sinh\left(\frac{x\xi}{2\kappa t}\right) d\xi$$

$$+ \frac{2}{\sqrt{\pi}} \int_{x/2\sqrt{\kappa t}}^\infty e^{-\eta^2} g\left(t - \frac{x^2}{4\kappa \eta^2}\right) d\eta .$$

(3.11)

From (3.11) it is clear that $T(0, t) = g(t)$.

Example 3.1 For the semi-infinite solid $0 \le x < \infty$ with constant initial temperature u_0 and prescribed zero temperature along the surface $x = 0$, show that the temperature (see Figure 3.1) is given by

$$u(x, t) = u_0 \operatorname{erf}\left(\frac{x}{2\sqrt{\kappa t}}\right),$$

where the error function $\operatorname{erf}(x)$ is defined by (2.11).

From (3.8) with $f(x) = u_0$ and $g(t)$ zero, we have

$$u(x, t) = \frac{u_0}{2\sqrt{\pi \kappa t}} \int_0^\infty \left\{\exp \frac{-(x - \xi)^2}{4\kappa t} - \exp \frac{-(x + \xi)^2}{4\kappa t}\right\} d\xi ,$$

and the respective substitutions

$$\xi = x + 2\eta\sqrt{\kappa t}, \quad \xi = -x + 2\eta\sqrt{\kappa t}, \tag{3.12}$$

in the first and second parts of this integral yield

$$u(x, t) = \frac{u_0}{\sqrt{\pi}} \int_{-x/2\sqrt{\kappa t}}^\infty e^{-\eta^2} d\eta - \frac{u_0}{\sqrt{\pi}} \int_{x/2\sqrt{\kappa t}}^\infty e^{-\eta^2} d\eta$$

$$= \frac{u_0}{\sqrt{\pi}} \int_{-x/2\sqrt{\kappa t}}^{x/2\sqrt{\kappa t}} e^{-\eta^2} d\eta$$

$$= \frac{2u_0}{\sqrt{\pi}} \int_0^{x/2\sqrt{\kappa t}} e^{-\eta^2} d\eta ,$$

and hence the given expression follows.

<u>Example 3.2</u> For the semi-infinite solid $0 \le x < \infty$ with zero initial temperature and prescribed constant temperature v_0 along the surface $x = 0$, show that the temperature is given by

$$v(x,t) = v_0 \operatorname{erfc}\left(\frac{x}{2\sqrt{\kappa t}}\right),$$

where the complementary error function $\operatorname{erfc}(x)$ is defined by (2.13).

From (3.8) with $f(x) \equiv 0$ and $g(t) = v_0$, we have on using the alternative form (3.11) for the second integral

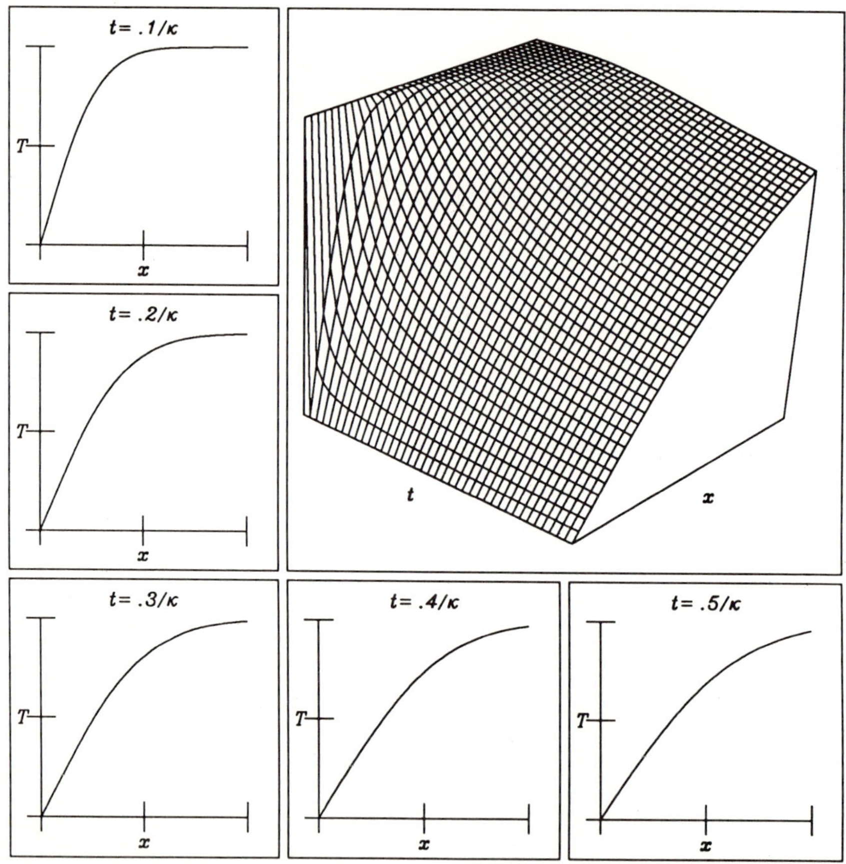

Figure 3.1. Time development of the solution of Example 3.1 with u_0 unity.

$$v(x,t) = \frac{2v_0}{\sqrt{\pi}} \int_{x/2\sqrt{\kappa t}}^{\infty} e^{-\eta^2} \, d\eta,$$

from which the given expression is immediately apparent. In particular we observe that the surface flux is inversely proportional to $\sqrt{t}$ since we have

$$\frac{\partial v}{\partial x}(0,t) = \frac{-v_0}{\sqrt{\pi \kappa t}}.$$

Finally we note that on inspection of (3.8), it is evident that the Green's function $G(x,\xi,t,\tau)$ for the problem (3.1), that is the Green's function which vanishes at $\xi = 0$ (and $\xi = \infty$) and has zero flux at $\xi = \infty$ is given by

$$G(x,\xi,t,\tau) = \frac{H(t-\tau)}{2\sqrt{\pi \kappa (t-\tau)}}\left(\exp \frac{-(x-\xi)^2}{4\kappa(t-\tau)} - \exp \frac{-(x+\xi)^2}{4\kappa(t-\tau)}\right),$$

where $H(x)$ is the Heaviside unit step function (1.15). We leave it to the reader to verify that (3.8) may be written as

$$T(x,t) = \int_0^{\infty} f(\xi)G(x,\xi,t,0)\,d\xi + \kappa \int_0^t g(\tau)\frac{\partial G}{\partial \xi}(x,0,t,\tau)\,d\tau.$$

The Green's function here consists of the difference of two parts, namely

$$H(t-\tau)[K_1(x-\xi,t-\tau) - K_1(x+\xi,t-\tau)],$$

where $K_1(x,t)$ is the one dimensional instantaneous unit point source solution. Note that the second of these terms arises from the first by replacing ξ by $-\xi$, that is by reflecting ξ in the line $\xi = 0$. Thus the Green's function arises as the difference of an instantaneous point source of unit strength located at ξ and an instantaneous point source of unit strength located at $-\xi$. Alternatively, we can think of this as the sum of an instantaneous point *source* of unit strength at ξ and an instantaneous point *sink* at $-\xi$. Since

$$G(x,-\xi,t,\tau) = -G(x,\xi,t,\tau),$$

it follows that

$$G(x,0,t,\tau) = 0.$$

This reflects the physical situation where the effects of the point source at ξ and the point sink at $-\xi$ cancel each other along the line of symmetry $\xi = 0$ (see Figure 3.2). Thus by the introduction of a fictitious point source at $-\xi$, outside the region of interest, a Green's function for the problem (3.1) can be found. The construction of a Green's function in this essentially geometric manner, by exploiting the symmetry of the problem at hand, is known as the method of images.

3.3 Surface $x = 0$ with prescribed flux $h(t)$

We now consider linear flow for the semi-infinite solid $0 \le x < \infty$ with an arbitrary initial temperature $f(x)$ and arbitrary prescribed flux on the surface $x = 0$. Thus we consider the problem

$$\left.\begin{array}{l}
\dfrac{\partial T}{\partial t} = \kappa \dfrac{\partial^2 T}{\partial x^2}, \quad 0 < x < \infty, \\[2ex]
T(x,0) = f(x), \quad 0 \le x < \infty, \\[2ex]
-k\dfrac{\partial T}{\partial x}(0,t) = h(t), \quad t > 0, \qquad \dfrac{\partial T}{\partial x}(x,t) \to 0 \text{ as } x \to \infty.
\end{array}\right\} \qquad (3.13)$$

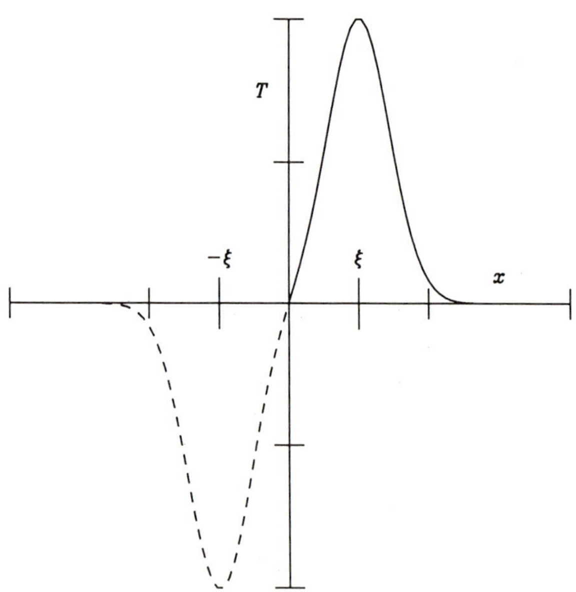

Figure 3.2. Source and sink for geometric derivation of Green's function by method of images (Section 3.2).

On taking the Laplace transform of (3.13) with respect to time we may again deduce $(3.2)_1$ and $(3.2)_3$, but in place of $(3.2)_2$ we have

$$-k\frac{\partial \hat{T}}{\partial x}(0,p) = \hat{h}(p).$$

Thus from this equation, $(3.2)_3$ and (3.6) we may deduce

$$C(p) - D(p) = \frac{\hat{h}(p)}{kq}, \qquad D(p) = \frac{1}{2\kappa q}\int_0^\infty e^{-q\xi}f(\xi)\,\mathrm{d}\xi,$$

and (3.6) becomes

$$\hat{T}(x,p) = \frac{e^{-qx}}{\kappa q}\int_0^x f(\xi)\cosh q\xi \,\mathrm{d}\xi + \frac{\cosh qx}{\kappa q}\int_x^\infty e^{-q\xi}f(\xi)\,\mathrm{d}\xi + \frac{\hat{h}(p)e^{-qx}}{kq}.$$

$$(3.14)$$

Again on using the table of Laplace transforms for

$$\frac{e^{-qx}}{q},$$

and the convolution theorem we may invert this expression to obtain

$$T(x,t) = \frac{1}{2\sqrt{\pi\kappa t}}\int_0^\infty f(\xi)\left\{\exp\frac{-(x-\xi)^2}{4\kappa t} + \exp\frac{-(x+\xi)^2}{4\kappa t}\right\}\mathrm{d}\xi$$

$$+ \frac{1}{k}\sqrt{\frac{\kappa}{\pi}}\int_0^t \frac{h(\tau)}{\sqrt{t-\tau}}\exp\frac{-x^2}{4\kappa(t-\tau)}\,\mathrm{d}\tau.$$

$$(3.15)$$

As in the previous section, we may confirm the infinite integral in (3.15) since

$$\int_0^\infty \frac{e^{-pt}}{2\sqrt{\pi\kappa t}}\int_0^\infty f(\xi)\left\{\exp\frac{-(x-\xi)^2}{4\kappa t} + \exp\frac{-(x+\xi)^2}{4\kappa t}\right\}\mathrm{d}\xi$$

$$= \frac{1}{2\kappa q}\int_0^\infty f(\xi)\left\{e^{-q|x-\xi|} + e^{-q(x+\xi)}\right\}\mathrm{d}\xi$$

$$= \frac{1}{2\kappa q}\left\{\int_0^x f(\xi)e^{-q(x-\xi)}\,\mathrm{d}\xi + \int_x^\infty f(\xi)e^{-q(\xi-x)}\,\mathrm{d}\xi\right.$$

$$\left. + \int_0^x f(\xi)e^{-q(x+\xi)}\,\mathrm{d}\xi + \int_x^\infty f(\xi)e^{-q(x+\xi)}\,\mathrm{d}\xi\right\}$$

$$= \frac{e^{-qx}}{\kappa q}\int_0^x f(\xi)\cosh q\xi \,\mathrm{d}\xi + \frac{\cosh qx}{\kappa q}\int_x^\infty e^{-q\xi}f(\xi)\,\mathrm{d}\xi.$$

Again it is apparent that (3.15) consists of two parts, namely that arising from an initial temperature $f(x)$ with zero prescribed flux on the surface $x = 0$ and that arising from zero initial temperature and prescribed flux at $x = 0$. Further, the first integral may be deduced from (2.9) by assuming that the solid is continued on the negative side of $x = 0$ such that the initial temperature at $-x$ is $f(x)$, so that $f(x)$ is extended to an even function. Thus from (2.9) we have the contribution

$$\frac{1}{2\sqrt{\pi \kappa t}} \left\{ \int_0^\infty f(\xi) \exp \frac{-(x - \xi)^2}{4\kappa t}\, \mathrm{d}\xi + \int_{-\infty}^0 f(-\xi) \exp \frac{-(x - \xi)^2}{4\kappa t}\, \mathrm{d}\xi \right\},$$

which is clearly the first integral in (3.15). Using the substitution (3.9) and (3.10) we observe that the second integral in (3.15) admits the alternative form

$$\frac{x}{k\sqrt{\pi}} \int_{x/2\sqrt{\kappa t}}^\infty e^{-\eta^2} h\!\left(t - \frac{x^2}{4\kappa \eta^2}\right) \frac{\mathrm{d}\eta}{\eta^2}.$$

On differentiating this with respect to x we have

$$\frac{1}{k\sqrt{\pi}} \int_{x/2\sqrt{\kappa t}}^\infty e^{-\eta^2} h\!\left(t - \frac{x^2}{4\kappa \eta^2}\right) \frac{\mathrm{d}\eta}{\eta^2}$$

$$-\frac{x^2}{2k\kappa\sqrt{\pi}} \int_{x/\sqrt{\kappa t}}^\infty e^{-\eta^2} h'\!\left(t - \frac{x^2}{4\kappa \eta^2}\right) \frac{\mathrm{d}\eta}{\eta^4} - \frac{2h(0)}{kx} \sqrt{\frac{\kappa t}{\pi}} \exp\!\left(\frac{-x^2}{4\kappa t}\right),$$

which on using integration by parts for the first integral gives simply

$$-\frac{2}{k\sqrt{\pi}} \int_{x/2\sqrt{\kappa t}}^\infty e^{-\eta^2} h\!\left(t - \frac{x^2}{4\kappa \eta^2}\right) \mathrm{d}\eta,$$

and, for any $t > 0$, in the limit as x tends to zero this gives $-h(t)/k$ as required. Thus with some simple rearrangement of the first integral (3.15) becomes

$$T(x, t) = \frac{1}{\sqrt{\pi \kappa t}} \exp\!\left(\frac{-x^2}{4\kappa t}\right) \int_0^\infty f(\xi) \exp\!\left(\frac{-\xi^2}{4\kappa t}\right) \cosh\!\left(\frac{x\xi}{2\kappa t}\right) \mathrm{d}\xi$$

$$+ \frac{x}{k\sqrt{\pi}} \int_{x/2\sqrt{\kappa t}}^\infty e^{-\eta^2} h\!\left(t - \frac{x^2}{4\kappa \eta^2}\right) \frac{\mathrm{d}\eta}{\eta^2}.$$

$$(3.16)$$

<u>Example 3.3</u> For the semi-infinite solid $0 \le x < \infty$ with zero initial temperature and prescribed constant flux w_0 at the surface $x = 0$, show that the temperature (see Figure 3.3) is given by

$$v(x,t) = \frac{w_0}{k}\left\{ 2\sqrt{\frac{\kappa t}{\pi}}\,\exp\left(\frac{-x^2}{4\kappa t}\right) - x\,\mathrm{erfc}\left(\frac{x}{2\sqrt{\kappa t}}\right)\right\}.$$

From (3.15) with $f(x) \equiv 0$ and $h(t) = w_0$ we have

$$T(x,t) = \frac{w_0}{k}\sqrt{\frac{\kappa}{\pi}}\int_0^t \frac{1}{\sqrt{t-\tau}}\,\exp\left(\frac{-x^2}{4\kappa(t-\tau)}\right)\mathrm{d}\tau = \frac{w_0 x}{k\sqrt{\pi}}\int_{x/2\sqrt{\kappa t}}^\infty e^{-\eta^2}\frac{\mathrm{d}\eta}{\eta^2},$$

and on using the alternative form (3.16) for this integral we have

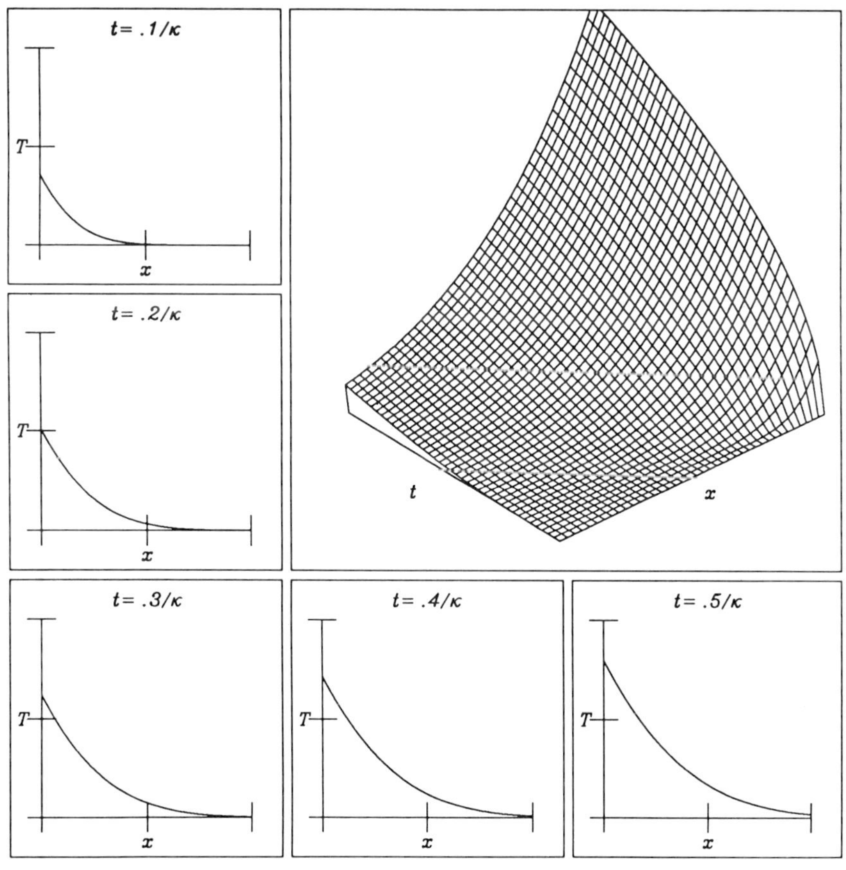

Figure 3.3. Time development of the solution of Example 3.3 with w_0 unity.

$$v(x,t) = \frac{w_0 x}{k\sqrt{\pi}} \int_{x/2\sqrt{\kappa t}}^{\infty} \frac{e^{-\eta^2}}{\eta^2}\, d\eta$$

$$= \frac{w_0 x}{k\sqrt{\pi}} \left\{ \frac{2\sqrt{\kappa t}}{x} \exp\left(\frac{-x^2}{4\kappa t}\right) - 2\int_{x/2\sqrt{\kappa t}}^{\infty} e^{-\eta^2}\, d\eta \right\}$$

$$= \frac{w_0}{k} \left\{ 2\sqrt{\frac{\kappa t}{\pi}} \exp\left(\frac{-x^2}{4\kappa t}\right) - x\,\mathrm{erfc}\left(\frac{x}{2\sqrt{\kappa t}}\right) \right\},$$

as required. In particular, we observe that the surface temperature $v(0,t)$ varies as $\sqrt{t}$, thus

$$v(0,t) = \frac{2w_0}{k}\sqrt{\frac{\kappa t}{\pi}}.$$

To construct a Green's function for the problem (3.13), that is a Green's function $G(x,\xi,t,\tau)$ which satisfies the homogeneous boundary condition at $\xi = 0$, namely

$$\frac{\partial G}{\partial \xi}(x,0,t,\tau) = 0,$$

we start with the fundamental one dimensional Green's function for the heat equation, that is the function

$$H(t-\tau)K_1(x-\xi,t-\tau),$$

where $K_1(x,t)$ is the unit source solution, and apply the method of images to produce a Green's function which is an even function in ξ. Specifically, if we introduce a fictitious unit point source at the point $-\xi$ (outside the region of interest) and time τ, namely

$$H(t-\tau)K_1(x+\xi,t-\tau),$$

and add this to the point source at ξ we obtain the function

$$G(x,\xi,t,\tau) = \frac{H(t-\tau)}{2\sqrt{\pi\kappa(t-\tau)}} \left\{ \exp\frac{-(x-\xi)^2}{4\kappa(t-\tau)} + \exp\frac{-(x+\xi)^2}{4\kappa(t-\tau)} \right\}.$$

By its construction, this function has the property (see Figure 3.4)

$$G(x,\xi,t,\tau) = G(x,-\xi,t,\tau),$$

so that in particular

$$\frac{\partial G}{\partial \xi}(x,0,t,\tau) = 0,$$

and therefore this is the required Green's function. Of course this Green's function is also apparent from inspection of (3.15). We leave it to the reader to verify that (3.15) can be written in terms of $G(x,\xi,t,\tau)$ as

$$T(x,t) = \int_0^\infty f(\xi)G(x,\xi,t,0)\,d\xi + \frac{\kappa}{k}\int_0^t h(\tau)G(x,0,t,\tau)\,d\tau.$$

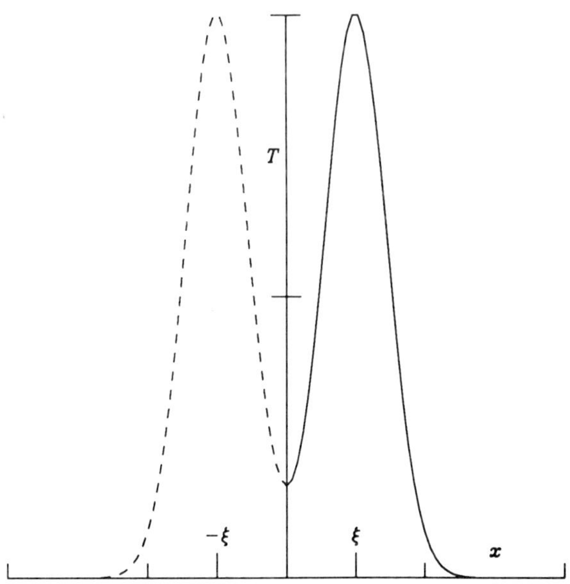

Figure 3.4. Source and sink for geometric derivation of Green's function by method of images (Section 3.3).

3.4 Newton cooling at surface $x = 0$ into a medium of prescribed temperature $T_0(t)$

In this section we consider linear flow for the semi-infinite solid $0 \le x < \infty$ with an arbitrary initial temperature $f(x)$ and Newton cooling at the surface $x = 0$ into a medium of prescribed temperature $T_0(t)$. We therefore consider the problem

$$\left. \begin{aligned} \frac{\partial T}{\partial t} &= \kappa \frac{\partial^2 T}{\partial x^2}, \quad 0 < x < \infty, \\[2mm] T(x,0) &= f(x), \quad 0 \le x < \infty, \\[2mm] -\frac{\partial T}{\partial x}(0,t) + \mu T(0,t) &= \mu T_0(t), \quad t > 0, \qquad \frac{\partial T}{\partial x}(x,t) \to 0 \text{ as } x \to \infty, \end{aligned} \right\} \tag{3.17}$$

where for convenience we have introduced $\mu = h/k$. Again, on taking the Laplace transform of (3.17) we obtain from (3.6) the following equations for $C(p)$ and $D(p)$

$$(q + \mu)C(p) - (q - \mu)D(p) = \mu \hat{T}_0(p), \quad D(p) = \frac{1}{2\kappa q} \int_0^\infty e^{-q\xi} f(\xi)\,d\xi,$$

so that (3.6) may be simplified to yield

$$\hat{T}(x,p) = \frac{e^{-qx}}{\kappa q} \int_0^x f(\xi) \cosh q\xi\,d\xi + \frac{\cosh qx}{\kappa q} \int_x^\infty e^{-q\xi} f(\xi)\,d\xi$$

$$- \frac{\mu e^{-qx}}{\kappa q(q + \mu)} \int_0^\infty e^{-q\xi} f(\xi)\,d\xi + \frac{\mu \hat{T}_0(p) e^{-qx}}{(q + \mu)}.$$

Using the table of Laplace transforms for

$$\frac{e^{-qx}}{q}, \quad \frac{e^{-qx}}{q(q + \mu)}, \quad \frac{e^{-qx}}{(q + \mu)},$$

and the convolution theorem, we may deduce

$$T(x,t) = \int_0^\infty \left\{ \frac{1}{2\sqrt{\pi \kappa t}} \left(\exp \frac{-(x - \xi)^2}{4\kappa t} + \exp \frac{-(x + \xi)^2}{4\kappa t} \right) \right.$$

$$\left. - \mu \exp\left[\mu(x + \xi) + \kappa \mu^2 t \right] \operatorname{erfc}\left[\frac{(x + \xi)}{2\sqrt{\kappa t}} + \mu\sqrt{\kappa t} \right] \right\} f(\xi)\,d\xi$$

$$+ \kappa\mu \int_0^t \left\{ \frac{1}{\sqrt{\pi \kappa (t - \tau)}} \exp\left(\frac{-x^2}{4\kappa (t - \tau)} \right) \right.$$

$$\left. - \mu \exp\left[\mu x + \kappa\mu^2(t - \tau) \right] \operatorname{erfc}\left[\frac{x}{2\sqrt{\kappa (t - \tau)}} + \mu\sqrt{\kappa (t - \tau)} \right] \right\} T_0(\tau)\,d\tau.$$

$$\tag{3.18}$$

An alternative method of solution of the above problem which produces another expression for the temperature is as follows. We introduce $S(x, t)$ defined by

$$S(x, t) = -\frac{\partial T}{\partial x}(x, t) + \mu T(x, t),$$

then on solving for $T(x, t)$ as a standard linear first order partial differential equation we have

$$T(x, t) = \int_0^\infty e^{-\mu(\xi - x)} S(\xi, t) \, d\xi + C(t) e^{\mu x},$$

where $C(t)$ denotes an arbitrary function of time. But since we require no flux at infinity, $C(t)$ must be identically zero and the substitution $\eta = \xi - x$ gives

$$T(x, t) = \int_0^\infty e^{-\mu\eta} S(x + \eta, t) \, d\eta. \tag{3.19}$$

Now, in terms of $S(x, t)$ the problem (3.17) becomes

$$\left. \begin{array}{l} \dfrac{\partial S}{\partial t} = \kappa \dfrac{\partial^2 S}{\partial x^2}, \quad 0 < x < \infty, \\[2mm] S(x, 0) = F(x), \quad 0 \le x < \infty, \\[2mm] S(0, t) = \mu T_0(t), \quad \dfrac{\partial S}{\partial x}(x, t) \to 0 \text{ as } x \to \infty, \end{array} \right\} \tag{3.20}$$

where $F(x)$ is defined by

$$F(x) = -f'(x) + \mu f(x). \tag{3.21}$$

But the solution of this problem is given in Section 3.2, by equation (3.8) which we take in the form

$$S(x, t) = \frac{1}{2\sqrt{\pi \kappa t}} \int_0^\infty F(\xi) \left\{ \exp \frac{-(x - \xi)^2}{4\kappa t} - \exp \frac{-(x + \xi)^2}{4\kappa t} \right\} d\xi$$

$$+ \frac{2\mu}{\sqrt{\pi}} \int_{x/2\sqrt{\kappa t}}^\infty e^{-\xi^2} T_0\!\left(t - \frac{x^2}{4\kappa \xi^2} \right) d\xi,$$

and on using (3.21) we have by integration by parts on the first integral

$$S(x, t) = \frac{1}{2\sqrt{\pi \kappa t}} \int_0^\infty e^{-\mu\xi} f(\xi) \frac{d}{d\xi} \left\{ e^{\mu\xi} \left[\exp \frac{-(x - \xi)^2}{4\kappa t} - \exp \frac{-(x + \xi)^2}{4\kappa t} \right] \right\} d\xi$$

$$+ \frac{2\mu}{\sqrt{\pi}} \int_{x/2\sqrt{\kappa t}}^\infty e^{-\xi^2} T_0\!\left(t - \frac{x^2}{4\kappa \xi^2} \right) d\xi. \tag{3.22}$$

We may reconcile the first integral of (3.18) with equations (3.19) and (3.22) as follows. We need the following integral

$$\int_0^\infty e^{-s\tau} e^{-a\tau^2} \, d\tau = \frac{1}{2}\sqrt{\frac{\pi}{a}} e^{s^2/4a} \, \text{erfc}\left(\frac{s}{2\sqrt{a}}\right),$$

(3.23)

for $a, s > 0$. On using this result we have

$$\int_0^\infty e^{-\mu\eta} \left\{ \exp\frac{-(x+\eta-\xi)^2}{4\kappa t} - \exp\frac{-(x+\eta+\xi)^2}{4\kappa t} \right\} d\xi$$

$$= \sqrt{\pi\kappa t}\, e^{\mu x + \kappa\mu^2 t} \left\{ e^{-\mu\xi} \, \text{erfc}\left[\frac{(x-\xi)}{2\sqrt{\kappa t}} + \mu\sqrt{\kappa t}\right] - e^{\mu\xi} \, \text{erfc}\left[\frac{(x+\xi)}{2\sqrt{\kappa t}} + \mu\sqrt{\kappa t}\right] \right\},$$

so that from (3.19) and (3.22) on interchanging orders of integration and performing the differentiation with respect to ξ in equation (3.22) we may show that the first integral of (3.22) generates the first integral of (3.18). Thus altogether the solution (3.18) of problem (3.17) with an alternative expression for the second integral of (3.18) becomes

$$T(x,t) = \int_0^\infty \left\{ \frac{1}{2\sqrt{\pi\kappa t}} \left[\exp\frac{-(x-\xi)^2}{4\kappa t} + \exp\frac{-(x+\xi)^2}{4\kappa t} \right] \right.$$

$$\left. - \mu \exp\left[\mu(x+\xi) + \kappa\mu^2 t\right] \text{erfc}\left[\frac{(x+\xi)}{2\sqrt{\kappa t}} + \mu\sqrt{\kappa t}\right] \right\} f(\xi)\, d\xi$$

(3.24)

$$+ \frac{2\mu}{\sqrt{\pi}} \int_0^\infty e^{-\mu\eta} \int_{(x+\eta)/2\sqrt{\kappa t}}^\infty e^{-\xi^2} T_0\left(t - \frac{x^2}{4\kappa\xi^2}\right) d\xi\, d\eta.$$

Example 3.4 For the semi-infinite solid $0 \le x < \infty$ with constant initial temperature u_0 and Newton cooling along the surface $x = 0$ into a medium at zero temperature, show that the temperature (see Figure 3.5) is given by

$$u(x,t) = u_0\left\{\text{erf}\left[\frac{x}{2\sqrt{\kappa t}}\right] + e^{\mu x + \kappa \mu^2 t}\,\text{erfc}\left[\frac{x}{2\sqrt{\kappa t}} + \mu\sqrt{\kappa t}\right]\right\},$$

where μ is the ratio of the surface heat transfer coefficient to the thermal conductivity (namely, h/k).

From either (3.18) or (3.24) with $f(x) = u_0$ and $T_0(t)$ zero we have

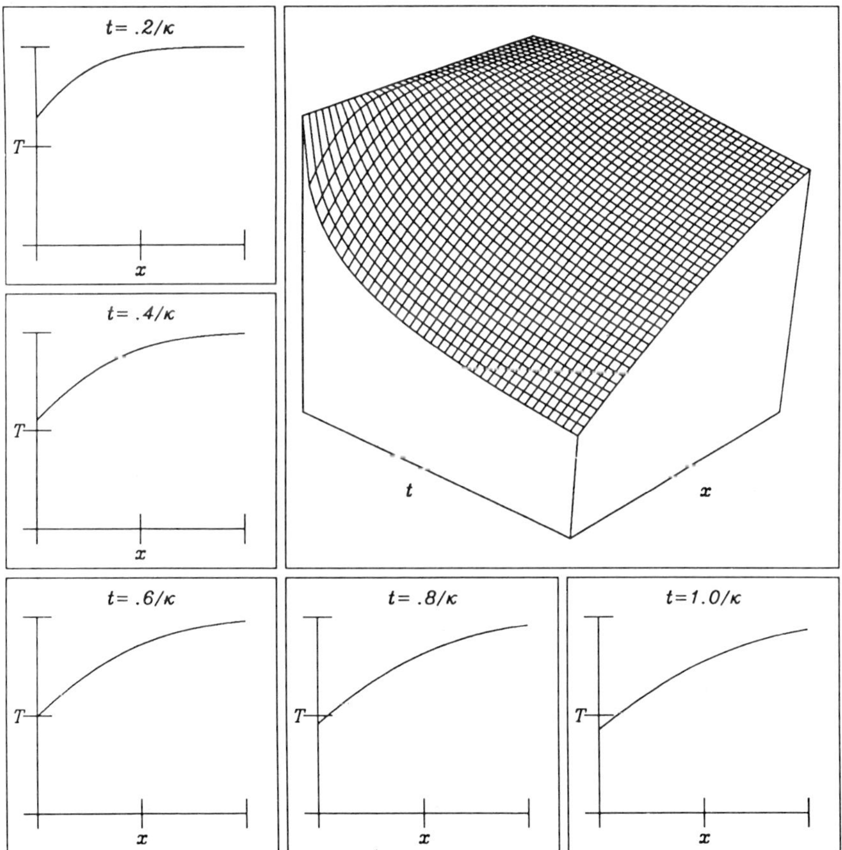

Figure 3.5. Time development of solution of Example 3.5 with u_0 and μ unity.

$$u(x,t) = u_0 \int_0^\infty \left\{ \frac{1}{2\sqrt{\pi \kappa t}} \left[\exp \frac{-(x-\xi)^2}{4\kappa t} + \exp \frac{-(x+\xi)^2}{4\kappa t} \right] \right.$$

$$\left. - \mu \exp\left[\mu(x+\xi) + \kappa\mu^2 t \right] \mathrm{erfc}\left[\frac{(x+\xi)}{2\sqrt{\kappa t}} + \mu\sqrt{\kappa t} \right] \right\} d\xi,$$

which on using the substitutions (3.12) in the first two parts of the integral simplifies to give

$$u(x,t) = u_0\left\{ 1 - \mu e^{\mu x + \kappa\mu^2 t} \int_0^\infty e^{\mu\xi}\, \mathrm{erfc}\left[\frac{(x+\xi)}{2\sqrt{\kappa t}} + \mu\sqrt{\kappa t} \right] d\xi \right\}.$$

Performing integration by parts on this integral yields

$$u(x,t) = u_0\left\{ 1 + e^{\mu x + \kappa\mu^2 t}\, \mathrm{erfc}\left[\frac{x}{2\sqrt{\kappa t}} + \mu\sqrt{\kappa t} \right] - \frac{1}{\sqrt{\pi \kappa t}} \int_0^\infty \exp \frac{-(x+\xi)^2}{4\kappa t}\, d\xi \right\},$$

and the desired result follows on making the substitution $(3.12)_2$ in the final term of this equation.

From inspection of (3.18) it is apparent that the Green's function $G(x,\xi,t,\tau)$ for the problem (3.17), that is the Green's function satisfying

$$-\frac{\partial G}{\partial \xi}(x,0,t,\tau) + \mu G(x,0,t,\tau) = 0,$$

(see Problem 2.12), is

$$G(x,\xi,t,\tau) = H(t-\tau)\left\{ \frac{1}{2\sqrt{\pi\kappa(t-\tau)}} \left[\exp \frac{-(x-\xi)^2}{4\kappa(t-\tau)} + \exp \frac{-(x+\xi)^2}{4\kappa(t-\tau)} \right] \right.$$

$$\left. - \mu \exp\left[\mu(x+\xi) + \kappa\mu^2(t-\tau) \right]\mathrm{erfc}\left[\frac{(x+\xi)}{2\sqrt{\kappa(t-\tau)}} + \mu\sqrt{\kappa(t-\tau)} \right] \right\}.$$

Unlike the previous cases of prescribed boundary temperature and prescribed rate of heat flow at the boundary, there is no simple method such as the method of images to construct $G(x,\xi,t,\tau)$. Construction of the general solution (3.18) is an effective means of finding $G(x,\xi,t,\tau)$. We leave it to the reader to verify that $G(x,\xi,t,\tau)$ does indeed satisfy the required boundary condition and that (3.18) can be written in the form

$$T(x,t) = \int_0^\infty f(\xi)G(x,\xi,t,0)\,d\xi + \mu\kappa \int_0^t T_0(\tau)G(x,0,t,\tau)\,d\tau.$$

3.5 Wedge initially at zero temperature and both faces at the constant temperature v_0

In this section we consider the problem of heat conduction in a semi-infinite wedge of angle θ_0 with faces $\theta = 0$ and $\theta = \theta_0$ where $\tan\theta = y/x$ (see Figure 3.6). We assume that the material in the wedge, $0 < \theta < \theta_0$, is initially at prescribed temperature zero and that both the faces $y = 0$ and $y = x\tan\theta_0$ are subjected to the same prescribed constant temperature v_0 from time zero onwards. Since thermal energy must be conserved there can be no heat flow into or out of infinity and hence the temperature inside the wedge as x and y tend to infinity must remain at its initial value of zero, except of course along the faces $y = 0$ and $y = x\tan\theta_0$ where the prescribed temperature condition applies. Thus we consider the following problem for the wedge temperature $v(x,y,t)$

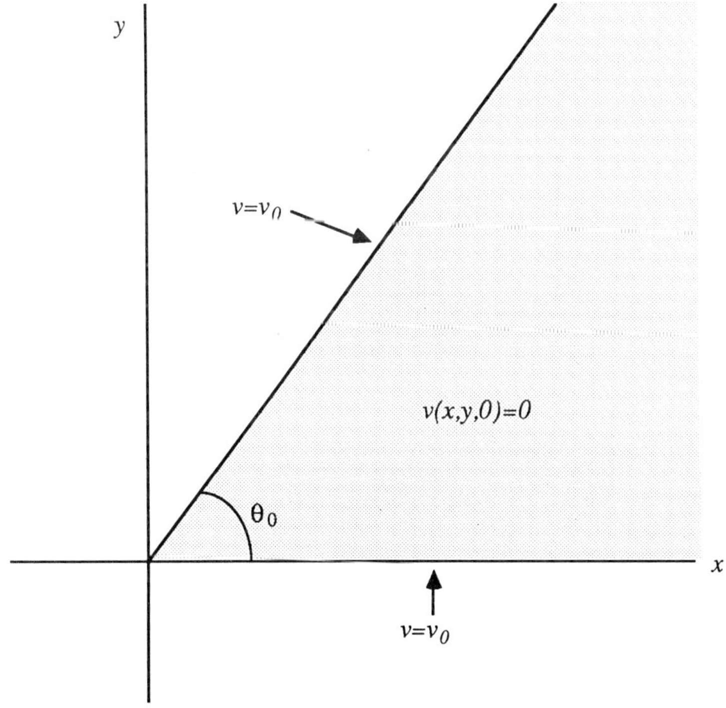

Figure 3.6. Semi-infinite wedge of angle θ_0.

$$\left.\begin{aligned}
\frac{\partial v}{\partial t} &= \kappa\left(\frac{\partial^2 v}{\partial x^2} + \frac{\partial^2 v}{\partial y^2}\right), \quad 0 < \theta < \theta_0, \\[2mm]
v(x,y,0) &= 0, \quad 0 \le \theta \le \theta_0, \\[2mm]
v(x,y,t) &= v_0 \text{ along } \theta = 0 \text{ and } \theta = \theta_0, \\[2mm]
v(x,y,t) &\to 0 \text{ as } |x|, |y| \to \infty.
\end{aligned}\right\}
\tag{3.25}$$

The characteristic feature of this problem is that it remains unchanged under the following stretching group of transformations

$$x^* = \lambda x, \quad y^* = \lambda y, \quad t^* = \lambda^2 t, \quad v^* = v.$$

Accordingly, on noting the relations

$$\frac{x^*}{\sqrt{t^*}} = \frac{x}{\sqrt{t}}, \quad \frac{y^*}{\sqrt{t^*}} = \frac{y}{\sqrt{t}}$$

we may deduce that the solution of problem (3.25) takes the form

$$v(x,y,t) = \psi\left(\frac{x}{\sqrt{t}}, \frac{y}{\sqrt{t}}\right).$$

Thus, this particular problem reduces to the determination of a function of two variables and on introducing

$$\zeta = \frac{x}{\sqrt{t}}, \quad \eta = \frac{y}{\sqrt{t}},$$

we observe that the partial differential equation $(3.25)_1$ becomes

$$\kappa\left(\frac{\partial^2 \psi}{\partial \zeta^2} + \frac{\partial^2 \psi}{\partial \eta^2}\right) + \frac{1}{2}\left(\zeta\frac{\partial \psi}{\partial \zeta} + \eta\frac{\partial \psi}{\partial \eta}\right) = 0.$$

In terms of the polar coordinates

$$R = \sqrt{\frac{\zeta^2 + \eta^2}{2\kappa}} = \sqrt{\frac{x^2 + y^2}{2\kappa t}},$$

$$\theta = \tan^{-1}\left(\frac{\zeta}{\eta}\right) = \tan^{-1}\left(\frac{y}{x}\right),$$

the wedge problem for

$$v(x,y,t) = \psi(\zeta,\eta) = \Psi(R,\theta),$$

becomes

$$\frac{\partial^2 \Psi}{\partial R^2} + (R + \frac{1}{R})\frac{\partial \Psi}{\partial R} + \frac{1}{R^2}\frac{\partial^2 \Psi}{\partial \theta^2} = 0, \quad 0 < R < \infty, \, 0 < \theta < \theta_0,$$

$$\Psi(\infty, \theta) = 0, \quad 0 < \theta < \theta_0,$$

$$\Psi(R, \theta) = v_0 \text{ for } \theta = 0 \text{ and } \theta = \theta_0. \tag{3.26}$$

We solve this problem by means of a Fourier sine series of the form

$$\Psi(R, \theta) = v_0 \left\{ 1 + \sum_{n=0}^{\infty} \Psi_n(R) \sin \lambda_n \theta \right\}, \tag{3.27}$$

where the constant λ_n must be chosen to satisfy the prescribed constant temperature requirements along $\theta = 0$ and $\theta = \theta_0$ as well as the condition of zero heat flux across $\theta = \theta_0/2$, that is

$$\frac{\partial \Psi}{\partial \theta}(R, \theta) = 0 \text{ for } \theta = \frac{\theta_0}{2},$$

which arises from physical symmetry considerations. After a little thought it is apparent that only the choice

$$\lambda_n = (2n + 1)\frac{\pi}{\theta_0}, \tag{3.28}$$

meets all these requirements. On substituting (3.27) into the partial differential equation $(3.26)_1$ we obtain the following ordinary differential equation for $\Psi_n(R)$,

$$\frac{d^2 \Psi_n}{dR^2} + \left(R + \frac{1}{R}\right)\frac{d\Psi_n}{dR} - \left(\frac{\lambda_n}{R}\right)^2 \Psi_n = 0. \tag{3.29}$$

Now the boundary conditions on $\Psi_n(R)$ are firstly that this function remain finite at the wedge vertex and that $\Psi_n(\infty)$ be determined as the Fourier coefficient in the Fourier sine series

$$\sum_{n=0}^{\infty} \Psi_n(\infty) \sin\left(\frac{(2n + 1)\pi\theta}{\theta_0}\right) = -1,$$

which ensures that $(3.26)_2$ is satisfied. Now from (2.56) for the Fourier sine coefficients we have

$$\Psi_n(\infty) = \frac{2}{\theta_0} \int_0^{\theta_0} (-1) \sin\left(\frac{(2n + 1)\pi\theta}{\theta_0}\right) d\theta,$$

from which we may readily deduce

$$\Psi_n(\infty) = \frac{-4}{(2n + 1)\pi}. \tag{3.30}$$

In order to solve (3.29) we use the *Frobenius series method*. We observe that equation (3.29) has a regular singularity at $R = 0$ and an irregular singularity at infinity. Thus, we assume that $\Psi_n(R)$ can be expressed as an infinite series

$$\Psi_n(R) = R^s \sum_{m=0}^{\infty} c_m R^m,$$

and on substituting this expression into (3.29) we obtain

$$R^s \left\{ \sum_{m=0}^{\infty} c_m (m+s)(m+s-1)R^{m-2} \right.$$

$$\left. + \sum_{m=0}^{\infty} c_m (m+s)\left(R^m + R^{m-2}\right) - \lambda_n^2 \sum_{m=0}^{\infty} c_m R^{m-2} \right\} = 0,$$

which on adjusting the summation indices becomes

$$\sum_{m=-2}^{\infty} c_{m+2}\left[(m+s+2)^2 - \lambda_n^2\right]R^m = - \sum_{m=0}^{\infty} c_m (m+s)R^m.$$

After equating like powers of R, we obtain the indicial equation

$$s^2 - \lambda_n^2 = 0,$$

for s and the recurrence relation

$$c_{m+2} = \frac{-(m+s)}{(m+s+2)^2 - \lambda_n^2} c_m, \quad m \geq 0,$$

for the coefficients c_m. Since our solution must be finite at the wedge vertex ($R = 0$), we choose the positive value of s, namely $s = \lambda_n$. The coefficients c_m are therefore determined by

$$c_{m+2} = \frac{-(m+\lambda_n)}{(m+2)(m+2+2\lambda_n)} c_m, \quad m \geq 0,$$

and the solution of (3.29) which remains finite at $R = 0$ may readily be seen to be

$$\Psi_n(R) = c_0 R^{\lambda_n}\left\{ 1 - \frac{\lambda_n}{(\lambda_n+1)}(R/2)^2 + \frac{\lambda_n(\lambda_n+2)}{(\lambda_n+1)(\lambda_n+2)}\frac{(R/2)^4}{2!} \right.$$

$$\left. - \frac{\lambda_n(\lambda_n+2)(\lambda_n+4)}{(\lambda_n+1)(\lambda_n+2)(\lambda_n+3)}\frac{(R/2)^6}{3!} + \cdots \right\}, \tag{3.31}$$

where c_0 denotes an arbitrary constant. The other solutions linearly independent to (3.31) either behave as $R^{-\lambda_n}$ for $R \to 0$ or contain a $\log R$ term which are not appropriate for our use if the temperature is to remain finite at the wedge vertex. In order to identify the arbitrary constant c_0 with the condition (3.30) we need to obtain a finite expression for the series (3.31).

We do not attempt to present the full details for the determination of such a finite expression. We simply note that on looking for a solution of (3.29) involving an integral representation in terms of a Mellin kernel we may deduce the following

$$\Psi_n(R) = c_0^* \int_0^\infty e^{-\xi^2/2} \frac{J_{\lambda_n}(\xi R)}{\xi} \, d\xi, \tag{3.32}$$

where c_0^* denotes another arbitrary constant, and $J_{\lambda_n}(z)$ is the usual Bessel function. On substituting the series for $J_\lambda(z)$, namely

$$J_\lambda(z) = \sum_{m=0}^\infty \frac{(-1)^m}{m!} \frac{(z/2)^{\lambda+2m}}{\Gamma(\lambda+m+1)},$$

it is not difficult to show that (3.32) becomes

$$\Psi_n(R) = \frac{c_0^*}{2} \sum_{m=0}^\infty \left\{ \frac{(-1)^m}{m!} \left(\frac{R^2}{2}\right)^{\frac{\lambda_n}{2}+m} \frac{\Gamma(\frac{\lambda_n}{2}+m)}{\Gamma(\lambda_n+m+1)} \right\}, \tag{3.33}$$

where $\Gamma(z)$ denotes the gamma function. On examination of (3.31) and (3.33) it is apparent that the two series are the same and the arbitrary constants are related by

$$c_0 = c_0^* \left(\frac{1}{2}\right)^{\frac{\lambda_n}{2}+1} \frac{\Gamma(\lambda_n/2)}{\Gamma(\lambda_n+1)}.$$

In order to determine c_0^* we make the substitution $\eta = \xi R$ so that (3.32) becomes

$$\Psi_n(R) = c_0^* \int_0^\infty \exp\left(\frac{-\eta^2}{2R^2}\right) \frac{J_{\lambda_n}(\eta)}{\eta} \, d\eta, \tag{3.34}$$

and therefore

$$\Psi_n(\infty) = c_0^* \int_0^\infty \frac{J_{\lambda_n}(\eta)}{\eta} \, d\eta = \frac{c_0^*}{\lambda_n}, \tag{3.35}$$

where have used the integral

$$\int_0^\infty \frac{J_\lambda(\eta)}{\eta} \, d\eta = \frac{1}{\lambda}. \tag{3.36}$$

From (3.30) and (3.35) and the definition (3.28) of λ_n we find that

$$c_0^* = \frac{-4}{\theta_0}.$$

Thus altogether the solution for a wedge initially at zero temperature and both faces subjected to the same constant temperature v_0 is given by

$$v(x,y,t) = v_0\left\{1 - \frac{4}{\theta_0}\sum_{n=0}^{\infty}\int_0^{\infty} e^{-\xi^2/2}\frac{J_{\lambda_n}(\xi r/\sqrt{2\kappa t})}{\xi}\,d\xi\,\sin\lambda_n\theta\right\}, \tag{3.37}$$

where $\lambda_n = (2n+1)\pi/\theta_0$ and (r,θ) denotes the usual plane cylindrical polar coordinates.

Example 3.5 By making use of the differential equation for $J_{\lambda_n}(z)$ verify that (3.32) is indeed a solution of (3.29). From (3.32) establish the alternative expression for $\Psi_n(R)$, namely

$$\Psi_n(R) = c_0^*\sqrt{\frac{\pi}{2}}\frac{R}{2\lambda_n}e^{-R^2/4}\left\{I_{\frac{\lambda_n+1}{2}}\left(\frac{R^2}{4}\right) + I_{\frac{\lambda_n-1}{2}}\left(\frac{R^2}{4}\right)\right\}, \tag{3.38}$$

where $I_\alpha(z)$ denotes the usual modified Bessel function of order α.

On differentiating (3.32) with respect to R we obtain

$$\frac{d\Psi_n}{dR} = c_0^*\int_0^{\infty} e^{-\xi^2/2}J_{\lambda_n}{}'(\xi R)\,d\xi,$$

where the prime denotes differentiation with respect to the argument. Further

$$\frac{d^2\Psi_n}{dR^2} = c_0^*\int_0^{\infty} e^{-\xi^2/2}J_{\lambda_n}{}''(\xi R)\xi\,d\xi.$$

Now, the differential equation for $J_{\lambda_n}(z)$ is

$$z^2 J_\lambda{}'' + z J_\lambda{}' + (z^2 - \lambda^2)J_\lambda = 0,$$

so that we have

$$R^2 \frac{\mathrm{d}^2 \Psi_n}{\mathrm{d}R^2} + (R + R^3)\frac{\mathrm{d}\Psi_n}{\mathrm{d}R} - \lambda_n^2 \Psi_n$$

$$= c_0^* \int_0^\infty \frac{e^{-\xi^2/2}}{\xi} \left\{ (\xi R)^2 \, \mathrm{J}_{\lambda_n}{}''(\xi R) + \left[(\xi R) + (\xi R^3) \right] \mathrm{J}_{\lambda_n}{}'(\xi R) - \lambda_n^2 \, \mathrm{J}_{\lambda_n}(\xi R) \right\} \mathrm{d}\xi$$

$$= c_0^* \int_0^\infty e^{-\xi^2/2} R^2 \left\{ R \, \mathrm{J}_{\lambda_n}{}'(\xi R) - \xi \, \mathrm{J}_{\lambda_n}(\xi R) \right\} \mathrm{d}\xi$$

$$= c_0^* R^2 \left\{ \left[e^{-\xi^2/2} \, \mathrm{J}_{\lambda_n}(\xi R) \right]_0^\infty + \int_0^\infty \xi e^{-\xi^2/2} \, \mathrm{J}_{\lambda_n}(\xi R) \, \mathrm{d}\xi - \int_0^\infty e^{-\xi^2/2} \xi \, \mathrm{J}_{\lambda_n}(\xi R) \, \mathrm{d}\xi \right\}$$

$$= 0,$$

and we have verified that (3.32) is indeed a solution of (3.29).

In order to establish the alternative expression (3.38) for $\Psi_n(R)$ we need the following formulae involving Bessel functions, namely

$$\mathrm{J}_\lambda(z) = \frac{z}{2\lambda}[\mathrm{J}_{\lambda+1}(z) + \mathrm{J}_{\lambda-1}(z)],$$

$$\int_0^\infty e^{-pt} \frac{\mathrm{J}_\lambda(a\sqrt{t})}{\sqrt{t}} \, \mathrm{d}t = \sqrt{\frac{\pi}{p}} e^{-a^2/8p} \, \mathrm{I}_{\frac{\lambda}{2}}\left(\frac{a^2}{8p}\right), \quad \lambda > -1,$$

Thus we have

$$\int_0^\infty e^{-\xi^2/2} \frac{\mathrm{J}_\lambda(\xi R)}{\xi} \, \mathrm{d}\xi = \frac{R}{2\lambda} \int_0^\infty e^{-\xi^2/2}[\mathrm{J}_{\lambda+1}(\xi R) + \mathrm{J}_{\lambda-1}(\xi R)] \, \mathrm{d}\xi$$

$$= \frac{R}{2\lambda} \sqrt{\frac{1}{2}} \int_0^\infty e^{-t} \left[\mathrm{J}_{\lambda+1}(R\sqrt{2t}) + \mathrm{J}_{\lambda-1}(R\sqrt{2t}) \right] \frac{\mathrm{d}t}{\sqrt{t}},$$

where we have made the substitution $\xi = \sqrt{2t}$ and the desired result follows immediately from the above Laplace transform with $p = 1$ and $a = \sqrt{2}R$.

It is of interest to note that from this alternative expression for $\Psi_n(R)$ and the asymptotic formulae

$$\mathrm{I}_\lambda(z) \sim \frac{e^z}{\sqrt{2\pi z}} \text{ as } z \to \infty,$$

we have immediately

$$\Psi_n(R) \sim \frac{c_0^*}{\lambda_n} \text{ as } R \to \infty,$$

which is in agreement with a previously obtained result and moreover, provides a derivation of the integral (3.36) obtained via (3.34) letting R tend to infinity.

3.6 Special wedge angles $\theta_0 = \pi/2$, π and 2π

In this section we examine the Fourier series solution (3.37) for three special wedge angles. For $\theta_0 = \pi/2$ and $\theta = \pi$ we show how the series solutions agree with known simple closed form expressions. Since the correspondence of (3.37) with these exact solutions is by no means obvious and hinges on Bessel function identities, it is therefore an instructive exercise in special functions. Although the case $\theta_0 = 2\pi$ does not admit a simple exact solution, the series (3.37) does nevertheless yield a closed expression and since this case represents important physical problems, the simplifications of (3.37) would seem worthwhile.

<u>Example 3.6</u> In the case of $\theta_0 = \pi$ show that the Fourier series (3.37) reduces to

$$v(x,y,t) = v_0\left\{1 - \text{erf}\left[\frac{y}{2\sqrt{\kappa t}}\right]\right\}, \tag{3.39}$$

which is the solution derived in Example 3.2.

In this case, the constants λ_n are simply $(2n + 1)$ and we may deduce (3.39) by either of two methods.

<u>Method 1</u> The first method utilizes the integral expression (3.34) for $\Psi_n(R)$ and makes use of the identity

$$\sin(r \sin \theta) = 2 \sum_{n=0}^{\infty} J_{2n+1}(r) \sin(2n + 1)\theta, \tag{3.40}$$

and the integral

$$\int_0^\infty e^{-\xi^2/2} \cos a\xi \, d\xi = \sqrt{\frac{\pi}{2}} e^{-a^2/2}. \tag{3.41}$$

Now

$$\sum_{n=0}^{\infty} \Psi_n(R)\sin(2n+1)\theta = \frac{-4}{\pi}\sum_{n=0}^{\infty}\int_0^{\infty} e^{-\xi^2/2}\frac{J_{2n+1}(\xi R)}{\xi}\,d\xi\,\sin(2n+1)\theta$$

$$= \frac{-2}{\pi}\int_0^{\infty}\frac{e^{-\xi^2/2}}{\xi}\sin(\xi R\sin\theta)\,d\xi$$

$$= \frac{-2}{\pi}\int_0^{R\sin\theta}\left(\int_0^{\infty} e^{-\xi^2/2}\cos\tau\xi\,d\xi\right)d\tau$$

$$= -\sqrt{\frac{2}{\pi}}\int_0^{R\sin\theta} e^{-\tau^2/2}\,d\tau$$

$$= \frac{-2}{\sqrt{\pi}}\int_0^{R(\sin\theta)/\sqrt{2}} e^{-\eta^2}\,d\eta,$$

where we have made the substitution $\tau = \sqrt{2}\eta$. But since

$$\frac{R\sin\theta}{\sqrt{2}} = \frac{r\sin\theta}{2\sqrt{\kappa t}} = \frac{y}{2\sqrt{\kappa t}},$$

the desired result follows immediately.

<u>Method 2</u> This method utilizes the alternative expression (3.38) for $\Psi_n(R)$ and makes use of the identity

$$e^{y\left(z + \frac{1}{z}\right)} = \sum_{n=-\infty}^{\infty} I_n(2y)z^n.$$

Now

$$\sum_{n=0}^{\infty}\Psi_n(R)\sin(2n+1)\theta$$

$$= -R\sqrt{\frac{2}{\pi}}e^{-R^2/4}\sum_{n=0}^{\infty}\left\{I_{n+1}(\frac{R^2}{4}) + I_n(\frac{R^2}{4})\right\}\frac{\sin(2n+1)\theta}{(2n+1)}$$

$$= -R\sqrt{\frac{2}{\pi}}e^{-R^2/4}\int_0^{\theta}\sum_{n=0}^{\infty}\left\{I_{n+1}(\frac{R^2}{4}) + I_n(\frac{R^2}{4})\right\}\cos(2n+1)\phi\,d\phi.$$

But on making the respective changes $m = n+1$ and $j = -m$ we have

$$\sum_{n=0}^{\infty} I_{n+1}(\tfrac{R^2}{4})\cos(2n+1)\phi = \sum_{m=1}^{\infty} I_m(\tfrac{R^2}{4})\cos(2m-1)\phi$$

$$= \sum_{j=-1}^{-\infty} I_{-j}(\tfrac{R^2}{4})\cos(2j+1)\phi$$

$$= \sum_{j=-1}^{-\infty} I_j(\tfrac{R^2}{4})\cos(2j+1)\phi ,$$

where we have used the facts that $\cos(-z) = \cos z$ and that $I_{-n}(z) = I_n(z)$. Thus altogether we have

$$\sum_{n=0}^{\infty} \Psi_n(R)\sin(2n+1)\theta = -R\sqrt{\tfrac{2}{\pi}}\,e^{-R^2/4}\int_0^{\theta}\sum_{n=-\infty}^{\infty} I_n(\tfrac{R^2}{4})\cos(2n+1)\phi\,\mathrm{d}\phi$$

$$= -R\sqrt{\tfrac{2}{\pi}}\,e^{-R^2/4}\mathrm{Re}\left\{\int_0^{\theta}\exp\left(\tfrac{R^2}{8}(e^{2i\phi}+e^{-2i\phi})\right)e^{i\phi}\,\mathrm{d}\phi\right\}$$

$$= -R\sqrt{\tfrac{2}{\pi}}\,e^{-R^2/4}\int_0^{\theta}\exp\left(\tfrac{R^2}{4}\cos 2\phi\right)\cos\phi\,\mathrm{d}\phi ,$$

so that, on using the identity $\cos 2\phi = 1 - 2\sin^2\phi$ in the final integral and then making the substitution $\eta = R\sin\phi/\sqrt{2}$ in the resulting integral, we again obtain the desired result.

Example 3.7 In the case of $\theta_0 = \pi/2$ verify directly that

$$v(x,y,t) = v_0\left\{1 - \mathrm{erf}\left[\frac{x}{2\sqrt{\kappa t}}\right]\mathrm{erf}\left[\frac{y}{2\sqrt{\kappa t}}\right]\right\}, \tag{3.42}$$

is a solution of (3.25) (see Figure 3.7) and show how this solution emerges from the Fourier series solution (3.37).

By direct differentiation we may verify that (3.42) is a solution of the partial differential equation $(3.25)_1$. Moreover, since erf(0) $= 0$ we see that the appropriate constant temperature

$$v(0,y,t) = v(x,0,t) = v_0, \quad x,y \geq 0, \quad t > 0,$$

is obtained along the positive x-axis and y-axis. Further, from the property erf(∞) $= 1$ we see that both the initial condition $(3.25)_2$ and the condition at infinity $(3.25)_4$ are also satisfied. We note that, due to the inconsistency between the initial and boundary conditions the temperature given by (3.42) is not continuous at the positive x-axis and y-axis at time zero.

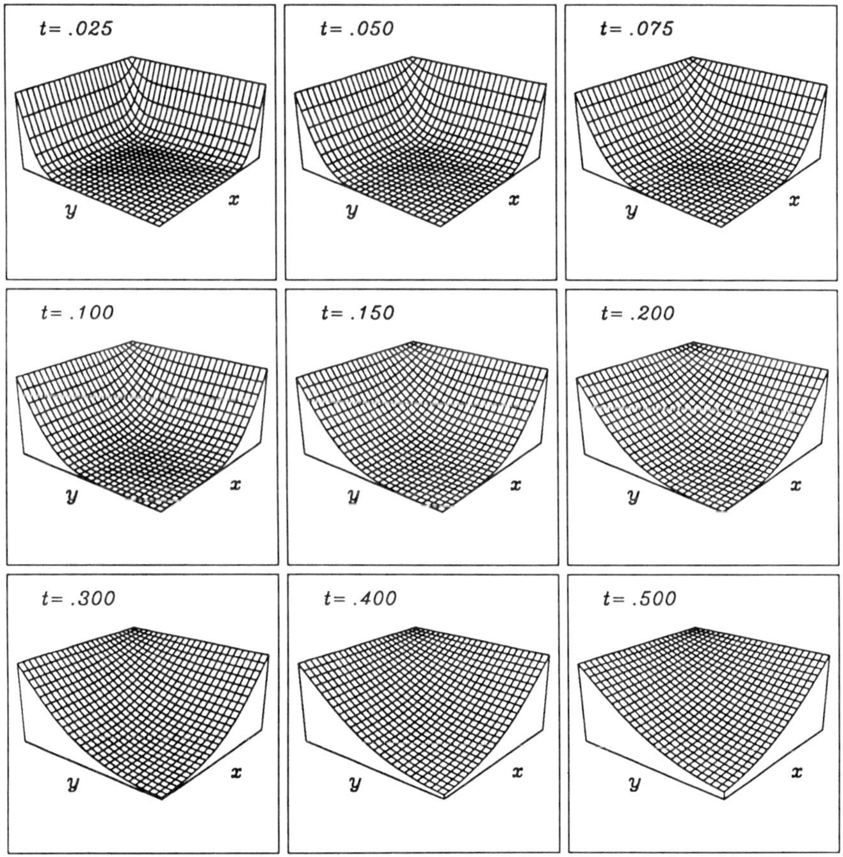

Figure 3.7. Two dimensional temperature profiles for Example 3.7 with v_0 unity, at nine values of time.

In order to show how (3.42) arises from the Fourier series solution (3.37) we exploit the formalism of Method 1 in the previous example but use the identity

$$\cos(r\cos\theta) = J_0(r) + 2\sum_{n=1}^{\infty}(-1)^n J_{2n}(r)\cos 2n\theta$$

$$= \sum_{n=0}^{\infty}(-1)^n \epsilon_n J_{2n}(r)\cos 2n\theta ,$$

(3.43)

where Neumann's factor ϵ_n is given by

$$\epsilon_n = \begin{cases} 1, & n = 0, \\ 2, & n \neq 0. \end{cases}$$

We also use the integral (3.41) and the integral

$$\int_0^{\pi}\frac{\sin m\psi}{\sin\psi}\,d\psi = \begin{cases} 0, & m \text{ even}, \\ \pi, & m \text{ odd}. \end{cases}$$

(3.44)

We proceed to show that

$$\text{erf}\left[\frac{x}{2\sqrt{\kappa t}}\right]\text{erf}\left[\frac{y}{2\sqrt{\kappa t}}\right] = \frac{8}{\pi}\sum_{n=0}^{\infty}\int_0^{\infty}e^{-\xi^2/2}\frac{J_{4n+2}\left(\frac{\xi r}{\sqrt{2\kappa t}}\right)}{\xi}\,d\xi\,\sin(4n+2)\theta .$$

Thus with $R = r/\sqrt{2\kappa t}$ we have

$$\text{erf}\left[\frac{x}{2\sqrt{\kappa t}}\right]\text{erf}\left[\frac{y}{2\sqrt{\kappa t}}\right] = \frac{2}{\pi}\int_0^{R\cos\theta}e^{-\xi^2/2}\,d\xi\int_0^{R\sin\theta}e^{-\eta^2/2}\,d\eta$$

$$= \frac{4}{\pi^2}\int_0^{R\cos\theta}\left(\int_0^{\infty}e^{-X^2/2}\cos\xi X\,dX\right)d\xi\int_0^{R\sin\theta}\left(\int_0^{\infty}e^{-Y^2/2}\cos\eta Y\,dY\right)d\eta$$

$$= \frac{4}{\pi^2}\int_0^{\infty}\frac{e^{-X^2/2}}{X}\sin(RX\cos\theta)\,dX\int_0^{\infty}\frac{e^{-Y^2/2}}{Y}\sin(RY\sin\theta)\,dY ,$$

which on introducing polar coordinates (ξ,ϕ) by

$$X = \xi\cos\phi, \quad Y = \xi\sin\phi ,$$

gives

$$\text{erf}\left[\frac{x}{2\sqrt{\kappa t}}\right]\text{erf}\left[\frac{y}{2\sqrt{\kappa t}}\right]$$

$$= \frac{4}{\pi^2}\int_0^{\infty}\frac{e^{-\xi^2/2}}{\xi}\left[\int_0^{\frac{\pi}{2}}\frac{\sin(\xi R\cos\theta\cos\phi)\sin(\xi R\sin\theta\sin\phi)}{\sin\phi\cos\phi}\,d\phi\right]d\xi .$$

But

$$\int_0^{\frac{\pi}{2}} \frac{\sin(\xi R \cos\theta \cos\phi)\sin(\xi R \sin\theta \sin\phi)}{\sin\phi \cos\phi}\, d\phi$$

$$= \int_0^{\frac{\pi}{2}} \frac{\{\cos[\xi R \cos(\theta + \phi)] - \cos[\xi R \cos(\theta - \phi)]\}}{\sin 2\phi}\, d\phi$$

$$= \sum_{m=0}^{\infty} (-1)^m\, J_{2m}(\xi R)\epsilon_m \int_0^{\frac{\pi}{2}} \frac{[\cos 2m(\theta + \phi) - \cos 2m(\theta - \phi)]}{\sin 2\phi}\, d\phi$$

$$= -2 \sum_{m=0}^{\infty} (-1)^m\, J_{2m}(\xi R)\epsilon_m \sin 2m\theta \int_0^{\frac{\pi}{2}} \frac{\sin 2m\phi}{\sin 2\phi}\, d\phi$$

$$= - \sum_{m=0}^{\infty} (-1)^m\, J_{2m}(\xi R)\epsilon_m \sin 2m\theta \int_0^{\pi} \frac{\sin m\psi}{\sin \psi}\, d\psi,$$

where we have made use of the identity (3.43). The required result now follows immediately on using the integral (3.44).

We observe that the integral (3.44) may be established by induction as follows. The cases m zero and m unity evidently immediately yield zero and π respectively. Now we have

$$\int_0^{\pi} \frac{\sin(m + 2)\psi}{\sin \psi}\, d\psi = \int_0^{\pi} \frac{[\sin m\psi \cos 2\psi + \cos m\psi \sin 2\psi]}{\sin \psi}\, d\psi$$

$$= \int_0^{\pi} \frac{\sin m\psi}{\sin \psi}\, d\psi + 2 \int_0^{\pi} [\cos m\psi \cos \psi - \sin m\psi \sin \psi]\, d\psi$$

$$= \int_0^{\pi} \frac{\sin m\psi}{\sin \psi}\, d\psi + 2 \int_0^{\pi} \cos(m + 1)\psi\, d\psi,$$

and since the second integral is zero the proof of (3.44) by induction is complete.

The special case of $\theta_0 = 2\pi$ is of particular interest since it provides the solution of two problems. On the one hand for the whole two dimensional plane, it represents a semi-infinite line source along the positive x-axis. On the other hand for the half-space $y > 0$ it represents the temperature distribution due to prescribed constant temperature along the positive x-axis and prescribed zero flux along the negative x-axis.

Example 3.8 In the case of $\theta_0 = 2\pi$ show that the Fourier series (3.37) reduces to

$$v(x,y,t) = v_0\left\{1 - \frac{2}{\pi} \int_0^{\infty} \frac{e^{-\xi^2/2}}{\xi}[G(\xi R, \theta) - G(\xi R, -\theta)]\, d\xi\right\},$$

where $R = r/\sqrt{2\kappa t}$ and the function $G(z,\theta)$ is defined by

$$G(z,\theta) = \frac{1}{\sqrt{\pi}} \int_0^{\sqrt{2z}\,\sin(\frac{\pi}{4} - \frac{\theta}{2})} \sin(\tau^2 + z\sin\theta - \tfrac{\pi}{4})\,d\tau.$$

For $\theta_0 = 2\pi$ we sum the series in (3.37) by making use of the formula

$$J_\lambda(z) = \frac{(z/2)^\lambda}{\sqrt{\pi}\,\Gamma(\lambda + \tfrac{1}{2})} \int_0^\pi e^{iz\cos\phi}\left(\sin^2\phi\right)^\lambda d\phi.$$

Since $\lambda_n = n + \tfrac{1}{2}$ we have

$$\sum_{n=0}^\infty J_{n+\frac{1}{2}}(z)\sin(n + \tfrac{1}{2})\theta$$

$$= \mathrm{Im}\left\{\sum_{n=0}^\infty J_{n+\frac{1}{2}}(z)e^{i(n+\frac{1}{2})\theta}\right\}$$

$$= \frac{1}{\sqrt{\pi}}\,\mathrm{Im}\left\{\sum_{n=0}^\infty \int_0^\pi e^{iz\cos\phi}\frac{(z/2)^{n+\frac{1}{2}}}{n!}(\sin^2\phi)^{n+\frac{1}{2}}e^{i(n+\frac{1}{2})\theta}\,d\phi\right\}$$

$$= \sqrt{\frac{z}{2\pi}}\,\mathrm{Im}\left\{\sum_{n=0}^\infty \int_{-1}^1 e^{iz\mu}\frac{(z/2)^n}{n!}(1-\mu^2)^n e^{i(n+\frac{1}{2})\theta}\,d\mu\right\},$$

where we have made the substitution $\mu = \cos\phi$. On performing the sum we obtain

$$\sum_{n=0}^\infty J_{n+\frac{1}{2}}(z)\sin(n + \tfrac{1}{2})\theta$$

$$= \sqrt{\frac{z}{2\pi}}\,\mathrm{Im}\left\{\int_{-1}^1 \exp\left\{z[i\mu + (1-\mu^2)\frac{e^{i\theta}}{2}] + \frac{i\theta}{2}\right\}d\mu\right\}$$

$$= \sqrt{\frac{z}{2\pi}}\,\mathrm{Im}\left\{\int_{-1}^1 \exp\left\{-\frac{z}{2}\left(\mu e^{i\frac{\theta}{2}} - ie^{-i\frac{\theta}{2}}\right)^2 + i\left(z\sin\theta + \frac{\theta}{2}\right)\right\}d\mu\right\}$$

$$= \mathrm{Im}\left\{\frac{e^{iz\sin\theta}}{\sqrt{\pi}} \int_{-(1+i)\sqrt{z/2}(\cos\frac{\theta}{2} + \sin\frac{\theta}{2})}^{(1-i)\sqrt{z/2}(\cos\frac{\theta}{2} - \sin\frac{\theta}{2})} e^{-\xi^2}\,d\xi\right\},$$

where we have made the substitution

$$\xi = \sqrt{\frac{z}{2}}\left(\mu e^{i\frac{\theta}{2}} - ie^{-i\frac{\theta}{2}}\right),$$

so that

$$d\xi = \sqrt{\frac{z}{2}}\, e^{i\frac{\theta}{2}}\, d\mu\, .$$

We now split the integral into two parts and make the respective substitutions

$$\xi = e^{-i\pi/4}\tau, \quad \xi = e^{i\pi/4}\tau,$$

as follows

$$\int_{-\sqrt{z}e^{\frac{i\pi}{4}}a(\theta)}^{\sqrt{z}e^{\frac{-i\pi}{4}}b(\theta)} e^{-\xi^2}\, d\xi = \int_{0}^{\sqrt{z}e^{\frac{-i\pi}{4}}b(\theta)} e^{-\xi^2}\, d\xi + \int_{0}^{\sqrt{z}e^{\frac{i\pi}{4}}a(\theta)} e^{-\xi^2}\, d\xi$$

$$= e^{\frac{-i\pi}{4}}\int_{0}^{\sqrt{z}b(\theta)} e^{i\tau^2}\, d\tau + e^{\frac{i\pi}{4}}\int_{0}^{\sqrt{z}a(\theta)} e^{-i\tau^2}\, d\tau,$$

where $a(\theta)$ and $b(\theta)$ are given by

$$a(\theta) = \cos\frac{\theta}{2} + \sin\frac{\theta}{2}, \quad b(\theta) = \cos\frac{\theta}{2} - \sin\frac{\theta}{2}.$$

Thus altogether on multiplying by $e^{iz\sin\theta}$ and taking the imaginary part we obtain

$$\sum_{n=0}^{\infty} J_{n+\frac{1}{2}}(z)\sin(n + \tfrac{1}{2})\theta = \frac{1}{\sqrt{\pi}} \int_{0}^{\sqrt{z}b(\theta)} \sin(\tau^2 + z\sin\theta - \tfrac{\pi}{4})\, d\tau$$

$$- \frac{1}{\sqrt{\pi}} \int_{0}^{\sqrt{z}a(\theta)} \sin(\tau^2 - z\sin\theta - \tfrac{\pi}{4})\, d\tau,$$

which gives the desired result on noting that

$$\sqrt{z}\left(\cos\frac{\theta}{2} - \sin\frac{\theta}{2}\right) = \sqrt{2z}\,\sin\left(\frac{\pi}{4} - \frac{\theta}{2}\right).$$

PROBLEMS

1. For the semi-infinite solid $0 \le x < \infty$ with initial temperature

$$f(x) = \alpha + \beta x,$$

where α and β are constants, and $x = 0$ maintained at zero temperature show that

$$u(x,t) = \alpha \operatorname{erf}\left[\frac{x}{2\sqrt{\kappa t}}\right] + \beta x.$$

2. For the semi-infinite solid $0 \le x < \infty$ with zero initial temperature and the surface $x = 0$ maintained at temperature

$$g(t) = \alpha \sqrt{t},$$

where α is constant, show that

$$v(x,t) = \frac{\alpha}{2}\sqrt{\frac{\pi}{\kappa}}\left\{2\sqrt{\frac{\kappa t}{\pi}}\exp\left(\frac{-x^2}{4\kappa t}\right) - x\operatorname{erfc}\left[\frac{x}{2\sqrt{\kappa t}}\right]\right\}.$$

$\left[\text{Hint, invert the Laplace transform }\hat{g}(p)e^{-qx}\text{ directly from the table.}\right]$

3. For the semi-infinite solid $0 \le x < \infty$ with zero initial temperature and the surface $x = 0$ maintained at the temperature

$$g(t) = \alpha e^{\beta t},$$

where α and β are constants, show that

$$v(x,t) = \frac{\alpha e^{\beta t}}{2}\left\{e^{x\sqrt{\beta/\kappa}}\operatorname{erfc}\left[\frac{x}{2\sqrt{\kappa t}} + \sqrt{\beta t}\right] + e^{-x\sqrt{\beta/\kappa}}\operatorname{erfc}\left[\frac{x}{2\sqrt{\kappa t}} - \sqrt{\beta t}\right]\right\}.$$

$\left[\text{Hint, invert the Laplace transform }\hat{g}(p)e^{-qx}\text{ directly from the table.}\right]$

4. For the semi-infinite solid $0 \le x < \infty$ with zero initial temperature and the surface $x = 0$ maintained at temperature

$$g(t) = \alpha \sin(\beta t + \gamma),$$

where α, β and γ are all constants, show that

$$v(x,t) = \alpha e^{-x\sqrt{\beta/2\kappa}}\sin\left[\beta t + \gamma - x\sqrt{\beta/2\kappa}\right]$$

$$+ \frac{2\alpha\kappa}{\pi}\int_0^\infty e^{-\kappa\xi^2 t}\,\xi\,\frac{\left(\beta\cos\gamma - \kappa\xi^2\sin\gamma\right)}{\left(\beta^2 + \kappa^2\xi^4\right)}\sin\xi x\,d\xi.$$

$\Big[$Hint, use the Laplace transform inversion theorem and contour integration (also, see

Carslaw and Jaeger (1980), page 319). $\Big]$

5. If $v_1(x, t)$ denotes the temperature in the semi-infinite region $0 \le x < \infty$ with zero initial temperature and constant unit surface temperature at $x = 0$, then

$$v_1(x, t) = \mathrm{erfc}\left(\frac{x}{2\sqrt{\kappa t}}\right).$$

Show that the temperature $v(x, t)$ for the semi-infinite region $0 \le x < \infty$ with zero initial temperature and arbitrary prescribed temperature $g(t)$ at $x = 0$ can be expressed in terms of $v_1(x, t)$ and $g(t)$ by the convolution

$$v(x, t) = \int_0^t g(\tau) \frac{\partial v_1}{\partial t}(x, t - \tau)\, d\tau.$$

$\Big[$Hint, either use Laplace transforms or show that the given expression reduces to (3.8)

with $f(x) = 0.$ $\Big]$

6. For the semi-infinite solid $0 \le x < \infty$ with zero initial temperature and the flux along the surface $x = 0$ prescribed as

$$h(t) = \alpha \sin(\beta t + \gamma),$$

where α, β and γ are all constants, show that

$$v(x, t) = \frac{\alpha}{k}\sqrt{\frac{\kappa}{\beta}}\, e^{-x\sqrt{\beta/2\kappa}} \sin\left[\beta t + \gamma - x\sqrt{\beta/2\kappa} - \frac{\pi}{4}\right]$$

$$+ \frac{2\alpha\kappa}{k\pi} \int_0^\infty e^{-\kappa\xi^2 t} \frac{\left(\beta \cos\gamma - \kappa\xi^2 \sin\gamma\right)}{\left(\beta^2 + \kappa^2\xi^4\right)} \cos\xi x\, d\xi.$$

$\Big[$Hint, use the Laplace transform inversion theorem and contour integration (also, see

Carslaw and Jaeger (1980), pages 64 and 76). $\Big]$

7. If $v_1(x,t)$ denotes the temperature in the semi-infinite region $0 \leq x < \infty$ with zero initial temperature and constant unit heat flux at the surface $x = 0$, then

$$v_1(x,t) = \frac{1}{k}\left\{2\sqrt{\frac{\kappa t}{\pi}}\exp\left(\frac{-x^2}{4\kappa t}\right) - x\,\mathrm{erfc}\left(\frac{x}{2\sqrt{\kappa t}}\right)\right\}.$$

Show that the temperature $v(x,t)$ in the semi-infinite region $0 \leq x < \infty$ with zero initial temperature and arbitrary prescribed heat flux $h(t)$ at $x = 0$ is given in terms of $v_1(x,t)$ and $h(t)$ by the convolution

$$v(x,t) = \int_0^t h(\tau)\frac{\partial v_1}{\partial t}(x, t - \tau)\,\mathrm{d}\tau.$$

$\Big[$ Hint, either use Laplace transforms or show that this integral and (3.15) are equivalent

if $f(x) = 0.$ $\Big]$

8. For the semi-infinite solid $0 \leq x < \infty$ with zero initial temperature and Newton heat loss at the surface $x = 0$ into a medium at constant temperature T_0, show that

$$v(x,t) = T_0\left\{\mathrm{erfc}\left[\frac{x}{2\sqrt{\kappa t}}\right] - e^{\mu x + \kappa\mu^2 t}\,\mathrm{erfc}\left[\frac{x}{2\sqrt{\kappa t}} + \mu\sqrt{\kappa t}\right]\right\},$$

where the constant μ is h/k.

9. For the semi-infinite solid $0 \leq x < \infty$ with zero initial temperature and Newton heat loss at the surface $x = 0$ into a medium at temperature $T_0(t)$ given by

$$T_0(t) = \alpha\sin(\beta t + \gamma),$$

where α, β and γ are all constants, show that

$$v(x,t) = \frac{\alpha\mu e^{-x\sqrt{\beta/2\kappa}}}{\sqrt{\left[\mu + \sqrt{\beta/2\kappa}\right]^2 + (\beta/2\kappa)}}\sin\left[\beta t + \gamma - x\sqrt{\beta/2\kappa} - \delta\right]$$

$$+ \frac{2\alpha\kappa\mu}{\pi}\int_0^\infty e^{-\kappa\xi^2 t}\,\xi\,\frac{\left(\beta\cos\gamma - \kappa\xi^2\sin\gamma\right)\left(\mu\sin\xi x + \xi\cos\xi x\right)}{\left(\beta^2 + \kappa^2\xi^4\right)\left(\mu^2 + \xi^2\right)}\,\mathrm{d}\xi,$$

where the constant δ is defined by

$$\delta = \tan^{-1}\left(\frac{\sqrt{\beta/2\kappa}}{\mu + \sqrt{\beta/2\kappa}}\right).$$

$\Big[$ Hint, use the Laplace transform inversion theorem and contour integration. $\Big]$

10. If $v_1(x,t)$ denotes the temperature in the semi-infinite region $0 \le x < \infty$ with zero initial temperature and Newton heat loss at the surface $x = 0$ into a medium at constant unit temperature then

$$v_1(x,t) = \operatorname{erfc}\left(\frac{x}{2\sqrt{\kappa t}}\right) - e^{\mu x + \kappa \mu^2 t}\operatorname{erfc}\left(\frac{x}{2\sqrt{\kappa t}} + \mu\sqrt{\kappa t}\right),$$

where $\mu = h/k$. Show that the temperature $v(x,t)$ in the semi-infinite region $0 \le x < \infty$ with zero initial temperature and Newton heat loss at the surface $x = 0$ into a medium at arbitrary temperature $T_0(t)$ is given in terms of $v_1(x,t)$ and $T_0(t)$ by the convolution

$$v(x,t) = \int_0^t T_0(\tau)\frac{\partial v_1}{\partial t}(x, t - \tau)\,d\tau.$$

$\Big[$Hint, either use Laplace transforms or show that this integral and (3.18) are equivalent if $f(x) = 0.$ $\Big]$

11. Using a Fourier sine series, show that the wedge of angle θ_0, initially at zero temperature and with faces $\theta = 0$ and $\theta = \theta_0$ held at constant temperature zero and v_0 respectively has temperature $v(r,\theta,t)$ given by

$$v(r,\theta,t) = \frac{v_0}{\theta_0}\left\{\theta + 2\sum_{n=1}^{\infty}\int_0^{\infty}e^{-\xi^2/2}\frac{J_{\lambda_n}(\xi R)}{\xi}\,d\xi\,(-1)^n\sin\lambda_n\theta\right\},$$

where $\lambda_n = n\pi/\theta_0$ and $R = r/\sqrt{2\kappa t}$.

12. Verify that the infinite eighth space $x,y,z \ge 0$, initially at zero temperature and with all faces held at a constant temperature v_0 has temperature $v(x,y,z,t)$ given by

$$v(x,y,z,t) = v_0\left\{1 - \operatorname{erf}\left[\frac{x}{\sqrt{2\kappa t}}\right]\operatorname{erf}\left[\frac{y}{\sqrt{2\kappa t}}\right]\operatorname{erf}\left[\frac{z}{\sqrt{2\kappa t}}\right]\right\}.$$

Chapter Four

Exact analytical solutions by Fourier series

4.1 Introduction

In this chapter we obtain analytical solutions of heat flow problems by means of separation of variables and Fourier series. We first consider linear flow of heat in the region $(0, \ell)$ given an arbitrary initial temperature distribution $f(x)$ and in the next three sections we consider the three general problems of prescribed surface temperatures, prescribed surface fluxes and Newton heat loss into media at prescribed temperatures. Thus, we consider the following three problems for $T(x, t)$

$$\left.\begin{array}{l} \dfrac{\partial T}{\partial t} = \kappa \dfrac{\partial^2 T}{\partial x^2}, \quad 0 < x < \ell, \\[3mm] T(x, 0) = f(x), \quad 0 \le x \le \ell, \end{array}\right\} \tag{4.1}$$

with one of the following sets of boundary conditions for $t > 0$ at $x = 0$ and $x = \ell$

(i) $\quad T(0, t) = g_1(t), \quad T(\ell, t) = g_2(t),$

(ii) $\quad -k\dfrac{\partial T}{\partial x}(0, t) = h_1(t), \quad k\dfrac{\partial T}{\partial x}(\ell, t) = h_2(t),$

(iii) $\quad -\dfrac{\partial T}{\partial x}(0, t) + \mu T(0, t) = \mu T_1(t), \quad \dfrac{\partial T}{\partial x}(\ell, t) + \mu T(\ell, t) = \mu T_2(t),$

where $g_i(t)$, $h_i(t)$ and $T_i(t)$ $(i = 1, 2)$ denote arbitrary prescribed functions of time and as usual μ denotes h/k where h is the surface conductance and k is the thermal conductivity.

We solve these problems by utilizing solutions obtained by *separation of variables*. Thus we first look for a separable solution of $(4.1)_1$ of the form

$$T(x, t) = A(x)B(t),$$

so that, on substituting into the heat equation, we obtain

$$B'(t)A(x) = \kappa A''(x)B(t),$$

where primes here denote differentiation with respect to the argument indicated. On rearranging this equation we obtain

$$\frac{B'(t)}{\kappa B(t)} = \frac{A''(x)}{A(x)},$$

and since x and t are independent variables, these expressions must be equal to a constant (which is negative, say $-\lambda^2$, to ensure that the temperature does not become exponentially unbounded as time tends to infinity). Thus the two governing equations for $A(x)$ and $B(t)$ are

$$A''(x) + \lambda^2 A(x) = 0, \quad B'(t) + \kappa\lambda^2 B(t) = 0,$$

and our separable solution of $(4.1)_1$ takes the form

$$T(x,t) = e^{-\kappa\lambda^2 t}(C_1 \cos\lambda x + C_2 \sin\lambda x),$$

where C_1 and C_2 denote arbitrary constants. As noted previously, any linear combination of such solutions is also a solution of $(4.1)_1$ and accordingly, we can form the general trigonometric series solution of (4.1)

$$T(x,t) = \frac{a_0}{2} + \sum_{n-1}^{\infty} e^{-\kappa\lambda_n^2 t}(a_n \cos\lambda_n x + b_n \sin\lambda_n x), \tag{4.2}$$

where the choice of λ_n is determined by the boundary conditions and the constants a_n and b_n are determined as the Fourier coefficients of the initial temperature given by $(4.1)_2$, that is

$$f(x) = \frac{a_0}{2} + \sum_{n=1}^{\infty}(a_n \cos\lambda_n x + b_n \sin\lambda_n x).$$

In the fifth section of this chapter we deal with purely radial flow of heat in infinite solid circular cylinders and in solid spheres. For radial heat flow in cylinders the temperature $T(r,t)$ satisfies

$$\frac{\partial T}{\partial t} = \kappa\left(\frac{\partial^2 T}{\partial r^2} + \frac{1}{r}\frac{\partial T}{\partial r}\right), \tag{4.3}$$

and we build up our solution of the various problems from separable solutions of the form

$$T(r,t) = A(r)B(t).$$

As described above, we may readily establish that the two governing equations for $A(r)$ and $B(t)$ are

$$A''(r) + \frac{1}{r}A'(r) + \lambda^2 A(r) = 0, \quad B'(t) + \kappa\lambda^2 B(t) = 0,$$

again employing $-\lambda^2$ as the separation constant. We observe that the function $A(r)$ satisfies Bessel's equation of order zero. Thus, as before, we may form the general solution of (4.3) by taking a linear combination of the individual separable solutions

$$T(r,t) = \sum_{n=1}^{\infty} a_n e^{-\kappa\lambda_n^2 t} J_0(\lambda_n r), \tag{4.4}$$

where the a_n are determined from the initial condition while the λ_n are determined by the surface condition. We observe that $J_0(z)$ is the usual Bessel function of order zero defined by

$$J_0(z) = \sum_{n=0}^{\infty} (-1)^n \frac{(z/2)^{2n}}{(n!)^2},$$

and that in (4.4) we have purposely excluded the second linearly independent solution $Y_0(\lambda_n r)$ since this has a logarithmic singularity at the origin and is therefore only used for heat flow problems involving circular cylindrical tubes.

For radial heat flow in a solid sphere the temperature $T(r,t)$ satisfies

$$\frac{\partial T}{\partial t} = \kappa\left(\frac{\partial^2 T}{\partial r^2} + \frac{2}{r}\frac{\partial T}{\partial r}\right),$$

which as previously noted (see Problem 1.6) may be reduced to a linear heat flow problem by means of the transformation

$$T(r,t) = \frac{w(r,t)}{r}. \tag{4.5}$$

Thus $w(r,t)$ satisfies

$$\frac{\partial w}{\partial t} = \kappa\frac{\partial^2 w}{\partial r^2},$$

and we may therefore exploit linear heat flow results for radial heat flow in a sphere. However, it is also necessary to apply the transformation (4.5) to any imposed boundary or initial conditions and this may change the functional form of the boundary conditions. In particular, we note that this transformation changes a prescribed temperature gradient boundary condition on $T(r,t)$ into a Newton cooling boundary condition on $w(r,t)$.

In the final section of the chapter we consider the two dimensional problem for the finite rectangle $0 \le x \le a$, $0 \le y \le b$, with arbitrary initial temperature and prescribed temperature about the perimeter of the rectangle. For this problem we look for separable solutions of the form

$$T(x,y,t) = A(x)B(y)C(t),$$

which on substitution into the two dimensional heat equation

$$\frac{\partial T}{\partial t} = \kappa\left(\frac{\partial^2 T}{\partial x^2} + \frac{\partial^2 T}{\partial y^2}\right),$$

yields

$$\frac{C'(t)}{C(t)} = \kappa\left[\frac{A''(x)}{A(x)} + \frac{B''(y)}{B(y)}\right].$$

As before, we can argue that each of the terms in this equation must be constant, so that

$$A''(x) + \lambda^2 A(x) = 0, \quad B''(y) + \nu^2 B(x) = 0, \quad C'(t) + \kappa(\lambda^2 + \nu^2)C(t) = 0,$$

for constants λ and ν. Thus a separable solution of the two dimensional heat equation has the form

$$T(x,t) = (A_1 \cos \lambda x + A_2 \sin \lambda x)(B_1 \cos \nu y + B_2 \sin \nu y)e^{-\kappa(\lambda^2 + \nu^2)t},$$

where A_1, A_2, B_1 and B_2 denote arbitrary constants. From such separable solutions we can construct a general trigonometric series solution for the two dimensional heat equation

$$T(x,y,t) =$$

$$\sum_{m=0}^{\infty} \sum_{n=0}^{\infty} (a_m \cos \lambda_m x + b_m \sin \lambda_m x)(c_n \cos \nu_n y + d_n \sin \nu_n y)e^{-\kappa(\lambda_m^2 + \nu_n^2)t}.$$

In practice, the values of λ_m and ν_n are determined by the boundary conditions imposed on the perimeter of the rectangle and the constants a_m, b_m, c_n and d_n are determined by the initial temperature, namely

$$T(x,y,0) = \sum_{m=0}^{\infty} \sum_{n=0}^{\infty} (a_m \cos \lambda_m x + b_m \sin \lambda_m x)(c_n \cos \nu_n y + d_n \sin \nu_n y).$$

4.2 Linear flow in $(0, \ell)$ with prescribed surface temperatures

In this section we consider linear flow of heat in the finite region $0 \le x \le \ell$ with the planes $x = 0$ and $x = \ell$ at prescribed temperatures $g_1(t)$ and $g_2(t)$ respectively. Thus, we consider the problem

$$
\left.
\begin{aligned}
\frac{\partial T}{\partial t} &= \kappa \frac{\partial^2 T}{\partial x^2}, \quad 0 < x < \ell, \\[2mm]
T(x,0) &= f(x), \quad 0 \le x \le \ell, \\[2mm]
T(0,t) &= g_1(t), \quad T(\ell,t) = g_2(t), \quad t > 0.
\end{aligned}
\right\}
\tag{4.6}
$$

A standard method in heat conduction is to decompose the temperature as the sum of two components, one arising from the initial temperature subject to zero boundary conditions and the other arising from the prescribed boundary data but with zero initial temperature. The linearity of the heat equation (and boundary conditions) ensures that this procedure is effective. Thus for problem (4.6) we write

$$
T(x,t) = u(x,t) + v(x,t),
\tag{4.7}
$$

where $u(x,t)$ is the solution of

$$
\left.
\begin{aligned}
\frac{\partial u}{\partial t} &= \kappa \frac{\partial^2 u}{\partial x^2}, \quad 0 < x < \ell, \\[2mm]
u(x,0) &= f(x), \quad 0 \le x \le \ell, \\[2mm]
u(0,t) &= u(\ell,t) = 0, \quad t > 0,
\end{aligned}
\right\}
\tag{4.8}
$$

while $v(x,t)$ is the solution of

$$
\left.
\begin{aligned}
\frac{\partial v}{\partial t} &= \kappa \frac{\partial^2 v}{\partial x^2}, \quad 0 < x < \ell, \\[2mm]
v(x,0) &= 0, \quad 0 \le x \le \ell, \\[2mm]
v(0,t) &= g_1(t), \quad v(\ell,t) = g_2(t), \quad t > 0,
\end{aligned}
\right\}
\tag{4.9}
$$

We first deduce the solution of problem (4.8). Since $u(0,t) = 0$ we may reasonably assume a Fourier sine series

$$
u(x,t) = \sum_{n=1}^{\infty} b_n e^{-\kappa \lambda_n^2 t} \sin \lambda_n x,
$$

and further, since $u(\ell, t) = 0$, we take $\lambda_n = n\pi/\ell$, in which case b_n are given simply as Fourier sine coefficients (see equation (2.56)). Thus, altogether the solution of problem (4.8) becomes

$$u(x, t) = \sum_{n=1}^{\infty} b_n e^{-\kappa n^2 \pi^2 t / \ell^2} \sin\left(\frac{n\pi x}{\ell}\right), \tag{4.10}$$

where the constants b_n are determined from

$$b_n = \frac{2}{\ell} \int_0^{\ell} f(\xi) \sin\left(\frac{n\pi\xi}{\ell}\right) d\xi. \tag{4.11}$$

<u>Example 4.1</u> For the finite region $0 \le x \le \ell$ with constant initial temperature u_0 and prescribed zero temperature along the planes $x = 0$ and $x = \ell$, show that the temperature is given by

$$u(x, t) = \frac{4u_0}{\pi} \sum_{n=0}^{\infty} \frac{e^{-\kappa(2n+1)^2 \pi^2 t / \ell^2}}{(2n+1)} \sin\left(\frac{(2n+1)\pi x}{\ell}\right).$$

In this case we have simply problem (4.8), with $f(x) = u_0$ (a constant). Thus from (4.11) the Fourier sine coefficients are

$$b_n = \frac{2u_0}{\ell} \int_0^{\ell} \sin\left(\frac{n\pi\xi}{\ell}\right) d\xi = \frac{2u_0}{n\pi}\left[-\cos\left(\frac{n\pi\xi}{\ell}\right)\right]_0^{\ell}$$

$$= \frac{2u_0}{n\pi}[1 - \cos n\pi] = \frac{2u_0}{n\pi}\left[1 - (-1)^n\right],$$

and therefore, b_n is zero if n is even and given by

$$b_n = \frac{4u_0}{\pi(2m+1)},$$

when $n = 2m + 1$, $(m = 0, 1, 2, \ldots)$. The given expression now follows immediately from this result and (4.10).

<u>Example 4.2</u> For the finite region $0 \le x \le \ell$ with initial temperature αx where α is constant and prescribed zero temperature along the planes $x = 0$ and $x = \ell$, show that the temperature is given by (see Figure 4.1)

$$u(x, t) = \frac{2\alpha\ell}{\pi} \sum_{n=1}^{\infty} \frac{(-1)^{n+1}}{n} e^{-\kappa n^2 \pi^2 t / \ell^2} \sin\left(\frac{n\pi x}{\ell}\right).$$

Again this simply is problem (4.8) with $f(x) = \alpha x$ and from (4.11) we readily obtain

$$b_n = \frac{2\alpha}{\ell} \int_0^\ell \xi \sin\left(\frac{n\pi\xi}{\ell}\right) d\xi = \frac{2\alpha}{n\pi}\left\{\left[-\xi\cos\left(\frac{n\pi\xi}{\ell}\right)\right]_0^\ell + \int_0^\ell \cos\left(\frac{n\pi\xi}{\ell}\right)d\xi\right\}$$

$$= -\frac{2\alpha\ell}{n\pi}\cos n\pi = -\frac{2\alpha\ell}{n\pi}(-1)^n = \frac{2\alpha\ell}{n\pi}(-1)^{n+1},$$

by integration by parts and the desired result follows immediately from (4.10).

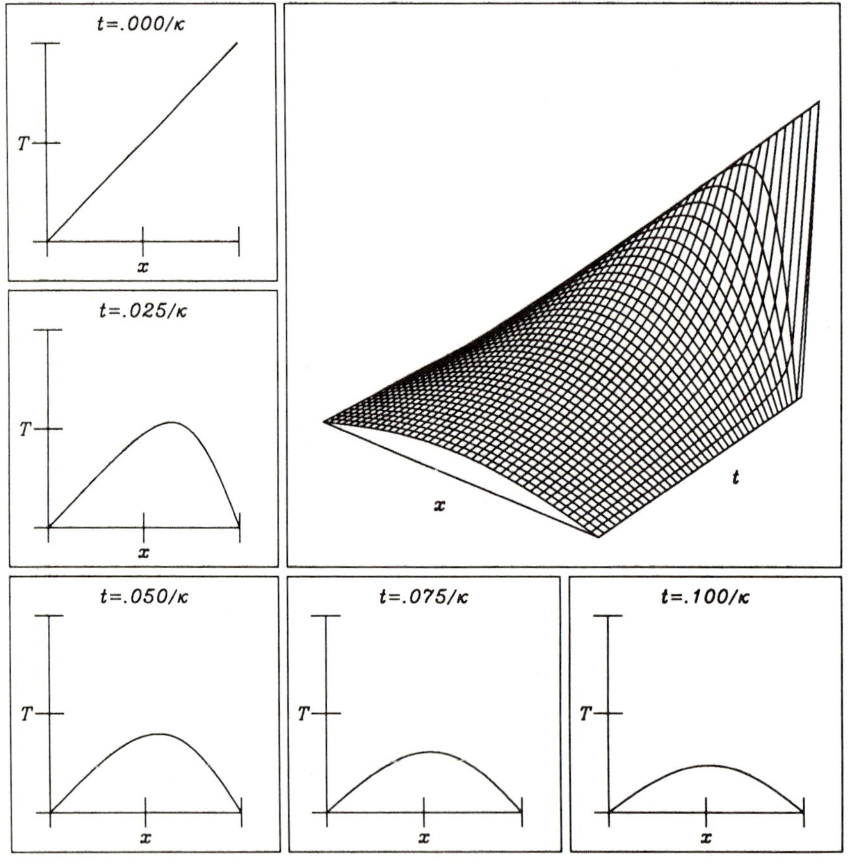

Figure 4.1. Time development of the solution of Example 4.2 with α unity.

Returning now to the solution of problem (4.9) we look for a solution in the form of a Fourier sine series with time dependent Fourier coefficients, that is

$$v(x,t) = \sum_{n=1}^{\infty} b_n(t) \sin\left(\frac{n\pi x}{\ell}\right), \tag{4.12}$$

so that formally we have

$$b_n(t) = \frac{2}{\ell} \int_0^\ell v(\xi,t) \sin\left(\frac{n\pi\xi}{\ell}\right) d\xi, \tag{4.13}$$

and, also on noting the zero initial condition $(4.9)_2$ we have

$$b_n(0) = 0. \tag{4.14}$$

In order to determine the functions $b_n(t)$ we differentiate (4.13) with respect to time and then, after performing two integrations by parts, we deduce a first order linear differential equation (4.15) which may be readily solved subject to (4.14). Thus, from (4.13) we have

$$\frac{db_n}{dt} = \frac{2}{\ell} \int_0^\ell \frac{\partial v}{\partial t}(\xi,t) \sin\left(\frac{n\pi\xi}{\ell}\right) d\xi$$

$$= \frac{2\kappa}{\ell} \int_0^\ell \frac{\partial^2 v}{\partial\xi^2}(\xi,t) \sin\left(\frac{n\pi\xi}{\ell}\right) d\xi$$

$$= \frac{2\kappa}{\ell} \left\{ \left[\frac{\partial v}{\partial\xi} \sin\left(\frac{n\pi\xi}{\ell}\right) \right]_0^\ell - \frac{n\pi}{\ell} \int_0^\ell \frac{\partial v}{\partial\xi}(\xi,t) \cos\left(\frac{n\pi\xi}{\ell}\right) d\xi \right\}$$

$$= \frac{-2\kappa n\pi}{\ell^2} \int_0^\ell \frac{\partial v}{\partial\xi}(\xi,t) \cos\left(\frac{n\pi\xi}{\ell}\right) d\xi$$

$$= \frac{-2\kappa n\pi}{\ell^2} \left\{ \left[v(\xi,t) \cos\left(\frac{n\pi\xi}{\ell}\right) \right]_0^\ell + \frac{n\pi}{\ell} \int_0^\ell v(\xi,t) \sin\left(\frac{n\pi\xi}{\ell}\right) d\xi \right\}.$$

From this equation and (4.13) we may deduce

$$\frac{db_n}{dt}(t) + \left(\frac{n\pi}{\ell}\right)^2 \kappa b_n(t) = \frac{2\kappa n\pi}{\ell^2}[v(0,t) - v(\ell,t)\cos n\pi],$$

which on using the surface conditions $(4.9)_3$ gives

$$\frac{db_n}{dt}(t) + \left(\frac{n\pi}{\ell}\right)^2 \kappa b_n(t) = \frac{2\kappa n\pi}{\ell^2}\left[g_1(t) - (-1)^n g_2(t)\right]. \tag{4.15}$$

On solving this first order linear ordinary differential equation subject to (4.14) we obtain

$$b_n(t) = \frac{2\kappa n\pi}{\ell^2} \int_0^t e^{-\kappa n^2\pi^2(t-\tau)/\ell^2}\Big[g_1(\tau) - (-1)^n g_2(\tau)\Big]\,d\tau, \qquad (4.16)$$

which, together with (4.12) constitutes the solution of problem (4.9).

We note that at first sight, on letting $x = 0$ or $x = \ell$, (4.12) appears not to assume the prescribed temperature values $(4.9)_3$. This is because, as noted in Section 2.5, a Fourier sine series necessarily converges to zero at the endpoints of the interval. If, however, we take the limit as $x \to 0$ or $x \to \ell$ of the sum (4.12) from within the interval $(0, \ell)$, as opposed to taking the sum of the limits, we recover $(4.9)_3$. We can see this by performing an integration by parts on (4.16) to obtain

$$b_n(t) = \frac{2}{n\pi}\Big[g_1(t) - (-1)^n g_2(t)\Big]$$

$$- \frac{2}{n\pi}\left\{\Big[g_1(0) - (-1)^n g_2(0)\Big] + \int_0^t \Big[g_1'(\tau) - (-1)^n g_2'(\tau)\Big]e^{\kappa\pi^2 n^2\tau/\ell^2}\,d\tau\right\}e^{-\kappa\pi^2 n^2 t/\ell^2},$$

so that (4.12) becomes

$$v(x,t) = \Big(1 - \frac{x}{\ell}\Big)g_1(t) + \frac{x}{\ell}g_2(t) - \frac{2}{\pi}\sum_{n=1}^{\infty}\frac{1}{n}\Big\{g_1(0) - (-1)^n g_2(0)$$

$$+ \int_0^t \Big[g_1'(\tau) - (-1)^n g_2'(\tau)\Big]e^{\kappa\pi^2 n^2\tau/\ell^2}\,d\tau\Big\}e^{-\kappa\pi^2 n^2 t/\ell^2}\sin\Big(\frac{n\pi x}{\ell}\Big),$$

where we have used the Fourier sine series

$$\Big(1 - \frac{x}{\ell}\Big) = \frac{2}{\pi}\sum_{n=1}^{\infty}\frac{1}{n}\sin\Big(\frac{n\pi x}{\ell}\Big),$$

$$\frac{x}{\ell} = \frac{2}{\pi}\sum_{n=1}^{\infty}\frac{(-1)^{n+1}}{n}\sin\Big(\frac{n\pi x}{\ell}\Big).$$

It can be shown, under quite mild restrictions on $g_i(t)$ and $g_i'(t)$ $(i = 1, 2)$ (see Problem 14), that the infinite sine series in this expression for $v(x,t)$ converges absolutely so that this series vanishes as $x \to 0$ and as $x \to \ell$ and the prescribed boundary values are recovered. Effectively we have removed from (4.12) the terms which do not converge uniformly on $[0, \ell]$ and replaced them by the actual functions instead of their Fourier series. This is essentially the process which occurs in the following examples as well.

<u>Example 4.3</u> For the finite region $0 \le x \le \ell$ with zero initial temperature and planes $x = 0$ and $x = \ell$ maintained at βt where β is a constant, show that the temperature (see Figure 4.2) is given by

$$v(x,t) = \beta t + \frac{\beta x(x - \ell)}{2\kappa} + \frac{4\beta \ell^2}{\kappa \pi^3} \sum_{n=1}^{\infty} \frac{e^{-\kappa(2n + 1)^2\pi^2 t/\ell^2}}{(2n + 1)^3} \sin\left(\frac{(2n + 1)\pi x}{\ell}\right).$$

This is merely problem (4.9) with $g_1(t) = g_2(t) = \beta t$. Before evaluating $b_n(t)$ from (4.16) we first note the Fourier sine coefficients of 1 and $x(x - \ell)$, thus

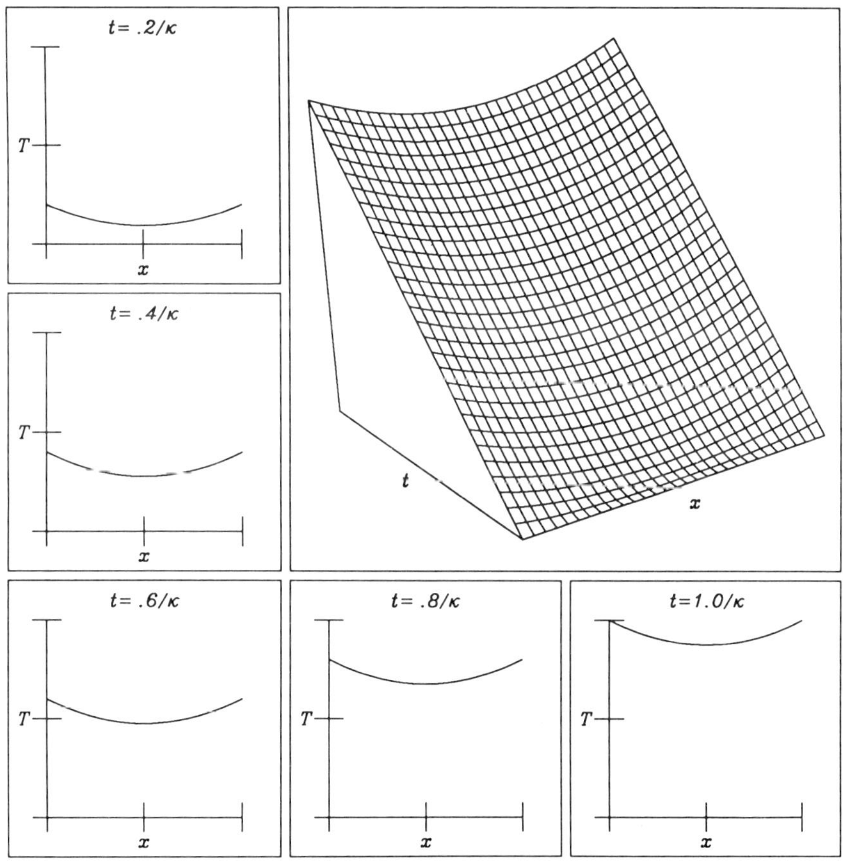

Figure 4.2. Time development of the solution of Example 4.3 with β unity.

$$\frac{2}{\ell} \int_0^\ell \sin\left(\frac{n\pi\xi}{\ell}\right) = \frac{2}{n\pi}\left[1 - (-1)^n\right],$$

$$\frac{2}{\ell} \int_0^\ell \xi(\xi - \ell)\sin\left(\frac{n\pi\xi}{\ell}\right)d\xi = \frac{-4\ell^2}{n^3\pi^3}\left[1 - (-1)^n\right],$$

where the latter result follows in a straightforward manner after two integrations by parts. Now from (4.16) with $g_1(t) = g_2(t) = \beta t$ we have

$$b_n(t) = \frac{2\beta\kappa n\pi}{\ell^2}\left[1 - (-1)^n\right]e^{-\kappa n^2\pi^2 t/\ell^2}\int_0^t \tau e^{\kappa n^2\pi^2\tau/\ell^2}\,d\tau$$

$$= \frac{2\beta\kappa n\pi}{\ell^2}\left[1 - (-1)^n\right]\left\{\frac{t\ell^2}{\kappa n^2\pi^2} - \frac{\ell^4}{\kappa^2 n^4\pi^4} + \frac{\ell^4 e^{-\kappa n^2\pi^2 t/\ell^2}}{\kappa^2 n^4\pi^4}\right\}$$

$$= 2\beta\left[1 - (-1)^n\right]\left\{\frac{t}{n\pi} - \frac{\ell^2}{\kappa n^3\pi^3} + \frac{\ell^2 e^{-\kappa n^2\pi^2 t/\ell^2}}{\kappa n^3\pi^3}\right\},$$

and on noting the above Fourier sine coefficients of 1 and $x(x - \ell)$ the desired result follows immediately.

Altogether from (4.7), (4.10), (4.11), (4.12) and (4.16) we may write the solution of the problem (4.6) with arbitrary initial temperature and prescribed surface temperature as

$$T(x,t) = \frac{2}{\ell}\sum_{n=1}^\infty\left(\int_0^\ell f(\xi)\sin\left(\frac{n\pi\xi}{\ell}\right)d\xi\right)e^{-\kappa n^2\pi^2 t/\ell^2}\sin\left(\frac{n\pi x}{\ell}\right)$$

$$+ \frac{2\kappa\pi}{\ell^2}\sum_{n=1}^\infty n\left(\int_0^t e^{-\kappa n^2\pi^2(t-\tau)/\ell^2}\left[g_1(\tau) - (-1)^n g_2(\tau)\right]d\tau\right)\sin\left(\frac{n\pi x}{\ell}\right). \tag{4.17}$$

Inspection of this result reveals that the Green's function $G(x,\xi,t,\tau)$ for problem (4.6) is

$$G(x,\xi,t,\tau) = \frac{2}{\ell}H(t - \tau)\sum_{n=1}^\infty \sin\left(\frac{n\pi\xi}{\ell}\right)\sin\left(\frac{n\pi x}{\ell}\right)e^{-\kappa n^2\pi^2(t-\tau)/\ell^2},$$

and (4.17) becomes

$$T(x,t) = \int_0^\ell G(x,\xi,t,0)f(\xi)\,d\xi$$

$$+ \kappa\int_0^t g_1(\tau)\frac{\partial G}{\partial\xi}(x,0,t,\tau)\,d\tau - \kappa\int_0^t g_2(\tau)\frac{\partial G}{\partial\xi}(x,\ell,t,\tau)\,d\tau.$$

<u>Example 4.4</u> For the finite region $0 \le x \le \ell$ with arbitrary initial temperature $f(x)$ and planes $x = 0$ and $x = \ell$ maintained at constant temperatures v_{10} and v_{20} respectively, show that the temperature is given by

$$T(x,t) = \frac{2}{\ell} \sum_{n=1}^{\infty} \left(\int_0^{\ell} f(\xi) \sin\left(\frac{n\pi\xi}{\ell}\right) d\xi \right) e^{-\kappa n^2 \pi^2 t/\ell^2} \sin\left(\frac{n\pi x}{\ell}\right)$$

$$+ v_{10} + (v_{20} - v_{10})\frac{x}{\ell} - \frac{2}{\pi} \sum_{n=1}^{\infty} \frac{\left[v_{10} - (-1)^n v_{20}\right]}{n} e^{-\kappa n^2 \pi^2 t/\ell^2} \sin\left(\frac{n\pi x}{\ell}\right).$$

This result evidently follows from (4.17) with $g_1(t) = v_{10}$ and $g_2(t) = v_{20}$ on noting that

$$\int_0^t e^{-\kappa n^2 \pi^2 (t-\tau)/\ell^2} \left[g_1(\tau) - (-1)^n g_2(\tau)\right] d\tau$$

$$= \frac{\ell^2}{\kappa n^2 \pi^2} \left[v_{10} - (-1)^n v_{20}\right]\left[1 - e^{-\kappa n^2 \pi^2 t/\ell^2}\right],$$

and that the Fourier sine coefficients of $v_{10} + (v_{20} - v_{10})x/\ell$ are

$$\frac{2}{\ell} \int_0^{\ell} \left\{v_{10} + (v_{20} - v_{10})\frac{\xi}{\ell}\right\} \sin\left(\frac{n\pi\xi}{\ell}\right) d\xi$$

$$= \frac{2}{n\pi} \left\{\left[-\left\{v_{10} + (v_{20} - v_{10})\frac{\xi}{\ell}\right\} \cos\left(\frac{n\pi\xi}{\ell}\right)\right]_0^{\ell} + \frac{(v_{20} - v_{10})}{\ell} \int_0^{\ell} \cos\left(\frac{n\pi\xi}{\ell}\right) d\xi\right\}$$

$$= \frac{2}{n\pi} \left[v_{10} - (-1)^n v_{20}\right].$$

The given result also follows by setting

$$T(x,t) = V(x) + U(x,t),$$

where $V(x)$ satisfies

$$\left.\begin{array}{l} \dfrac{d^2 V}{dx^2} = 0, \quad 0 < x < \ell, \\[2mm] V(0) = v_{10}, \quad V(\ell) = v_{20}, \end{array}\right\}$$

and $U(x,t)$ satisfies problem (4.8) but with initial condition

$$U(x,0) = f(x) - V(x).$$

A quick calculation shows that

$$V(x) = v_{10} + (v_{20} - v_{10})\frac{x}{\ell},$$

and the given expression then follows on using (4.10) and (4.11) for the solution for $U(x,t)$.

4.3 Linear flow in $(0, \ell)$ with prescribed surface fluxes

In this section we consider linear flow of heat in the finite region $0 \leq x \leq \ell$ with prescribed surface fluxes $h_1(t)$ and $h_2(t)$ across the planes $x = 0$ and $x = \ell$ respectively. We therefore consider the problem

$$\left.\begin{aligned}
\frac{\partial T}{\partial t} &= \kappa \frac{\partial^2 T}{\partial x^2}, \quad 0 < x < \ell, \\
T(x,0) &= f(x), \quad 0 \leq x \leq \ell, \\
-k\frac{\partial T}{\partial x}(0,t) &= h_1(t), \quad k\frac{\partial T}{\partial x}(\ell,t) = h_2(t), \quad t > 0.
\end{aligned}\right\} \tag{4.18}$$

As in the previous section we decompose the solution into two components, one arising from the initial temperature and zero surface fluxes and the other arising from the prescribed surface fluxes and zero initial temperature. Thus $T(x,t) = u(x,t) + v(x,t)$ where $u(x,t)$ satisfies

$$\left.\begin{aligned}
\frac{\partial u}{\partial t} &= \kappa \frac{\partial^2 u}{\partial x^2}, \quad 0 < x < \ell, \\
u(x,0) &= f(x), \quad 0 \leq x \leq \ell, \\
\frac{\partial u}{\partial x}(0,t) &= \frac{\partial u}{\partial x}(\ell,t) = 0, \quad t > 0,
\end{aligned}\right\} \tag{4.19}$$

and $v(x,t)$ satisfies

$$\left.\begin{aligned}
\frac{\partial v}{\partial t} &= \kappa \frac{\partial^2 v}{\partial x^2}, \quad 0 < x < \ell, \\
v(x,0) &= 0, \quad 0 \leq x \leq \ell, \\
-k\frac{\partial v}{\partial x}(0,t) &= h_1(t), \quad k\frac{\partial v}{\partial x}(\ell,t) = h_2(t), \quad t > 0.
\end{aligned}\right\} \tag{4.20}$$

Since the flux vanishes at both $x = 0$ and $x = \ell$, the solution of (4.19) is reasonably apparent as the Fourier cosine series

$$u(x,t) = \frac{a_0}{2} + \sum_{n=1}^{\infty} a_n e^{-\kappa n^2 \pi^2 t / \ell^2} \cos\left(\frac{n\pi x}{\ell}\right), \tag{4.21}$$

where the Fourier coefficients are determined from the initial condition $(4.19)_2$, namely

$$a_n = \frac{2}{\ell} \int_0^{\ell} f(\xi) \cos\left(\frac{n\pi\xi}{\ell}\right) d\xi. \tag{4.22}$$

<u>Example 4.5</u> Show that if the finite region $0 \le x \le \ell$ has constant initial temperature u_0 and insulated surfaces $x = 0$ and $x = \ell$ then the temperature remains at u_0 for all time.

This is immediately apparent from (4.21) and (4.22) since for $n \ge 1$ a_n is zero while $a_0 = 2u_0$ from which it follows that the temperature is u_0 for all time. Of course such a result should be expected, for with a zero initial temperature gradient and no heat flow at either end of the bar there is no heat flow at all, and therefore the temperature is necessarily constant.

<u>Example 4.6</u> For the finite region $0 \le x \le \ell$ with initial temperature αx where α is constant and insulated surfaces $x = 0$ and $x = \ell$, show that the temperature (see Figure 4.3) is

$$u(x,t) = \frac{\alpha\ell}{2} - \frac{4\alpha\ell}{\pi^2} \sum_{n=0}^{\infty} \frac{e^{-\kappa(2n+1)^2\pi^2 t/\ell^2}}{(2n+1)^2} \cos\left(\frac{(2n+1)\pi x}{\ell}\right).$$

From (4.22) with $f(x) = \alpha x$ we have for $n \ge 1$

$$a_n = \frac{2\alpha}{\ell} \int_0^{\ell} \xi \cos\left(\frac{n\pi\xi}{\ell}\right) d\xi$$

$$= \frac{2\alpha}{n\pi} \left\{ \left[\xi \sin\left(\frac{n\pi\xi}{\ell}\right)\right]_0^{\ell} - \int_0^{\ell} \sin\left(\frac{n\pi\xi}{\ell}\right) d\xi \right\}$$

$$= \frac{2\alpha\ell}{n^2\pi^2} \left[\cos\left(\frac{n\pi\xi}{\ell}\right)\right]_0^{\ell}$$

$$= \frac{2\alpha\ell}{n^2\pi^2} \left[(-1)^n - 1\right],$$

so that a_n is zero for n even and

$$a_n = \frac{-4\alpha\ell}{\pi^2(2m+1)^2},$$

for $n = 2m + 1$ ($m = 0, 1, 2, \ldots$). Further for n zero a_0 is $\alpha\ell$ and the desired result follows immediately.

Following the procedure described in the previous section we look for a solution of problem (4.20) as a Fourier cosine series with time dependent coefficients, thus

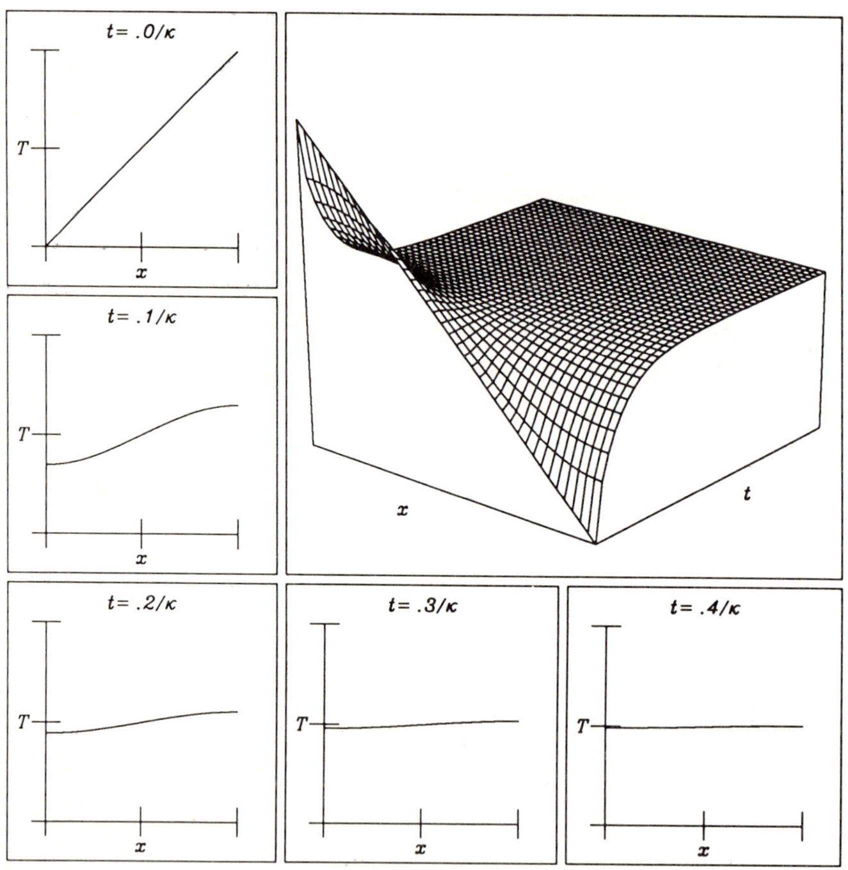

Figure 4.3. Time development of the solution of Example 4.6 with α unity.

$$v(x, t) = \frac{a_0(t)}{2} + \sum_{n=1}^{\infty} a_n(t) \cos\left(\frac{n\pi x}{\ell}\right), \tag{4.23}$$

so that formally we have

$$a_n(t) = \frac{2}{\ell} \int_0^{\ell} v(\xi, t) \cos\left(\frac{n\pi\xi}{\ell}\right) d\xi. \tag{4.24}$$

On differentiating this equation with respect to time and performing two integrations by parts we obtain

$$\frac{da_n}{dt} = \frac{2\kappa}{\ell} \int_0^{\ell} \frac{\partial^2 v}{\partial \xi^2}(\xi, t) \cos\left(\frac{n\pi\xi}{\ell}\right) d\xi$$

$$= \frac{2\kappa}{\ell} \left\{ \left[\frac{\partial v}{\partial \xi}(\xi, t) \cos\left(\frac{n\pi\xi}{\ell}\right)\right]_0^{\ell} + \frac{n\pi}{\ell} \int_0^{\ell} \frac{\partial v}{\partial \xi}(\xi, t) \sin\left(\frac{n\pi\xi}{\ell}\right) d\xi \right\}$$

$$= \frac{2\kappa}{\ell} \left\{ \frac{1}{k}\left[h_1(t) + (-1)^n h_2(t)\right] + \frac{n\pi}{\ell}\left[v(\xi, t) \sin\left(\frac{n\pi\xi}{\ell}\right)\right]_0^{\ell} \right.$$

$$\left. - \left(\frac{n\pi}{\ell}\right)^2 \int_0^{\ell} v(\xi, t) \cos\left(\frac{n\pi\xi}{\ell}\right) d\xi \right\},$$

from which we may readily deduce

$$\frac{da_n}{dt}(t) + \left(\frac{n\pi}{\ell}\right)^2 \kappa a_n(t) = \frac{2\kappa}{k\ell}\left[h_1(t) + (-1)^n h_2(t)\right].$$

Since v is zero initially we have $a_n(0) = 0$ and accordingly we may solve this equation to obtain

$$a_n(t) = \frac{2\kappa}{k\ell} \int_0^t e^{-\kappa n^2 \pi^2 (t-\tau)/\ell^2}\left[h_1(\tau) + (-1)^n h_2(\tau)\right] d\tau, \tag{4.25}$$

which, together with (4.23) constitutes the solution of problem (4.20). Again we note that after differentiating (4.23) with respect to x and letting $x = 0$ or $x = \ell$, it is not obvious at first sight that the prescribed flux conditions are satisfied. For reasons similar to those given in the previous section we find that as we approach $x = 0$ or $x = \ell$ from within the open interval $(0, \ell)$, the prescribed flux conditions emerge. However, if we simply put $x = 0$ or $x = \ell$ in the derivative of (4.23), that is if we interchange the order of limit and summation, then we appear to obtain zero flux at the boundaries because of the generally discontinuous nature of the Fourier series at the endpoints $x = 0$ and $x = \ell$.

<u>Example 4.7</u> For the finite region $0 \le x \le \ell$ with zero initial temperature, no flow of heat along $x = 0$ and constant flux w_0 across $x = \ell$, show that the temperature (see Figure 4.4) is given by

$$v(x,t) = \frac{w_0}{k\ell}\left\{\kappa t + \frac{(3x^2 - \ell^2)}{6} - \frac{2\ell^2}{\pi^2} \sum_{n=1}^{\infty} \frac{(-1)^n}{n^2} e^{-\kappa n^2 \pi^2 t/\ell^2} \cos\left(\frac{n\pi x}{\ell}\right)\right\}.$$

This result follows from (4.23) and (4.25) with $h_1(t) = 0$ and $h_2(t) = w_0$ and noting that the Fourier cosine coefficients for $3x^2 - \ell^2$ are

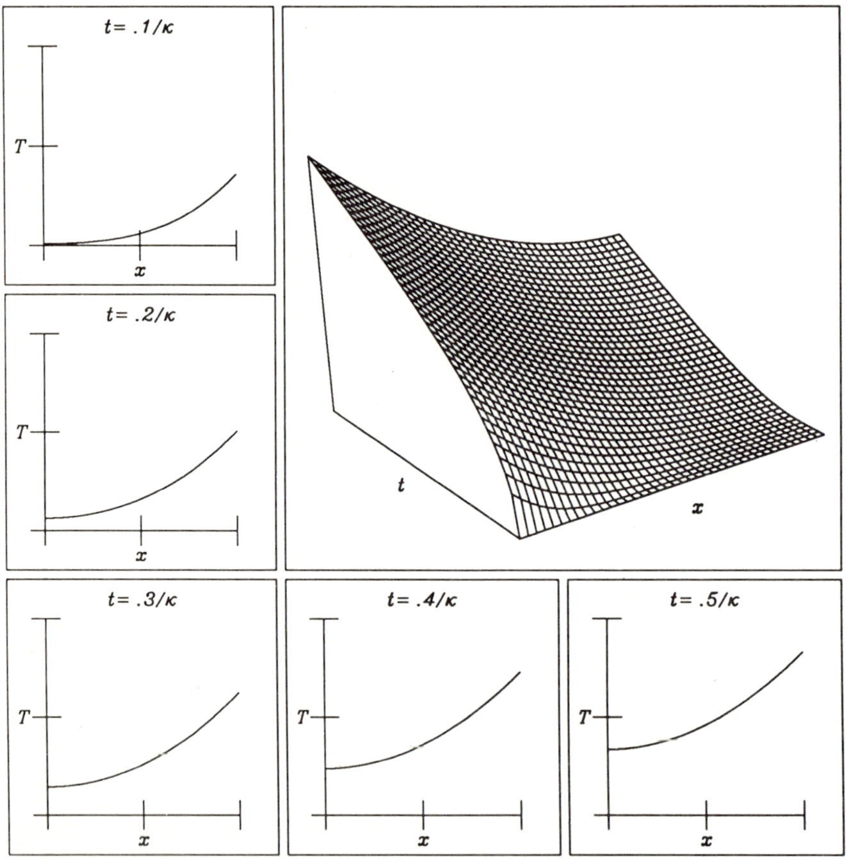

Figure 4.4. Time development of the solution of Example 4.7 with w_0 and k unity.

$$a_0 = 0, \quad a_n = \frac{12\ell^2(-1)^n}{n^2\pi^2}, \quad n \geq 1.$$

Further, from (4.25) we have

$$a_0(t) = \frac{2w_0\kappa t}{k\ell}, \quad a_n(t) = \frac{2w_0\ell(-1)^n}{kn^2\pi^2}\left(1 - e^{-\kappa n^2\pi^2 t/\ell^2}\right), \quad n \geq 1,$$

and the desired result is immediate using (4.23).

Altogether from (4.21), (4.22), (4.23), (4.25) and noting that $T = u + v$, the solution of the problem (4.18) becomes

$$T(x,t) = \frac{1}{\ell}\int_0^\ell f(\xi)\,d\xi + \frac{\kappa}{k\ell}\int_0^t [h_1(\tau) + h_2(\tau)]\,d\tau$$

$$+ \frac{2}{\ell}\sum_{n=1}^\infty \left(\int_0^\ell f(\xi)\cos\left(\frac{n\pi\xi}{\ell}\right)d\xi\right)e^{-\kappa n^2\pi^2 t/\ell^2}\cos\left(\frac{n\pi x}{\ell}\right) \qquad (4.26)$$

$$+ \frac{2\kappa}{k\ell}\sum_{n=1}^\infty \left(\int_0^t e^{-\kappa n^2\pi^2(t-\tau)/\ell^2}\left[h_1(\tau) + (-1)^n h_2(\tau)\right]d\tau\right)\cos\left(\frac{n\pi x}{\ell}\right).$$

Thus we see that the Green's function for problem (4.18) is

$$G(x,\xi,t,\tau) = H(t - \tau)\left[\frac{1}{\ell} + \frac{2}{\ell}\sum_{n=1}^\infty \cos\left(\frac{n\pi\xi}{\ell}\right)\cos\left(\frac{n\pi x}{\ell}\right)e^{-\kappa n^2\pi^2(t-\tau)/\ell^2}\right],$$

and (4.26) becomes

$$T(x,t) = \int_0^\ell G(x,\xi,t,0)f(\xi)\,d\xi$$

$$+ \frac{\kappa}{k}\int_0^t h_1(\tau)G(x,0,t,\tau)\,d\tau + \frac{\kappa}{k}\int_0^t h_2(\tau)G(x,\ell,t,\tau)\,d\tau.$$

4.4 Linear flow in $(0, \ell)$ with Newton heat loss into media at prescribed temperatures

For the problem of linear flow of heat in the finite region $0 \leq x \leq \ell$ with Newton heat loss across the surfaces $x = 0$ and $x = \ell$, the technique of solution is precisely as described in the two previous sections except that the details are considerably more complicated. The problem to solve is

$$\frac{\partial T}{\partial t} = \kappa \frac{\partial^2 T}{\partial x^2}, \quad 0 < x < \ell,$$

$$T(x,0) = f(x), \quad 0 \le x \le \ell,$$

$$-\frac{\partial T}{\partial x}(0,t) + \mu T(0,t) = \mu T_1(t), \quad \frac{\partial T}{\partial x}(\ell,t) + \mu T(\ell,t) = \mu T_2(t), \quad t > 0,$$

$$\tag{4.27}$$

where $\mu = h/k$. Again with $T = u + v$ we first solve the following problem for $u(x,t)$

$$\frac{\partial u}{\partial t} = \kappa \frac{\partial^2 u}{\partial x^2}, \quad 0 < x < \ell,$$

$$u(x,0) = f(x), \quad 0 \le x \le \ell, \tag{4.28}$$

$$-\frac{\partial u}{\partial x}(0,t) + \mu u(0,t) = \frac{\partial u}{\partial x}(\ell,t) + \mu u(\ell,t) = 0, \quad t > 0,$$

and then we solve the following problem for $v(x,t)$

$$\frac{\partial v}{\partial t} = \kappa \frac{\partial^2 v}{\partial x^2}, \quad 0 < x < \ell,$$

$$v(x,0) = 0, \quad 0 \le x \le \ell, \tag{4.29}$$

$$-\frac{\partial v}{\partial x}(0,t) + \mu v(0,t) = \mu T_1(t), \quad \frac{\partial v}{\partial x}(\ell,t) + \mu v(\ell,t) = \mu T_2(t), \quad t > 0.$$

We attempt to solve (4.28) by utilizing separable solutions of the form

$$u(x,t) = e^{-\kappa \lambda_n^2 t}(C_{1n} \cos \lambda_n x + C_{2n} \sin \lambda_n x), \tag{4.30}$$

where C_{1n}, C_{2n} and λ_n are all constants which have yet to be determined. The two homogeneous boundary conditions $(4.28)_3$ yield two equations for C_{1n} and C_{2n}, namely

$$\mu C_{1n} - \lambda_n C_{2n} = 0,$$

$$(\mu \cos \lambda_n \ell - \lambda_n \sin \lambda_n \ell)C_{1n} + (\mu \sin \lambda_n \ell + \lambda_n \cos \lambda_n \ell)C_{2n} = 0,$$

which yield non-trivial solutions only if the system of equations is singular. Thus

$$\mu(\mu \sin \lambda_n \ell + \lambda_n \cos \lambda_n \ell) + \lambda_n(\mu \cos \lambda_n \ell - \lambda_n \sin \lambda_n \ell) = 0,$$

which simplifies to give

$$\left(\lambda_n^2 - \mu^2\right) \sin \lambda_n \ell = 2\mu \lambda_n \cos \lambda_n \ell.$$

Thus we may conclude that the separable solution (4.30) satisfies the two homogeneous boundary conditions (4.28)₃ provided that λ_n is a root of the transcendental equation

$$\tan \lambda_n \ell = \frac{2\mu \lambda_n}{\left(\lambda_n^2 - \mu^2\right)}, \tag{4.31}$$

and if this is the case then

$$C_{2n} = \frac{\mu}{\lambda_n} C_{1n}.$$

It is apparent from graphical considerations that the transcendental equation (4.31) has infinitely many positive solutions (see Figure 4.5). Accordingly we are led to look for a solution of problem (4.28) of the form

$$u(x, t) = \sum_{n=1}^{\infty} a_n e^{-\kappa \lambda_n^2 t} X_n(x), \tag{4.32}$$

where the functions $X_n(x)$ are defined by

$$X_n(x) = \cos \lambda_n x + \frac{\mu}{\lambda_n} \sin \lambda_n x, \tag{4.33}$$

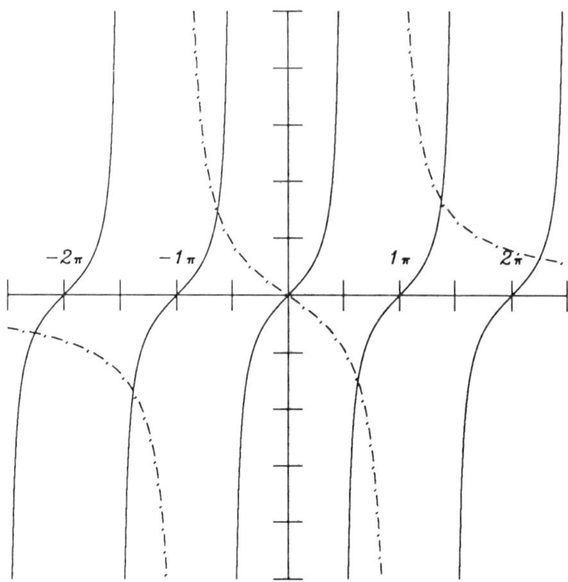

Figure 4.5. Graphical solution of the transcendental equation (4.31) with $\ell = 1$, $\tan(x)$ (—) and $2\mu x/(x^2 - \mu^2)$ ($-\cdot-$).

and the coefficients a_n are determined so that the initial condition $(4.28)_2$ is satisfied, thus

$$f(x) = \sum_{n=1}^{\infty} a_n X_n(x). \tag{4.34}$$

If for the time being we assume that

$$\int_0^{\ell} X_m(\xi) X_n(\xi)\, d\xi = 0, \tag{4.35}$$

for $m \neq n$ then it follows immediately from (4.34) that the coefficients a_n are given by

$$a_n = \frac{\displaystyle\int_0^{\ell} f(\xi) X_n(\xi)\, d\xi}{\displaystyle\int_0^{\ell} X_n^2(\xi)\, d\xi}, \tag{4.36}$$

and therefore we need to determine an expression for the integral occurring in the denominator, which we evaluate directly, thus

$$\int_0^{\ell} X_n(\xi)^2\, d\xi = \int_0^{\ell} \left(\cos \lambda_n \xi + \frac{\mu}{\lambda_n} \sin \lambda_n \xi\right)^2 d\xi$$

$$= \int_0^{\ell} \left\{ 1 + \frac{1}{2}\left(\frac{\mu^2}{\lambda_n^2} - 1\right)(1 - \cos 2\lambda_n \xi) + \frac{\mu}{\lambda_n} \sin 2\lambda_n \xi \right\} d\xi$$

$$= \frac{\ell}{2}\left(\frac{\mu^2}{\lambda_n^2} + 1\right) + \frac{\mu}{\lambda_n^2} \sin^2\left\{\lambda_n \ell\left(1 + \frac{(\lambda_n^2 - \mu^2)}{2\mu \lambda_n \tan \lambda_n \ell}\right)\right\}.$$

On using (4.31) this expression simplifies considerably to give

$$\int_0^{\ell} X_n(\xi)^2\, d\xi = \frac{\left[(\lambda_n^2 + \mu^2)\ell + 2\mu\right]}{2\lambda_n^2}, \tag{4.37}$$

and therefore equation (4.36) for the Fourier coefficient becomes

$$a_n = \frac{2\lambda_n^2}{\left[(\lambda_n^2 + \mu^2)\ell + 2\mu\right]} \int_0^{\ell} f(\xi) X_n(\xi)\, d\xi. \tag{4.38}$$

In order to prove (4.35) we observe that $X_n(x)$ defined by (4.33) is the solution of

$$\left.\begin{aligned}
\frac{d^2 X_n}{dx^2} + \lambda_n^2 X_n &= 0, \\[2mm]
-\frac{dX_n}{dx}(0) + \mu X_n(0) = 0, \qquad \frac{dX_n}{dx}(\ell) + \mu X_n(\ell) &= 0,
\end{aligned}\right\} \tag{4.39}$$

and so for $m \neq n$ we have

$$X_m \frac{d^2 X_n}{dx^2} - X_n \frac{d^2 X_m}{dx^2} + \left(\lambda_n^2 - \lambda_m^2\right) X_n X_m = 0,$$

from which it follows that

$$\begin{aligned}
\left(\lambda_n^2 - \lambda_m^2\right) \int_0^\ell X_n(\xi) X_m(\xi)\, d\xi &= \int_0^\ell \left\{ X_m(\xi) \frac{d^2 X_n}{d\xi^2}(\xi) - X_n(\xi) \frac{d^2 X_m}{d\xi^2}(\xi) \right\} d\xi \\[2mm]
&= \left[X_m(\xi) \frac{dX_n}{d\xi}(\xi) - X_n(\xi) \frac{dX_m}{d\xi}(\xi) \right]_0^\ell \\[2mm]
&= 0,
\end{aligned}$$

on using the boundary conditions $(4.39)_2$.

In order to solve problem (4.29) we again assume a Fourier series with time dependent coefficients, thus

$$v(x,t) = \sum_{n=1}^{\infty} a_n(t) X_n(x), \tag{4.40}$$

where $X_n(x)$ are again defined by (4.33) and $a_n(t)$ are formally given by

$$a_n(t) = \frac{2\lambda_n^2}{\left[(\lambda_n^2 + \mu^2)\ell + 2\mu\right]} \int_0^\ell v(\xi, t) X_n(\xi)\, d\xi.$$

Again on differentiating with respect to time and performing two integrations by parts we have

$$\frac{\mathrm{d}a_n}{\mathrm{d}t}$$

$$= \frac{2\lambda_n^2 \kappa}{\left[(\lambda_n^2 + \mu^2)\ell + 2\mu\right]} \int_0^\ell \frac{\partial^2 v}{\partial \xi^2}(\xi, t) X_n(\xi)\,\mathrm{d}\xi$$

$$= \frac{2\lambda_n^2 \kappa}{\left[(\lambda_n^2 + \mu^2)\ell + 2\mu\right]} \left\{ \left[\frac{\partial v}{\partial \xi}(\xi, t) X_n(\xi)\right]_0^\ell - \int_0^\ell \frac{\partial v}{\partial \xi}(\xi, t)\frac{\mathrm{d}X_n}{\mathrm{d}\xi}(\xi)\,\mathrm{d}\xi \right\}$$

$$= \frac{2\lambda_n^2 \kappa}{\left[(\lambda_n^2 + \mu^2)\ell + 2\mu\right]} \left\{ \left[\frac{\partial v}{\partial \xi}(\xi, t) X_n(\xi) - v(\xi, t)\frac{\mathrm{d}X_n}{\mathrm{d}\xi}(\xi)\right]_0^\ell - \int_0^\ell v(\xi, t)\lambda_n^2 X_n(\xi)\,\mathrm{d}\xi \right\},$$

on using $(4.39)_1$. On using the boundary conditions $(4.29)_3$ and $(4.39)_2$ and noting that $X_n(0) = 1$, this expression simplifies to give

$$\frac{\mathrm{d}a_n}{\mathrm{d}t}(t) + \kappa\lambda_n^2 a_n(t) = \frac{2\kappa\lambda_n^2\mu[T_1(t) + X_n(\ell)T_2(t)]}{\left[(\lambda_n^2 + \mu^2)\ell + 2\mu\right]}.$$

The solution of this equation subject to $a_n(0) = 0$ is simply

$$a_n(t) = \frac{2\kappa\lambda_n^2\mu}{\left[(\lambda_n^2 + \mu^2)\ell + 2\mu\right]} \int_0^t e^{-\kappa\lambda_n^2(t-\tau)}[T_1(\tau) + X_n(\ell)T_2(\tau)]\,\mathrm{d}\tau. \qquad (4.41)$$

Thus altogether from (4.32), (4.38), (4.40) and (4.41) the solution of problem (4.27) becomes

$$T(x, t) = \sum_{n=1}^\infty \frac{2\lambda_n^2}{\left[(\lambda_n^2 + \mu^2)\ell + 2\mu\right]} \left(\int_0^\ell f(\xi) X_n(\xi)\,\mathrm{d}\xi \right) X_n(x) e^{-\kappa\lambda_n^2 t}$$

$$+ \sum_{n=1}^\infty \frac{2\kappa\lambda_n^2\mu}{\left[(\lambda_n^2 + \mu^2)\ell + 2\mu\right]} \left(\int_0^t e^{-\kappa\lambda_n^2(t-\tau)}[T_1(\tau) + X_n(\ell)T_2(\tau)]\,\mathrm{d}\tau \right) X_n(x),$$

$$(4.42)$$

where $X_n(x)$ are given by (4.33) and λ_n is assumed to be the nth positive root of the transcendental equation (4.31). In this case we find that the Green's function

$G(x, \xi, t, \tau)$ for problem (4.27) is given by

$$G(x, \xi, t, \tau) = H(t - \tau) \sum_{n=1}^{\infty} \frac{2\lambda_n^2}{\left[(\lambda_n^2 + \mu^2)\ell + 2\mu\right]} X_n(\xi) X_n(x) e^{-\kappa \lambda_n^2 (t - \tau)},$$

and (4.42) becomes

$$T(x, t) = \int_0^{\ell} G(x, \xi, t, 0) f(\xi)\, d\xi$$

$$+ \kappa\mu \int_0^t T_1(\tau) G(x, 0, t, \tau)\, d\tau + \kappa\mu \int_0^t T_2(\tau) G(x, \ell, t, \tau)\, d\tau.$$

<u>Example 4.8</u> For the finite region $0 < x < \ell$ with arbitrary initial temperature $f(x)$ and Newton heat loss from surfaces $x = 0$ and $x = \ell$ into media at constant temperatures T_{10} and T_{20} respectively, show that the temperature is given by

$$T(x, t) = Ax + B + \sum_{n=1}^{\infty} \frac{2\lambda_n^2 e^{-\kappa\lambda_n^2 t}}{\left[(\lambda_n^2 + \mu^2)\ell + 2\mu\right]} \left(\int_0^{\ell} [f(\xi) - A\xi - B] X_n(\xi)\, d\xi \right) X_n(x),$$

where $X_n(x)$ is defined by (4.33) and the constants A and B are given respectively by

$$A = \frac{\mu(T_{20} - T_{10})}{(2 + \mu\ell)}, \quad B = \frac{T_{20} + (1 + \mu\ell)T_{10}}{(2 + \mu\ell)}.$$

This result follows from (4.42) with $T_1(t)$ and $T_2(t)$ constants but is most easily verified by setting

$$T(x, t) = V(x) + U(x, t),$$

where $V(x)$ satisfies

$$\frac{d^2 V}{dx^2} = 0,$$

$$-\frac{dV}{dx}(0) + \mu V(0) = \mu T_{10}, \quad \frac{dV}{dx}(\ell) + \mu V(\ell) = \mu T_{20},$$

and where $U(x, t)$ satisfies problem (4.28) but with initial condition

$$U(x, 0) = f(x) - V(x).$$

We may readily confirm that the solution for $V(x)$ is simply $Ax + B$ where the constants A and B are as given. On using (4.32) and (4.38) for the solution of $U(x, t)$ the desired result follows immediately.

4.5 Radial flow in solid circular cylinders and spheres

In this section we deal with heat flow in infinite solid circular cylinders and solid spheres. Thus, for cylinders and spheres of radius a we consider the problem for $T(r,t)$

$$\left.\begin{aligned}\frac{\partial T}{\partial t} &= \kappa\left(\frac{\partial^2 T}{\partial r^2} + \frac{m}{r}\frac{\partial T}{\partial r}\right), \quad 0 < r < a, \\[2mm] T(r,0) &= f(r), \quad 0 \le r \le a,\end{aligned}\right\} \tag{4.43}$$

subject to one of the following surface conditions for $t > 0$

$$\left.\begin{aligned} &\textbf{(i)} \quad T(a,t) = g(t), \\[2mm] &\textbf{(ii)} \quad k\frac{\partial T}{\partial r}(a,t) = h(t), \\[2mm] &\textbf{(iii)} \quad \frac{\partial T}{\partial r}(a,t) + \mu\,T(a,t) = \mu\,T_0(t), \end{aligned}\right\} \tag{4.44}$$

where as before $\mu = h/k$ and $m = 1$ for cylinders and $m = 2$ for spheres. In the usual way we solve these problems by assuming $T = u + v$ where $u(r,t)$ satisfies the initial condition and zero surface condition, while $v(r,t)$ is zero initially and satisfies one of the surface conditions (4.44).

For the cylinder we first consider the problem for $u(r,t)$, which as described in the introduction to this chapter has a series solution of the form (4.4), namely

$$u(r,t) = \sum_{n=1}^{\infty} a_n e^{-\kappa\lambda_n^2 t}\, \mathrm{J}_0(\lambda_n r), \tag{4.45}$$

which satisfies the appropriate surface condition, provided that λ_n is a positive root of one of

$$\left.\begin{aligned} &\textbf{(i)} \quad \mathrm{J}_0(\lambda_n a) = 0, \\[2mm] &\textbf{(ii)} \quad \mathrm{J}_0{}'(\lambda_n a) = 0, \\[2mm] &\textbf{(iii)} \quad \lambda_n \mathrm{J}_0{}'(\lambda_n a) + \mu\,\mathrm{J}_0(\lambda_n a) = 0. \end{aligned}\right\} \tag{4.46}$$

Now if for the time being we assume

$$\int_0^a \xi\,\mathrm{J}_0(\lambda_n \xi)\,\mathrm{J}_0(\lambda_m \xi)\,\mathrm{d}\xi = 0, \tag{4.47}$$

for $m \ne n$ then it follows that (4.45) satisfies the initial condition $(4.43)_2$ with the constants a_n given by

$$a_n = \frac{\int_0^a \xi f(\xi) \, J_0(\lambda_n \xi) \, d\xi}{\int_0^a \xi \, J_0(\lambda_n \xi)^2 \, d\xi}. \tag{4.48}$$

But on making use of Bessel's equation for $J_0(z)$, namely

$$J_0''(z) + \frac{1}{z} J_0'(z) + J_0(z) = 0, \tag{4.49}$$

it is a simple matter to verify by direct differentiation the standard Bessel function integral

$$\int_0^r \xi \, J_0(\lambda_n \xi)^2 \, d\xi = \frac{r^2}{2} \Big[J_0(\lambda_n r)^2 + J_0'(\lambda_n r)^2 \Big],$$

so that on setting $r = a$ we may deduce from (4.46) the following values of the integral appearing in the denominator of (4.48), thus

$$
\left.
\begin{aligned}
\textbf{(i)} \quad & \int_0^a \xi \, J_0(\lambda_n \xi)^2 \, d\xi = \frac{a^2}{2} J_0'(\lambda_n a)^2, \\[2ex]
\textbf{(ii)} \quad & \int_0^a \xi \, J_0(\lambda_n \xi)^2 \, d\xi = \frac{a^2}{2} J_0(\lambda_n a)^2, \\[2ex]
\textbf{(iii)} \quad & \int_0^a \xi \, J_0(\lambda_n \xi)^2 \, d\xi = \frac{a^2}{2}\left(1 + \frac{\mu^2}{\lambda_n^2}\right) J_0(\lambda_n a)^2.
\end{aligned}
\right\} \tag{4.50}
$$

In order to establish (4.47) we observe that $A_n(r) = J_0(\lambda_n r)$ satisfies

$$\frac{d^2 A_n}{dr^2} + \frac{1}{r}\frac{dA_n}{dr} + \lambda_n^2 A_n = 0,$$

and therefore

$$(\lambda_n^2 - \lambda_m^2) r A_n A_m = A_n \frac{d}{dr}\left(r \frac{dA_m}{dr}\right) - A_m \frac{d}{dr}\left(r \frac{dA_n}{dr}\right),$$

$$= \frac{d}{dr}\left\{ r\left(A_n \frac{dA_m}{dr} - A_m \frac{dA_n}{dr}\right)\right\},$$

and on integrating we have

$$(\lambda_n^2 - \lambda_m^2) \int_0^a \xi \, J_0(\lambda_n \xi) \, J_0(\lambda_m \xi) \, d\xi = \lambda_n a \Big[J_0(\lambda_n a) \, J_0'(\lambda_m a) - J_0(\lambda_m a) \, J_0'(\lambda_n a) \Big].$$

This expression is clearly zero provided λ_n are the roots of any one of (4.46) and therefore we have established (4.47) for $n \neq m$.

<u>Example 4.9</u> For the infinite solid circular cylinder $0 \le r \le a$ with zero surface temperature and constant initial temperature u_0 show that the temperature (see Figure 4.6) is given by

$$u(r,t) = \frac{-2u_0}{a} \sum_{n=1}^{\infty} e^{-\kappa \lambda_n^2 t} \frac{J_0(\lambda_n r)}{\lambda_n J_0'(\lambda_n a)},$$

where λ_n denote the positive roots of $J_0(\lambda_n a) = 0$.

The required solution is given by (4.45) with coefficients determined by (4.48). On using Bessel's equation for the following integral we have

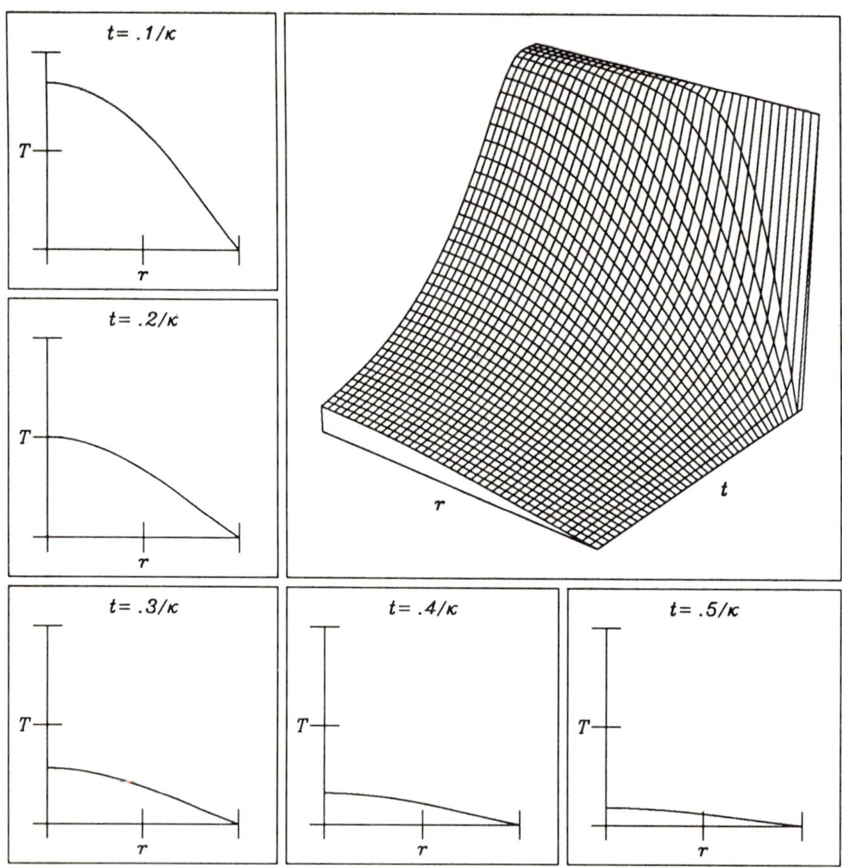

Figure 4.6. Time development of the solution of Example 4.9 with u_0 and a unity.

$$\int_0^a \xi\, J_0(\lambda_n \xi)\, d\xi \;=\; \frac{-1}{\lambda_n} \int_0^a \left\{ \lambda_n \xi\, J_0''(\lambda_n \xi) + J_0'(\lambda_n \xi) \right\} d\xi$$

$$= \frac{-1}{\lambda_n^2} \int_0^a \frac{d}{d\xi}\left[\lambda_n \xi\, J_0'(\lambda_n \xi) \right] d\xi$$

$$= \frac{-a}{\lambda_n}\, J_0'(\lambda_n a),$$

so that from (4.48) and on using (4.50)$_1$ we find

$$a_n \;=\; \frac{-2u_0}{\lambda_n a}\, \frac{1}{J_0'(\lambda_n a)},$$

and the desired result follows immediately from (4.45).

Thus we might summarize the solution for $u(r, t)$ by

$$u(r, t) \;=\; \sum_{n=1}^{\infty} \frac{\left(\int_0^a \xi f(\xi)\, J_0(\lambda_n \xi)\, d\xi \right)}{\left(\int_0^a \xi\, J_0(\lambda_n \xi)^2\, d\xi \right)}\, e^{-\kappa \lambda_n^2 t}\, J_0(\lambda_n r), \tag{4.51}$$

where λ_n are the positive roots of one of (4.46) with corresponding values of the integral in the denominator of (4.51) given by (4.50). We observe however that since in case **(ii)** $\lambda_0 \equiv 0$ is a root of $J_0'(\lambda_n a) = 0$ the solution for this case takes the form

$$u(r, t) \;=\; \frac{2}{a^2} \int_0^a \xi f(\xi)\, d\xi \;+\; \frac{2}{a^2} \sum_{n=1}^{\infty} \left(\int_0^a \xi f(\xi)\, J_0(\lambda_n \xi)\, d\xi \right) e^{-\kappa \lambda_n^2 t}\, \frac{J_0(\lambda_n r)}{J_0(\lambda_n a)^2}, \tag{4.52}$$

where we have utilized the integral (4.50)$_2$.

<u>Example 4.10</u> For the infinite solid circular cylinder $0 \le r \le a$ with zero surface flux and constant initial temperature u_0 show that the temperature remains at u_0 for all time.

This result is apparent from the above equation (4.52) on noting the integrals

$$\int_0^a \xi\, d\xi \;=\; \frac{a^2}{2}, \qquad \int_0^a \xi\, J_0(\lambda_n \xi)\, d\xi \;=\; \frac{-a}{\lambda_n}\, J_0'(\lambda_n a),$$

and recalling that in this case λ_n are the roots of $J_0'(\lambda_n a) = 0$. This result should of course be obvious, as with no initial temperature gradient and no heat flow at the surface of the cylinder, there is no heat flow at all within the cylinder so the temperature is necessarily constant.

In order to determine the solution $v(r,t)$ which satisfies the surface condition but has zero initial temperature, we seek a solution of the form

$$v(r,t) = \sum_{n=1}^{\infty} a_n(t)\, J_0(\lambda_n r), \tag{4.53}$$

where the λ_n are still given by the appropriate expression in (4.46) and where the time dependent coefficients are given formally by

$$a_n(t) = \frac{\displaystyle\int_0^a \xi\, v(\xi,t)\, J_0(\lambda_n \xi)\, \mathrm{d}\xi}{\displaystyle\int_0^a \xi\, J_0(\lambda_n \xi)^2\, \mathrm{d}\xi}. \tag{4.54}$$

On differentiating (4.54) with respect to time, using the heat equation (4.3) and performing two integrations by parts we may deduce

$$\frac{\mathrm{d}a_n}{\mathrm{d}t}(t) + \kappa\lambda_n^2 a_n(t) = \frac{\kappa a\left\{ J_0(\lambda_n a)\dfrac{\partial v}{\partial r}(a,t) - \lambda_n J_0{}'(\lambda_n a) v(a,t)\right\}}{\displaystyle\int_0^a \xi\, J_0(\lambda_n \xi)^2\, \mathrm{d}\xi}.$$

On integrating this equation using $a_n(0) = 0$ we may deduce from (4.53) the following solutions for $v(r,t)$,

$$\textbf{(i)} \qquad v(r,t) = \frac{-2\kappa}{a} \sum_{n=1}^{\infty} \frac{\lambda_n J_0(\lambda_n r)}{J_0{}'(\lambda_n a)} \int_0^t e^{-\kappa\lambda_n^2(t-\tau)} g(\tau)\, \mathrm{d}\tau, \tag{4.55}$$

where λ_n are the positive roots of $J_0(\lambda_n a) = 0$.

$$\textbf{(ii)} \quad v(r,t) = \frac{2\kappa}{ka}\left\{ \int_0^t h(\tau)\, \mathrm{d}\tau + \sum_{n=1}^{\infty} \frac{J_0(\lambda_n r)}{J_0(\lambda_n a)} \int_0^t e^{-\kappa\lambda_n^2(t-\tau)} h(\tau)\, \mathrm{d}\tau \right\}, \tag{4.56}$$

where λ_n are the positive roots of $J_0{}'(\lambda_n a) = 0$.

$$\textbf{(iii)} \quad v(r,t) = \frac{2\kappa\mu}{a} \sum_{n=1}^{\infty} \frac{\lambda_n^2 J_0(\lambda_n r)}{(\lambda_n^2 + \mu^2) J_0(\lambda_n a)} \int_0^t e^{-\kappa\lambda_n^2(t-\tau)} T_0(\tau)\, \mathrm{d}\tau, \tag{4.57}$$

where λ_n are the positive roots of $\lambda_n J_0{}'(\lambda_n a) + \mu J_0(\lambda_n a) = 0$.

<u>Example 4.11</u> For the infinite solid circular cylinder $0 \le r \le a$ with constant surface flux w_0 and zero initial temperature show that the temperature is given by

$$v(r,t) = \frac{2w_0}{ka}\left\{ \kappa t + \frac{r^2}{4} - \frac{a^2}{8} - \sum_{n=1}^{\infty} e^{-\kappa\lambda_n^2 t} \frac{J_0(\lambda_n r)}{\lambda_n^2 J_0(\lambda_n a)} \right\},$$

where λ_n are the positive roots of $J_0{}'(\lambda_n a) = 0$.

This follows immediately from (4.56) on noting the result

$$\frac{r^2}{4} - \frac{a^2}{8} = \sum_{n=1}^{\infty} \frac{J_0(\lambda_n r)}{\lambda_n^2 \, J_0(\lambda_n a)},$$

which may be verified using (4.48) as follows

$$\frac{\int_0^a \frac{\xi}{4}\left(\xi^2 - \frac{a^2}{2}\right) J_0(\lambda_n \xi)\, d\xi}{\int_0^a \xi \, J_0(\lambda_n \xi)^2 \, d\xi}$$

$$= \frac{1}{2a^2 \, J_0(\lambda_n a)^2} \left\{ \left[-\left(\xi^2 - \frac{a^2}{2}\right)\frac{\xi \, J_0'(\lambda_n \xi)}{\lambda_n} \right]_0^a + \frac{2}{\lambda_n} \int_0^a \xi^2 \, J_0'(\lambda_n \xi)\, d\xi \right\}$$

$$= \frac{1}{\lambda_n a^2 \, J_0(\lambda_n a)^2} \int_0^a \xi^2 \, J_0'(\lambda_n \xi)\, d\xi$$

$$= \frac{1}{\lambda_n a^2 \, J_0(\lambda_n a)^2} \left\{ \left[\xi^2 \frac{J_0(\lambda_n \xi)}{\lambda_n} \right]_0^a - \frac{2}{\lambda_n} \int_0^a \xi \, J_0(\lambda_n \xi)\, d\xi \right\}$$

$$= \frac{1}{\lambda_n^2 \, J_0(\lambda_n a)},$$

since the integral appearing in the previous equation is zero.

For radial heat flow in a solid sphere we introduce $w(r, t)$ defined by (4.5) and the problems (4.43) (with $m = 2$) and (4.44) become

$$\left.\begin{aligned}
\frac{\partial w}{\partial t} &= \kappa \frac{\partial^2 w}{\partial r^2}, \quad 0 < r < a, \\
w(r, 0) &= rf(r), \quad 0 \leq r \leq a, \\
w(0, t) &= 0, \quad t > 0,
\end{aligned}\right\} \tag{4.58}$$

subject to one of the following surface conditions for $t > 0$

$$\left.\begin{aligned}
&\textbf{(i)} \quad w(a, t) = ag(t), \\
&\textbf{(ii)} \quad \frac{\partial w}{\partial r}(a, t) - \frac{1}{a} w(a, t) = \frac{a}{k} h(t), \\
&\textbf{(iii)} \quad \frac{\partial w}{\partial r}(a, t) + \left(\mu - \frac{1}{a}\right) w(a, t) = \mu a T_0(t).
\end{aligned}\right\} \tag{4.59}$$

We observe that the condition $w(0, t) = 0$ arises since we require that the temperature $T(0, t)$ be finite. Further we note that the prescribed flux case is essentially a special case of the Newton heat loss boundary condition, and accordingly we give only the results for cases **(i)** and **(iii)**.

For case **(i)** we may write down the solution of (4.58) and (4.59)$_1$ immediately from equation (4.17) with $g_1(t) = 0$ and $g_2(t) = ag(t)$, thus

$$T(r, t) = \frac{2}{ra} \sum_{n=1}^{\infty} \left(\int_0^a \xi f(\xi) \sin\left(\frac{n\pi\xi}{a}\right) d\xi \right) e^{-\kappa n^2 \pi^2 t/a^2} \sin\left(\frac{n\pi r}{a}\right)$$

$$- \frac{2\kappa\pi}{ra} \sum_{n=1}^{\infty} (-1)^n n \left(\int_0^t e^{-\kappa n^2 \pi^2 (t-\tau)/a^2} g(\tau) d\tau \right) \sin\left(\frac{n\pi r}{a}\right). \tag{4.60}$$

For case **(iii)** the solution $u(r, t)$ for an arbitrary initial condition and Newton heat loss at the surface into a medium at zero temperature the solution of (4.58) and (4.59)$_3$ becomes (see Problem 6)

$$u(r, t) = \frac{2}{r} \sum_{n=1}^{\infty} \frac{(\tilde{\lambda}_n^2 + \tilde{\mu}^2)e^{-\kappa\tilde{\lambda}_n^2 t}}{\left[(\tilde{\lambda}_n^2 + \tilde{\mu}^2)a + \tilde{\mu}\right]} \left(\int_0^a \xi f(\xi) \sin\tilde{\lambda}_n \xi \, d\xi \right) \sin\tilde{\lambda}_n r, \tag{4.61}$$

where $\tilde{\mu} = \mu - 1/a$ and $\tilde{\lambda}_n$ are the positive roots of

$$\tilde{\lambda}_n \cot\tilde{\lambda}_n a + \tilde{\mu} = 0. \tag{4.62}$$

For the solution $v(r, t)$ which is zero initially but satisfies Newton heat loss at the surface into a medium at temperature $T_0(t)$ we proceed in the usual way with

$$w(r, t) = \sum_{n=1}^{\infty} a_n(t) \sin\tilde{\lambda}_n r, \tag{4.63}$$

where formally we have

$$a_n(t) = \frac{\int_0^a w(\xi, t) \sin \tilde\lambda_n \xi \, d\xi}{\int_0^a (\sin \tilde\lambda_n \xi)^2 \, d\xi}.$$

Following our established procedure we obtain

$$\frac{da_n}{dt}(t) + \kappa \tilde\lambda_n^2 a_n(t) = \frac{\kappa \mu a \, T_0(t) \sin \tilde\lambda_n a}{\int_0^a (\sin \tilde\lambda_n \xi)^2 \, d\xi},$$

which we solve subject to $a_n(0) = 0$, thus

$$a_n(t) = \frac{\kappa \mu a \sin \tilde\lambda_n a}{\int_0^a (\sin \tilde\lambda_n \xi)^2 \, d\xi} \int_0^t e^{-\kappa \tilde\lambda_n^2 (t - \tau)} T_0(\tau) \, d\tau. \tag{4.64}$$

In order to evaluate the integral in the denominator we use integration by parts

$$\int_0^a (\sin \tilde\lambda_n \xi)^2 \, d\xi = \left[- \sin \tilde\lambda_n \xi \, \frac{\cos \tilde\lambda_n \xi}{\tilde\lambda_n} \right]_0^a + \int_0^a (\cos \tilde\lambda_n \xi)^2 \, d\xi$$

$$= - \frac{1}{\tilde\lambda_n} \sin \tilde\lambda_n a \cos \tilde\lambda_n a + \int_0^a \left[1 - (\sin \tilde\lambda_n \xi)^2 \right] d\xi,$$

From which we may deduce

$$\int_0^a (\sin \tilde\lambda_n \xi)^2 \, d\xi = \frac{a}{2} \left(1 - \frac{\sin \tilde\lambda_n a \cos \tilde\lambda_n a}{\tilde\lambda_n a} \right). \tag{4.65}$$

Thus from (4.63), (4.64) and (4.65) we may deduce the following solution for $v(r, t)$

$$v(r, t) = \frac{2\kappa\mu}{r} \sum_{n=1}^{\infty} \frac{\tilde\lambda_n a \sin \tilde\lambda_n a \sin \tilde\lambda_n r}{\tilde\lambda_n a - \sin \tilde\lambda_n a \cos \tilde\lambda_n a} \int_0^t e^{-\kappa \tilde\lambda_n^2 (t - \tau)} T_0(\tau) \, d\tau, \tag{4.66}$$

where $\tilde\lambda_n$ are the positive roots of (4.62). Finally the solution of (4.43) (with $m = 2$) and $(4.44)_3$ may be obtained from (4.61) and (4.66) as $T(r, t) = u(r, t) + v(r, t)$.

4.6 Flow in a rectangle with prescribed boundary temperature

In this section we show how the methods for the one dimensional heat equation in a finite domain can be extended to the heat equation in two (or more) spatial dimensions. By way of illustration we consider the problem of heat flow in the rectangle $0 \le x \le a$, $0 \le y \le b$ with arbitrary initial temperature and prescribed temperature along the perimeter of the rectangle. Specifically we consider the problem

$$
\left.
\begin{aligned}
\frac{\partial T}{\partial t} &= \kappa\left(\frac{\partial^2 T}{\partial x^2} + \frac{\partial^2 T}{\partial y^2}\right), \quad 0 < x < a, \, 0 < y < b, \\[2mm]
T(x,y,0) &= f(x,y), \quad 0 \le x \le a, \, 0 \le y \le b, \\[2mm]
T(0,y,t) &= g_1(y,t), \quad T(a,y,t) = g_3(y,t), \quad 0 < y < b, \, t > 0, \\[2mm]
T(x,0,t) &= g_2(x,t), \quad T(x,b,t) = g_4(x,t), \quad 0, < x < a, \, t > 0,
\end{aligned}
\right\} \tag{4.67}
$$

for arbitrary initial temperature $f(x,y)$ and boundary temperatures $g_1(y,t)$, $g_2(x,t)$, $g_3(y,t)$ and $g_4(x,t)$ (see Figure 4.7). As before, our strategy is to express the temperature as the sum $T(x,y,t) = u(x,y,t) + v(x,y,t)$ where $u(x,y,t)$ is the solution of

$$
\left.
\begin{aligned}
\frac{\partial u}{\partial t} &= \kappa\left(\frac{\partial^2 u}{\partial x^2} + \frac{\partial^2 u}{\partial y^2}\right), \quad 0 < x < a, \, 0 < y < b, \\[2mm]
u(x,y,0) &= f(x,y), \quad 0 \le x \le a, \, 0 \le y \le b, \\[2mm]
u(0,y,t) &= u(x,0,t) = u(a,y,t) = u(x,b,t) = 0, \, t > 0,
\end{aligned}
\right\} \tag{4.68}
$$

and $v(x,y,t)$ satisfies

$$
\left.
\begin{aligned}
\frac{\partial v}{\partial t} &= \kappa\left(\frac{\partial^2 v}{\partial x^2} + \frac{\partial^2 v}{\partial y^2}\right), \quad 0 < x < a, \, 0 < y < b, \\[2mm]
v(x,y,0) &= 0, \quad 0 \le x \le a, \, 0 \le y \le b, \\[2mm]
v(0,y,t) &= g_1(y,t), \quad v(a,y,t) = g_3(y,t), \quad 0 < y < b, \, t > 0, \\[2mm]
v(x,0,t) &= g_2(x,t), \quad v(x,b,t) = g_4(x,t), \quad 0 < x < a, \, t > 0.
\end{aligned}
\right\} \tag{4.69}
$$

In view of the boundary conditions $(4.68)_3$ imposed on $u(x,y,t)$, it is reasonable to assume that

$$u(x,y,t) = \sum_{m=1}^{\infty} \sum_{n=1}^{\infty} A_{mn} \sin\left(\frac{m\pi x}{a}\right) \sin\left(\frac{n\pi y}{b}\right) e^{-\kappa\pi^2\lambda_{mn}t}, \tag{4.70}$$

where λ_{mn} is defined to be

$$\lambda_{mn} = \left(\frac{m^2}{a^2} + \frac{n^2}{b^2}\right), \tag{4.71}$$

and the coefficients A_{mn} are determined by the initial condition $(4.68)_2$, namely

$$f(x,y) = \sum_{m=1}^{\infty} \sum_{n=1}^{\infty} A_{mn} \sin\left(\frac{m\pi x}{a}\right) \sin\left(\frac{n\pi y}{b}\right). \tag{4.72}$$

To expand $f(x,y)$ as a double Fourier sine series in this way we proceed as follows. Firstly we assume that for any fixed y in the range $0 \leq y \leq b$ it is possible to expand

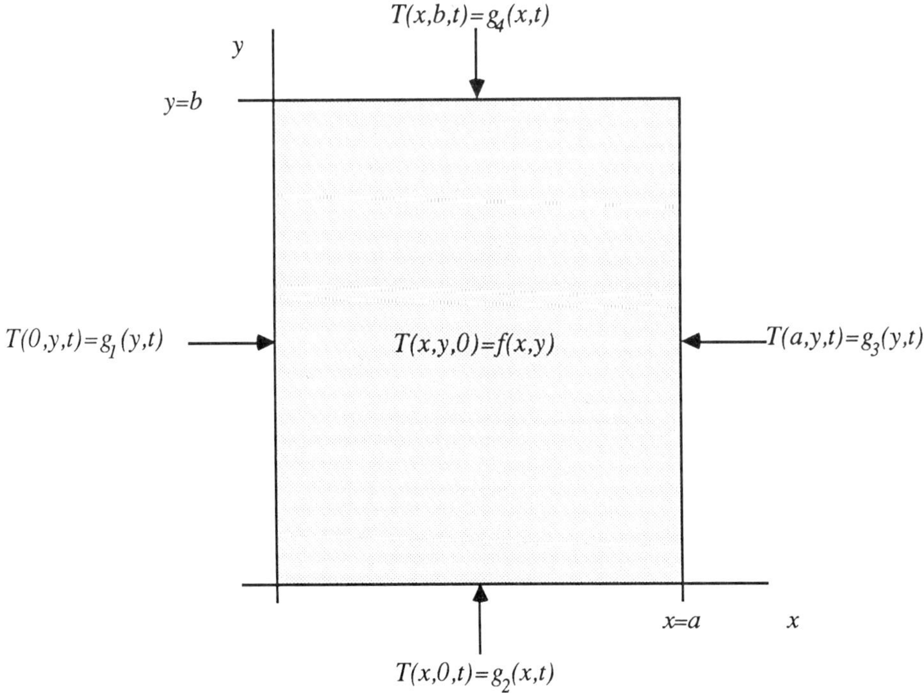

Figure 4.7. Rectangular heat flow problem.

$f(x, y)$ as a Fourier sine series in x, so that

$$f(x, y) = \sum_{m=1}^{\infty} a_m(y) \sin\left(\frac{m\pi x}{a}\right), \qquad (4.73)$$

where the functions $a_m(y)$ are determined by

$$a_m(y) = \frac{2}{a} \int_0^a f(\xi, y) \sin\left(\frac{m\pi\xi}{a}\right) d\xi.$$

If we now suppose that each function $a_m(y)$ can be expressed as a Fourier sine series in y we have

$$a_m(y) = \sum_{n=1}^{\infty} A_{mn} \sin\left(\frac{n\pi y}{b}\right),$$

where the constants A_{mn} are determined by

$$A_{mn} = \frac{2}{b} \int_0^b a_m(\eta) \sin\left(\frac{n\pi\eta}{b}\right) d\eta,$$

so that we recover (4.72) from (4.73) and find that the constants A_{mn} are given by

$$A_{mn} = \frac{4}{ab} \int_0^b \int_0^a f(\xi, \eta) \sin\left(\frac{m\pi\xi}{a}\right) \sin\left(\frac{n\pi\eta}{b}\right) d\xi \, d\eta. \qquad (4.74)$$

Provided that $f(x, y)$ is an integrable function of x and y, so that in particular the value of A_{mn} is independent of the order of integration in (4.74), and is such that

$$\int_0^b \int_0^a |f(\xi, \eta)|^2 \, d\xi \, d\eta < \infty,$$

then the series (4.72) with coefficients given by (4.74) converges to $f(x, y)$ almost everywhere. In particular the series (4.72) converges to $f(x, y)$ at any point (x, y) inside the rectangle $0 < x < a$, $0 < y < b$ at which $f(x, y)$ is continuous. Thus, in summary the solution of (4.68) is given by (4.70), where A_{mn} are determined by (4.74).

Example 4.12 For the finite rectangle $0 \leq x \leq a$, $0 \leq y \leq b$ with constant initial temperature u_0 and prescribed zero temperature on the faces $x = 0$, $x = a$, $y = 0$ and $y = b$, show that the temperature is given by (see Figure 4.8)

$$u(x, y, t) =$$

$$\frac{16u_0}{\pi^2} \sum_{m=0}^{\infty} \sum_{n=0}^{\infty} \frac{e^{-\kappa\pi^2\lambda_{(2m+1)(2n+1)}t}}{(2m+1)(2n+1)} \sin\left(\frac{(2m+1)\pi x}{a}\right) \sin\left(\frac{(2n+1)\pi y}{b}\right),$$

where λ_{mn} is given by (4.71).

Here we have simply $f(x, y) = u_0$, and thus from (4.74) we have

$$A_{mn} = \frac{4u_0}{ab} \int_0^a \int_0^b \sin\left(\frac{m\pi\xi}{a}\right) \sin\left(\frac{n\pi\eta}{b}\right) d\eta \, d\xi$$

$$= \frac{4u_0}{ab} \left(\int_0^a \sin\left(\frac{m\pi\xi}{a}\right) d\xi \right) \left(\int_0^b \sin\left(\frac{n\pi\eta}{b}\right) d\eta \right)$$

$$= \frac{4u_0}{\pi^2 mn}[1 - \cos m\pi][1 - \cos n\pi] = \frac{4u_0}{\pi^2 mn}\left[1 - (-1)^m\right]\left[1 - (-1)^n\right],$$

so that A_{mn} is zero if one or both of m and n is even, and

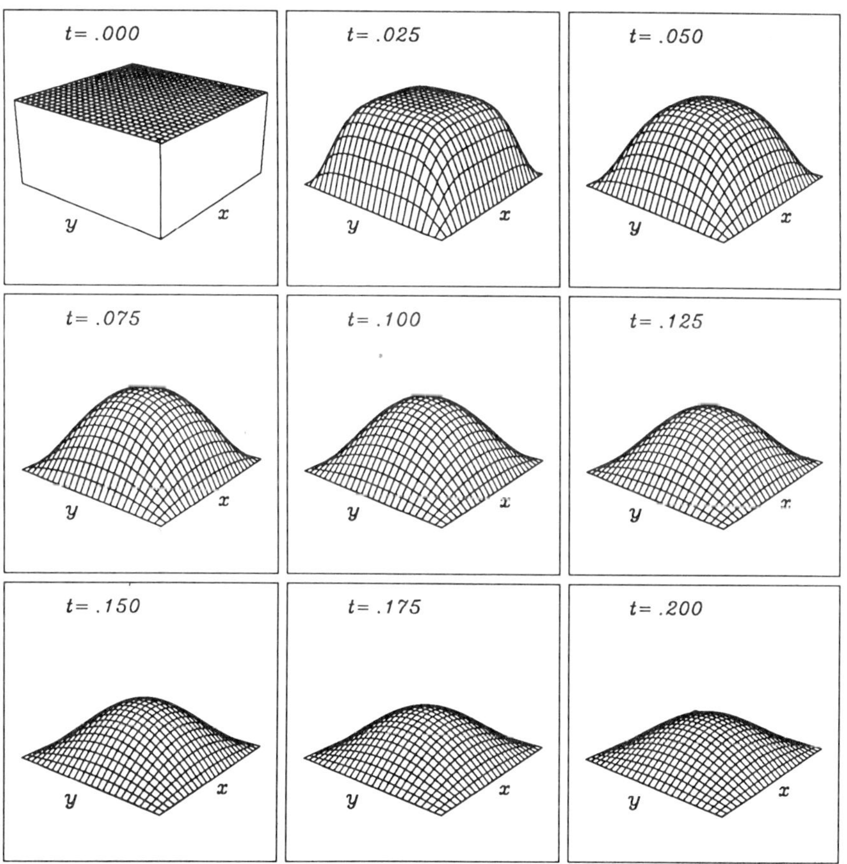

Figure 4.8. Two dimensional temperature profiles for Example 4.12 with u_0 unity at 9 time values.

$$A_{mn} = \frac{16u_0}{\pi^2(2i + 1)(2j + 1)},$$

when both m and n are odd and given by $m = (2i + 1)$, $n = (2j + 1)$ for $i, j = 0, 1, 2, \ldots$. The given solution now follows immediately from (4.70). We observe that we can write the solution to this problem in the form

$$u(x,y,t) = u_0\left\{\frac{4}{\pi}\sum_{m=0}^{\infty}\frac{e^{-\kappa(2m+1)^2\pi^2t/a^2}}{(2m+1)}\sin\left(\frac{(2m+1)\pi x}{a}\right)\right\} \times$$

$$\left\{\frac{4}{\pi}\sum_{n=0}^{\infty}\frac{e^{-\kappa(2n+1)^2\pi^2t/b^2}}{(2n+1)}\sin\left(\frac{(2n+1)\pi y}{b}\right)\right\},$$

which is simply a product of solutions obtained in Example 4.1.

To solve (4.69) for $v(x,y,t)$ we assume that

$$v(x,y,t) = \sum_{m=1}^{\infty}\sum_{n=1}^{\infty} A_{mn}(t)\sin\left(\frac{m\pi x}{a}\right)\sin\left(\frac{n\pi y}{b}\right), \tag{4.75}$$

so that, formally

$$A_{mn}(t) = \frac{4}{ab}\int_0^a\int_0^b v(\xi,\eta,t)\sin\left(\frac{m\pi\xi}{a}\right)\sin\left(\frac{n\pi\eta}{b}\right)d\eta\,d\xi, \tag{4.76}$$

and, as $v(x,y,t)$ vanishes at time zero

$$A_{mn}(0) = 0. \tag{4.77}$$

Proceeding exactly as in previous sections, we deduce a first order ordinary differential equation for $A_{mn}(t)$ by differentiating (4.76) with respect to time so as to obtain

$$\frac{dA_{mn}}{dt}(t) = \frac{4}{ab}\int_0^a\int_0^b\frac{\partial v}{\partial t}(\xi,\eta,t)\sin\left(\frac{m\pi\xi}{a}\right)\sin\left(\frac{n\pi\eta}{b}\right)d\eta\,d\xi$$

$$= \frac{4}{ab}\int_0^a\int_0^b\kappa\left(\frac{\partial^2 v}{\partial\xi^2} + \frac{\partial^2 v}{\partial\eta^2}\right)\sin\left(\frac{m\pi\xi}{a}\right)\sin\left(\frac{n\pi\eta}{b}\right)d\eta\,d\xi$$

$$= \frac{4\kappa}{ab}\left\{\int_0^a\left(\int_0^b\frac{\partial^2 v}{\partial\eta^2}\sin\left(\frac{n\pi\eta}{b}\right)d\eta\right)\sin\left(\frac{m\pi\xi}{a}\right)d\xi\right.$$

$$\left. + \int_0^b\left(\int_0^a\frac{\partial^2 v}{\partial\xi^2}\sin\left(\frac{m\pi\xi}{a}\right)d\xi\right)\sin\left(\frac{n\pi\eta}{b}\right)d\eta\right\},$$

and since, as before

$$\int_0^a \frac{\partial^2 v}{\partial \xi^2}(\xi,\eta)\sin\left(\frac{m\pi\xi}{a}\right)d\xi = \left(\frac{m\pi}{a}\right)\left[v(0,\eta,t) - (-1)^m v(a,\eta,t)\right]$$
$$- \left(\frac{m\pi}{a}\right)^2 \int_0^a v(\xi,\eta,t)\sin\left(\frac{m\pi\xi}{a}\right)d\xi,$$

$$\int_0^b \frac{\partial^2 v}{\partial \eta^2}(\xi,\eta)\sin\left(\frac{n\pi\eta}{b}\right)d\eta = \left(\frac{n\pi}{b}\right)\left[v(\xi,0,t) - (-1)^n v(\xi,b,t)\right]$$
$$- \left(\frac{n\pi}{b}\right)^2 \int_0^b v(\xi,\eta,t)\sin\left(\frac{n\pi\eta}{b}\right)d\eta,$$

we have, altogether

$$\frac{dA_{mn}}{dt}(t) + \kappa\pi^2\lambda_{mn}A_{mn} = \frac{4\pi\kappa}{ab}\left\{\frac{n}{b}\int_0^a \left[v(\xi,0,t) - (-1)^n v(\xi,b,t)\right]\sin\left(\frac{m\pi\xi}{a}\right)d\xi\right.$$
$$\left. + \frac{m}{a}\int_0^b \left[v(0,\eta,t) - (-1)^m v(a,\eta,t)\right]\sin\left(\frac{n\pi\eta}{b}\right)d\eta,\right\}.$$
$$(4.78)$$

Thus, solving (4.78) subject to the initial condition (4.77) we find

$$A_{mn}(t) = \int_0^t F_{mn}(\tau)e^{-\kappa\pi^2\lambda_{mn}(t-\tau)}\,d\tau,\qquad (4.79)$$

where λ_{mn} is defined by (4.71) and the functions $F_{mn}(t)$ are given by

$$F_{mn}(t) = \frac{4\pi\kappa}{ab}\left\{\frac{n}{b}\int_0^a \left[g_2(\xi,t) - (-1)^n g_4(\xi,t)\right]\sin\left(\frac{m\pi\xi}{a}\right)d\xi\right.$$
$$(4.80)$$
$$\left. + \frac{m}{a}\int_0^b \left[g_1(\eta,t) - (-1)^m g_3(\eta,t)\right]\sin\left(\frac{n\pi\eta}{b}\right)d\eta\right\},$$

where we have used the prescribed boundary data $(4.69)_3$ and $(4.69)_4$ for $v(x,y,t)$ on the perimeter of the rectangle.

<u>Example 4.13</u> For the finite rectangle $0 \le x \le a$, $0 \le y \le b$ with zero initial temperature and sides maintained at the constant temperature v_0 show that the temperature is given by

$$v(x,y,t) =$$

$$\frac{16v_0}{\pi^2}\sum_{m=0}^\infty \sum_{n=0}^\infty \frac{\left[1 - e^{-\kappa\pi^2\lambda_{(2m+1)(2n+1)}t}\right]}{(2m+1)(2n+1)}\sin\left(\frac{(2m+1)\pi x}{a}\right)\sin\left(\frac{(2n+1)\pi y}{b}\right).$$

Here we must solve problem (4.69) with $g_1(y,t) = g_2(x,t) = g_3(y,t) = g_4(x,t) = v_0$. Thus from (4.80) we find that

$$F_{mn}(t)$$

$$= \frac{4\pi\kappa v_0}{ab}\left\{\frac{n}{b}\int_0^a\left[1 - (-1)^n\right]\sin\left(\frac{m\pi\xi}{a}\right)d\xi + \frac{m}{a}\int_0^b\left[1 - (-1)^m\right]\sin\left(\frac{n\pi\eta}{b}\right)d\eta\right\}$$

$$= \frac{4\kappa v_0}{ab}\left\{\frac{-na}{mb}\left[1 - (-1)^n\right]\left[\cos\left(\frac{m\pi\xi}{a}\right)\right]_0^a - \frac{mb}{na}\left[1 - (-1)^m\right]\left[\cos\left(\frac{n\pi\eta}{b}\right)\right]_0^b\right\}$$

$$= \frac{4\kappa v_0}{mn}\lambda_{mn}\left[1 - (-1)^n\right]\left[1 - (-1)^m\right],$$

which is zero if either of n and m is even. Thus

$$F_{mn}(t) = \frac{16\kappa v_0}{mn}\lambda_{mn},$$

if m and n are both odd and zero otherwise. So for odd m and n we have

$$A_{mn}(t) = \frac{16\kappa v_0}{mn}\lambda_{mn}\int_0^t e^{-\kappa\pi^2\lambda_{mn}(t-\tau)}\,d\tau$$

$$= \frac{16v_0}{\pi^2 mn}\left\{1 - e^{-\kappa\pi^2\lambda_{mn}t}\right\},$$

and otherwise $A_{mn}(t) = 0$. The given solution now follows immediately from (4.75). On expanding the constant function $f(x,y) = 1$ as a double Fourier series over the rectangle $0 \leq x \leq a$, $0 \leq y \leq b$ (see (4.74)) we find

$$1 = \frac{16}{\pi^2}\sum_{m=0}^{\infty}\sum_{n=0}^{\infty}\frac{1}{(2m+1)(2n+1)}\sin\left(\frac{(2m+1)\pi x}{a}\right)\sin\left(\frac{(2n+1)\pi y}{b}\right),$$

so that the given solution can also be written in the form

$$T(x,y,t) =$$

$$v_0\left\{1 - \frac{16}{\pi^2}\sum_{m=0}^{\infty}\sum_{n=0}^{\infty}\frac{e^{-\kappa\pi^2\lambda_{(2m+1)(2n+1)}t}}{(2m+1)(2n+1)}\sin\left(\frac{(2m+1)\pi x}{a}\right)\sin\left(\frac{(2n+1)\pi y}{b}\right)\right\}.$$

Finally putting $T(x,y,t) = u(x,y,t) + v(x,y,t)$ we find from (4.70), (4.74), (4.75) and (4.79) that the general solution of problem (4.67) is given by

$$T(x,y,t) = \frac{4}{ab} \sum_{m=1}^{\infty} \sum_{n=1}^{\infty} \left\{ \int_0^a \int_0^b f(\xi,\eta) \sin\left(\frac{m\pi\xi}{a}\right) \sin\left(\frac{n\pi\eta}{b}\right) d\eta\, d\xi \right\}$$

$$\times \sin\left(\frac{m\pi x}{a}\right) \sin\left(\frac{n\pi y}{b}\right) e^{-\kappa\pi^2\lambda_{mn}t}$$

$$+ \frac{4\pi\kappa}{ab} \sum_{m=1}^{\infty} \sum_{n=1}^{\infty} \left\{ \int_0^t e^{-\kappa\pi^2\lambda_{mn}(t-\tau)} \left[\frac{m}{a} \int_0^b \left(g_1(\eta,\tau) - (-1)^m g_3(\eta,\tau) \right) \sin\left(\frac{n\pi\eta}{b}\right) d\eta \right.\right.$$

$$\left.\left. + \frac{n}{b} \int_0^a \left(g_2(\xi,\tau) - (-1)^n g_4(\xi,\tau) \right) \sin\left(\frac{m\pi\xi}{a}\right) d\xi \right] d\tau \right\} \sin\left(\frac{m\pi x}{a}\right) \sin\left(\frac{n\pi y}{b}\right).$$

This can be written more compactly in terms of a Green's function as

$$T(x,y,t) = \int_0^a \int_0^b G(x,y,\xi,\eta,t,0) f(\xi,\eta)\, d\xi\, d\eta$$

$$- \kappa \int_0^t \oint_S \frac{\partial G}{\partial n}(x,y,\xi,\eta,t,\tau) g(\xi,\eta,\tau)\, dS\, d\tau$$

where $G(x,y,\xi,\eta,t,\tau)$ is given by

$$G(x,y,\xi,\eta,t,\tau) = H(t-\tau) \times$$

$$\frac{4}{ab} \sum_{m=1}^{\infty} \sum_{n=1}^{\infty} \left\{ \sin\left(\frac{m\pi\xi}{a}\right) \sin\left(\frac{n\pi\eta}{b}\right) \sin\left(\frac{m\pi x}{a}\right) \sin\left(\frac{n\pi y}{b}\right) e^{-\kappa\pi^2\lambda_{mn}(t-\tau)} \right\},$$

and $g(x,y,t)$ is defined to be

$$g(x,y,t) = \begin{cases} g_1(y,t), & x = 0, \\ g_2(x,t), & y = 0, \\ g_3(y,t), & x = a, \\ g_4(x,t), & y = b, \end{cases}$$

and the normal derivative and surface integral are taken with respect to the perimeter of the rectangle $0 \le \xi \le a$, $0 \le \eta \le b$ in (ξ,η) space. We note that the Green's function $G(x,y,\xi,\eta,t,\tau)$ is simply the product of two Green's functions for problem (4.6).

PROBLEMS

1. In this question we outline the main steps involved in a rigorous proof that the series solutions given in this chapter are in fact the exact solutions of the heat equation and satisfy the appropriate boundary and initial conditions. Specifically, consider the solution given in Example 4.2, namely

$$u(x,t) = \frac{2\alpha\ell}{\pi} \sum_{n=1}^{\infty} \frac{(-1)^{n+1}}{n} \sin\left(\frac{n\pi x}{\ell}\right) e^{-\kappa n^2 \pi^2 t/\ell^2}.$$

(i) Show that for all $t > 0$ this series converges absolutely, so that in particular

$$u(0,t) = u(\ell,t) = 0, \quad t > 0.$$

$$\left[\text{Hint, use the fact that } \left|\frac{(-1)^{n+1}}{n} \sin\left(\frac{n\pi x}{\ell}\right) e^{-\kappa n^2 \pi^2 t/\ell^2}\right| \le e^{-\kappa n^2 \pi^2 t/\ell^2}.\right]$$

(ii) Show that the series for $\frac{\partial u}{\partial t}$ and $\frac{\partial^2 u}{\partial x^2}$ which arise by termwise differentiation of the series for $u(x,t)$ are also absolutely convergent for $t > 0$ (so that the termwise differentiation is valid) and hence that $u(x,t)$ satisfies the heat equation for $t > 0$.

(iii) Show that the series

$$u(x,0) = \frac{2\alpha\ell}{\pi} \sum_{n=1}^{\infty} \frac{(-1)^{n+1}}{n} \sin\left(\frac{n\pi x}{\ell}\right),$$

does indeed converge to the function αx for $0 < x < \ell$.

$$\left[\text{Hint, use the results given in Section 2.5.}\right]$$

2. For the finite region $0 \le x \le \ell$ with arbitrary initial temperature $f(x)$, no flow of heat at $x = 0$ and prescribed zero temperature at $x = \ell$, show that the temperature is given by

$$u(x,t) = \frac{2}{\ell} \sum_{n=0}^{\infty} \left(\int_0^\ell f(\xi) \cos \lambda_n \xi \, d\xi\right) e^{-\kappa \lambda_n^2 t} \cos \lambda_n x,$$

where $\lambda_n = (2n + 1)\pi/2\ell$. Show that as t tends to infinity, $u(x,t)$ tends to zero. Why is it physically obvious that this should happen?

3. **Continuation.** For the finite region $0 \le x \le \ell$ with zero initial temperature, no flow of heat at $x = 0$ and prescribed temperature $g(t)$ at $x = \ell$, show that the temperature is given by

$$v(x,t) = \frac{2\kappa}{\ell} \sum_{n=0}^{\infty} (-1)^n \lambda_n \left(\int_0^t e^{-\kappa\lambda_n^2(t-\tau)} g(\tau)\, d\tau \right) \cos \lambda_n x,$$

where $\lambda_n = (2n+1)\pi/2\ell$.

4. Continuation. For the finite region $0 \le x \le \ell$ with constant initial temperature, u_0, no flow of heat at $x = 0$ and prescribed temperature $u_0 e^{\gamma t}$ at $x = \ell$ where γ is a constant, show that the temperature is given by

$$T(x,t) = \frac{2u_0}{\ell} \sum_{n=0}^{\infty} \frac{(-1)^n}{\lambda_n} \left(\frac{\gamma e^{-\kappa\lambda_n^2 t} + \kappa\lambda_n^2 e^{\gamma t}}{\gamma + \kappa\lambda_n^2} \right) \cos \lambda_n x,$$

where $\lambda_n = (2n+1)\pi/2\ell$. Observe that this solution can be rewritten as

$$T(x,t) = u_0 e^{\gamma t} + \frac{2\gamma u_0}{\ell} \sum_{n=0}^{\infty} \frac{(-1)^n \left(e^{-\kappa\lambda_n^2 t} - e^{\gamma t} \right)}{\lambda_n \left(\gamma + \kappa\lambda_n^2 \right)} \cos \lambda_n x,$$

and therefore $T(x,t) = u_0$ in the special case $\gamma = 0$.

5. For the finite region $0 \le x \le \ell$ with zero initial temperature, $x = 0$ maintained at zero and $x = \ell$ has constant flux w_0 across it, show that

$$v(x,t) = \frac{w_0}{k} \left\{ x - \frac{2}{\ell} \sum_{n=0}^{\infty} \frac{(-1)^n}{\lambda_n^2} e^{-\kappa\lambda_n^2 t} \sin \lambda_n x \right\},$$

where $\lambda_n = (2n+1)\pi/2\ell$. Notice that as t tends to infinity $v(x,t)$ tends to $w_0 x/k$. Why is it physically obvious that this should happen?

6. For the finite region $0 \le x \le \ell$ with arbitrary initial temperature $f(x)$, the surface $x = 0$ maintained at zero temperature while there is Newton heat loss across the surface $x = \ell$ into a medium at zero temperature, show that the temperature is given by

$$u(x,t) = \sum_{n=1}^{\infty} \frac{2(\lambda_n^2 + \mu^2)e^{-\kappa\lambda_n^2 t}}{\left[(\lambda_n^2 + \mu^2)\ell + \mu \right]} \left(\int_0^\ell f(\xi) \sin \lambda_n \xi \, d\xi \right) \sin \lambda_n x,$$

where $\mu = h/k$ and λ_n are the positive roots of

$$\lambda_n \cot \lambda_n \ell + \mu = 0.$$

7.   For the finite region $0 \leq x \leq \ell$ with arbitrary initial temperature f(x), no flow of heat across $x = 0$ while the surface $x = \ell$ has Newton heat loss into a medium at zero temperature, show that the temperature is given by

$$u(x,t) = \sum_{n=1}^{\infty} \frac{2(\lambda_n^2 + \mu^2)e^{-\kappa\lambda_n^2 t}}{\left[(\lambda_n^2 + \mu^2)\ell + \mu\right]} \left(\int_0^{\ell} f(\xi) \cos \lambda_n \xi \, d\xi\right) \cos \lambda_n x,$$

where $\mu = h/k$ and λ_n are the positive roots of

$$\lambda_n \tan \lambda_n \ell - \mu = 0.$$

8.   For the infinite solid circular cylinder $0 \leq r \leq a$ with zero surface temperature and parabolic initial temperature

$$f(r) = \alpha - \beta r^2, \quad 0 \leq r \leq a,$$

where α and β are constants, show that the temperature is given by

$$u(r,t) = \frac{2}{a} \sum_{n=1}^{\infty} e^{-\kappa\lambda_n^2 t} \frac{\left[\lambda_n(\alpha - \beta a^2) J_1(\lambda_n a) + 2\beta a J_2(\lambda_n a)\right]}{\lambda_n^2 J_1(\lambda_n a)^2} J_0(\lambda_n r),$$

where λ_n denote the positive roots of $J_0(\lambda_n a) = 0$.

9.   For the solid sphere $0 \leq r \leq a$ with constant initial temperature u_0 and zero surface temperature show that the temperature is given by

$$u(r,t) = \frac{-2a u_0}{\pi r} \sum_{n=1}^{\infty} \frac{(-1)^n}{n} e^{-\kappa n^2 \pi^2 t / a^2} \sin\left(\frac{n\pi r}{a}\right).$$

Show that as t tends to infinity $u(r,t)$ tends to zero. Why is it physically obvious that this should be the case?

10.   For the solid sphere $0 \leq r \leq a$ with zero surface temperature and initial temperature

$$f(r) = u_0\left(1 - \frac{r}{a}\right), \quad 0 \leq r \leq a,$$

where u_0 is a constant, show that the temperature is given by

$$u(r,t) = \frac{8a u_0}{\pi^3 r} \sum_{n=0}^{\infty} \frac{e^{-\kappa(2n+1)^2\pi^2 t / a^2}}{(2n+1)^3} \sin\left(\frac{(2n+1)\pi r}{a}\right).$$

11. For the solid sphere $0 \leq r \leq a$ with zero initial temperature and constant surface temperature v_0 show that

$$v(r,t) = v_0 \left\{ 1 + \frac{2a}{\pi r} \sum_{n=1}^{\infty} \frac{(-1)^n}{n} e^{-\kappa n^2 \pi^2 t / a^2} \sin\left(\frac{n \pi r}{a}\right) \right\}.$$

12. For the solid sphere $0 \leq r \leq a$ with zero initial temperature and surface temperature βt where β is a constant, show that

$$v(r,t) = \beta \left\{ t + \frac{(r^2 - a^2)}{6\kappa} - \frac{2k a^3}{\kappa \pi^3 r} \sum_{n=1}^{\infty} \frac{(-1)^n}{n^3} e^{-\kappa n^2 \pi^2 t / a^2} \sin\left(\frac{n \pi r}{a}\right) \right\}.$$

13. For the solid sphere $0 \leq r \leq a$ initially at zero temperature and has Newton heat loss into a medium at temperature βt where β is a constant, show that

$$v(r,t) =$$

$$\beta \left\{ t + \frac{1}{6\kappa\mu} \left[\mu r^2 - (2 + \mu a)a \right] + \frac{2\mu a^2}{\kappa r} \sum_{n=1}^{\infty} \frac{e^{-\kappa \tilde{\lambda}_n^2 t} \sin \tilde{\lambda}_n r}{\tilde{\lambda}_n^2 \left[a^2 \tilde{\lambda}_n^2 + \mu a(\mu a - 1) \right] \sin \tilde{\lambda}_n a} \right\},$$

where $\tilde{\lambda}_n$ are the positive roots of

$$\tilde{\lambda}_n a \cot \tilde{\lambda}_n a + \mu a - 1 = 0.$$

14. Show that the series

$$\sum_{n=1}^{\infty} \frac{1}{n} \left\{ g_1(0) - (-1)^n g_2(0) + \int_0^t \left[g_1'(\tau) - (-1)^n g_2'(\tau) \right] e^{\kappa \pi^2 n^2 \tau / \ell^2} \, d\tau \right\}$$

$$\times e^{-\kappa \pi^2 n^2 t / \ell^2} \sin\left(\frac{n \pi x}{\ell}\right),$$

converges absolutely for $t > 0$ and $0 \leq x \leq \ell$ provided that $g_i(t)$ and $g_i'(t)$ ($i = 1, 2$) are bounded. $\Big[$ Hint, consider the sums containing $g_1(0)$, $g_2(0)$, $g_1'(\tau)$ and $g_2'(\tau)$ separately and show that each of these series converge absolutely by construction of suitable bounds. $\Big]$

Chapter Five

Approximate analytical solutions by heat-balance

5.1 Introduction

In this chapter we consider the approximation of solutions of linear heat conduction problems by the *heat-balance* integral method. For solvable and unsolvable problems alike there exist a multitude of possible approximate solutions all of which exhibit some desirable and some undesirable features. Indeed, there are very few analytical approximations to exact solutions which are fully robust in the sense that meaningful results are generated for all situations or all parameter values. It is certainly the case that the heat-balance method generates meaningful results for many problems while for others there is a complete mathematical breakdown of the procedure. The heat-balance integral method is widely employed for problems in diffusion and heat conduction. Its main merits are that, firstly, it frequently provides reasonable results quickly and, secondly, it applies equally well to non-linear problems where Laplace transforms and Fourier series are not effective. Its main disadvantages, apart from the occasional complete mathematical breakdown, are that for some problems (usually non-planar problems) it becomes technically extremely complicated while for other problems the approximation simply does not exist (for example see (5.33) and subsequently). Thus the approximation obtained for the specific problem at hand must be taken on its particular merits.

In general terms the heat-balance method is as follows:

(i)   Assume a simple approximation for the temperature profile. This is usually a polynomial expression in the spatial variable with time dependent coefficients.

(ii)   Introduce the notion of a *penetration depth* $\rho(t)$, beyond which there is no heat flow so that the prescribed initial condition still applies.

(iii)   Using the prescribed boundary conditions and the assumed conditions beyond the penetration depth $\rho(t)$, determine the time dependent coefficients in the assumed temperature profile solely as functions of $\rho(t)$.

(iv)   Integrate the governing equation across the penetration distance and from the assumed temperature profile determine $\rho(t)$ explicitly from the integrated form of the governing equation. Thus the pointwise partial differential equation is satisfied in an 'average' or integral sense only.

The astute reader will readily agree that a good deal of this procedure is somewhat arbitrary and that the final approximation is by no means unique. Firstly, in step **(i)**, with respect to the actual choice of the approximating expression for the temperature profile, essentially any simple functions may be used. Secondly, in step **(iii)**, with respect to the actual conditions used to determine the time dependent coefficients, there is no unique procedure. With regard to this matter, the fundamental rule which we adopt is to select the constraints involving the lower order derivatives and if there is no unique way of doing this, then the choice is arbitrary. This rule arises because in approximating the temperature profile the priorities at the end points of the penetration depth are first to obtain agreement with the temperature values themselves, then the first derivatives, then the second derivatives and so on. This aspect is discussed further in subsequent sections in the context of specific problems. A completely alternative strategy is to make the approximate temperature profile as smooth as possible at the penetration depth, subject to the imposed boundary conditions of course, by matching as many derivatives at the penetration depth as is possible. Finally in step **(iv)**, the integral form of the governing partial differential equation is clearly not unique since prior to integration we may multiply the equation by any function. The significance of all these issues will be clarified with reference to specific problems in the following sections.

In the following three sections we consider in detail, as simple illustrative examples of the method, the three semi-infinite media problems discussed previously in Chapter 3. However, in order to keep the details to a minimum we only consider the cases of zero initial temperature and constant surface conditions. These semi-infinite media approximations are also relevant in the context of finite media problems, since the latter problems may be viewed in two stages. Firstly, an initial period during which semi-infinite media approximations apply (until the penetration fronts advancing from each end reach opposite ends, that is until the penetration depths exceed the width and the effect of the heat flow from one boundary is felt at the other boundary) and secondly a subsequent period which must be treated separately. The extension of the heat-balance method to finite media planar problems is considered in Section 5.5 and in the final section of the chapter we deal with cylindrical and spherical problems.

5.2 Semi-infinite media with $x = 0$ at constant temperature v_0

In this section we consider the semi-infinite one dimensional region $0 \le x < \infty$ which is initially at zero temperature. At time $t = 0$ and subsequently the surface $x = 0$ is maintained at constant temperature v_0. Thus we consider the special case of the problem given by (3.1), namely

$$
\left.
\begin{aligned}
\frac{\partial v}{\partial t} &= \kappa \frac{\partial^2 v}{\partial x^2}, \quad 0 < x < \infty, \\[2mm]
v(x,0) &= 0, \quad 0 \le x < \infty, \\[2mm]
v(0,t) = v_0, \quad t > 0, \qquad &\frac{\partial v}{\partial x}(x,t) \to 0 \text{ as } x \to \infty,
\end{aligned}
\right\}
\tag{5.1}
$$

the solution of which is given in Example 3.2 as

$$
v(x,t) = v_0 \operatorname{erfc}\left(\frac{x}{2\sqrt{\kappa t}}\right),
$$

so that, in particular, we have

$$
\frac{\partial v}{\partial x}(0,t) = \frac{-v_0}{\sqrt{\pi \kappa t}},
\tag{5.2}
$$

for the temperature gradient at the origin.

We assume there exists a penetration depth $x = \rho(t)$ such that effectively heat conduction only occurs for $0 \le x \le \rho(t)$ while in the region $x > \rho(t)$ we have no heat flow at all, so that

$$
v(x,t) = \frac{\partial v}{\partial x}(x,t) = 0, \quad x > \rho(t).
$$

This is not an unreasonable assumption, since physically heat cannot be transmitted at infinite speed, and moreover it can be reconciled with the exact solution by imagining $x = \rho(t)$ to be the point beyond which the temperature and heat flux is negligible (see Problem 1). At $x = \rho(t)$ we have

$$
v(\rho(t),t) = \frac{\partial v}{\partial x}(\rho(t),t) = 0,
\tag{5.3}
$$

since on the most basic of thermodynamic grounds both the temperature and heat flux must be continuous (the zeroth law of thermodynamics and the conservation of energy). We utilize (5.3) to determine the approximate temperature profiles. On integrating the heat equation $(5.1)_1$ from $x = 0$ to $x = \rho(t)$ we obtain

$$\int_0^{\rho(t)} \frac{\partial v}{\partial t}(x, t)\, dx = \kappa\left(\frac{\partial v}{\partial x}(\rho(t), t) - \frac{\partial v}{\partial x}(0, t)\right),$$

which on employing both conditions in (5.3) becomes

$$\frac{d\theta}{dt} = -\kappa \frac{\partial v}{\partial x}(0, t), \tag{5.4}$$

where the heat-balance integral $\theta(t)$ is defined by

$$\theta(t) = \int_0^{\rho(t)} v(x, t)\, dx. \tag{5.5}$$

Physically $\theta(t)$ is proportional to the total heat content of the region between the origin and penetration depth, and (5.4) expresses the fact that the rate at which the total heat content varies is equal to the rate at which heat flows into the region across the surface $x = 0$. Thus the heat equation is satisfied in an average sense over the region $[0, \rho(t)]$. We now derive three approximating temperature profiles by assuming quadratic and cubic expressions for the temperature profile.

Quadratic temperature profile

Here we approximate $v(x, t)$ by an expression of the form

$$v(x, t) = a(t) + b(t)x + c(t)x^2, \tag{5.6}$$

where $a(t)$, $b(t)$ and $c(t)$ denote functions of time, which are determined from the three equations $(5.1)_3$ and $(5.3)_{1,2}$. From $(5.1)_3$ we have immediately

$$a(t) = v_0,$$

while from (5.3) and (5.6) we obtain

$$v_0 + b(t)\rho(t) + c(t)\rho(t)^2 = b(t) + 2c(t)\rho(t) = 0,$$

so that

$$b(t) = \frac{-2v_0}{\rho(t)}, \quad c(t) = \frac{v_0}{\rho(t)^2}.$$

Altogether we have

$$v(x, t) = v_0\left(1 - \frac{x}{\rho(t)}\right)^2. \tag{5.7}$$

On substituting this expression into (5.5) we find

$$\theta(t) = \frac{v_0}{3}\rho(t),$$

and since from (5.7) we have

$$\frac{\partial v}{\partial x}(0, t) = \frac{-2v_0}{\rho(t)}, \tag{5.8}$$

it follows that (5.4) becomes

$$\frac{d\rho}{dt}(t) = \frac{6\kappa}{\rho(t)}.$$

Thus noting that $\rho(0) = 0$ we readily deduce

$$\rho(t) = 2\sqrt{3\kappa t},$$

and (5.8) becomes

$$\frac{\partial v}{\partial x}(0, t) = \frac{-v_0}{\sqrt{3\kappa t}}. \tag{5.9}$$

On comparing (5.9) with the exact expression (5.2) we see that the two results are identical except that the constant $\sqrt{\pi}$ in (5.2) is replaced by $\sqrt{3}$ in (5.9). In the derivation of (5.7) we have made no attempt to accommodate the initial condition $(5.1)_2$, and indeed (5.7) does not vanish as time tends to zero. However, (5.7) represents the temperature between the origin and the penetration depth $\rho(t)$ only, and outside this region we take the temperature to be zero. In particular, at time zero the penetration depth is also zero so that we recover our zero initial temperature condition.

<u>Cubic temperature profile</u>

Here we assume that $v(x, t)$ can be approximated by an expression of the form

$$v(x, t) = a(t) + b(t)x + c(t)x^2 + d(t)x^3, \tag{5.10}$$

where the four functions of time are determined from $(5.1)_3$, $(5.3)_{1,2}$ and an additional equation. The extra equation is obtained by differentiating either $v(0, t) = v_0$ or $v(\rho(t), t) = 0$ with respect to time and using the governing partial differential equation $(5.1)_1$. Thus we obtain either

$$\frac{\partial^2 v}{\partial x^2}(0, t) = 0, \tag{5.11}$$

or

$$\frac{\partial^2 v}{\partial x^2}(\rho(t), t) = 0, \tag{5.12}$$

and we note that in deriving the latter we have used $(5.3)_2$ as well. We consider the two possibilities separately.

<u>Method 1</u> Suppose that (5.11) holds, then from (5.10), $(5.1)_3$ and (5.11) we may deduce immediately

$$a(t) = v_0, \quad c(t) = 0.$$

Further from $(5.3)_{1,2}$ we obtain

$$v_0 + b(t)\rho(t) + d(t)\rho(t)^3 = b(t) + 3d(t)\rho(t)^2 = 0,$$

and therefore

$$b(t) = \frac{-3v_0}{2\rho(t)}, \quad d(t) = \frac{v_0}{2\rho(t)^3}.$$

Thus, altogether we have

$$v(x,t) = \frac{v_0}{2}\left(2 + \frac{x}{\rho(t)}\right)\left(1 - \frac{x}{\rho(t)}\right)^2, \tag{5.13}$$

so that in this case we obtain

$$\theta(t) = \frac{3v_0}{8}\rho(t), \quad \frac{\partial v}{\partial x}(0,t) = \frac{3v_0}{2\rho(t)}.$$

Using these equations the heat-balance equation (5.4) eventually yields

$$\rho(t) = 2\sqrt{2\kappa t},$$

so that

$$\frac{\partial v}{\partial x}(0,t) = \frac{-3v_0}{4\sqrt{\kappa t}},$$

and therefore the factor of $\sqrt{\pi}$ in the exact expression (5.2) is approximated by $\sqrt{32/9}$.

<u>Method 2</u> Suppose that (5.12) holds. Then in this case we have $a(t) = v_0$ while the remaining three functions are determined from the equations

$$b(t)\rho(t) + c(t)\rho(t)^2 + d(t)\rho(t)^3 = -v_0,$$

$$b(t) + 2c(t)\rho(t) + 3d(t)\rho(t)^2 = 0,$$

$$c(t) + 3d(t)\rho(t) = 0.$$

Solving these equations in a routine manner we obtain

$$b(t) = \frac{-3v_0}{\rho(t)}, \quad c(t) = \frac{3v_0}{\rho(t)^2}, \quad d(t) = \frac{-v_0}{\rho(t)^3},$$

and therefore the approximating temperature profile is given by

$$v(x,t) = v_0\left(1 - \frac{x}{\rho(t)}\right)^3.$$

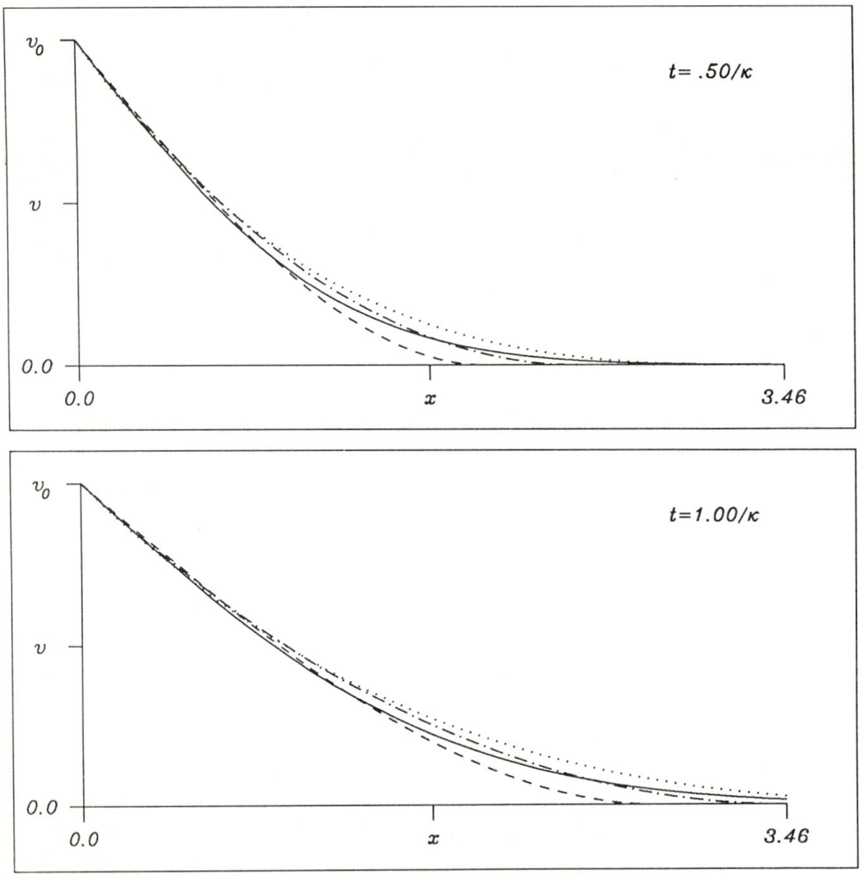

Figure 5.1. Comparison of heat-balance approximations for problem (5.1) with the exact solution (exact solution (—), the quadratic approximation ($-\cdot-$), cubic method 1 ($-\,-$) and cubic method 2 ($\cdots$)).

From this expression we obtain

$$\theta(t) = \frac{v_0}{4}\rho(t), \qquad \frac{\partial v}{\partial x}(0, t) = \frac{-3v_0}{\rho(t)},$$

which on using (5.4) gives

$$\rho(t) = 2\sqrt{6\kappa t},$$

so that

$$\frac{\partial v}{\partial x}(0, t) = \frac{-3v_0}{2\sqrt{6\kappa t}},$$

and therefore $\sqrt{\pi}$ in the exact expression (5.2) is in this instance approximated by $\sqrt{8/3}$.

From this example it is apparent that increasing the order of the polynomial does not necessarily increase the accuracy of the approximation. Indeed if we take a quartic approximation to the temperature profile (see Problem 2) we obtain the approximation $\sqrt{10/3}$ for $\sqrt{\pi}$ occurring in (5.2). Thus in terms of the four approximations considered, the simple quadratic expression (5.6) yields the best result. The various heat-balance approximations for the problem given by (5.1) are compared graphically with the exact solution in Figure 5.1.

5.3 Semi-infinite media with constant flux w_0 along $x = 0$

Here we consider the semi-infinite one dimensional region $0 \leq x < \infty$ which is initially at zero temperature. At time $t = 0$ and subsequently a constant flux w_0 is maintained across the surface $x = 0$. The equations for this situation are a special case of (3.13), namely

$$\left. \begin{aligned} \frac{\partial v}{\partial t} &= \kappa \frac{\partial^2 v}{\partial x^2}, \quad 0 < x < \infty, \\ v(x, 0) &= 0, \quad 0 \leq x < \infty, \\ -k \frac{\partial v}{\partial x}(0, t) &= w_0, \quad t > 0, \qquad \frac{\partial v}{\partial x} \to 0 \text{ as } x \to \infty, \end{aligned} \right\} \tag{5.14}$$

the solution of which is given in Example 3.3 as

$$v(x, t) = \frac{w_0}{k}\left\{ 2\sqrt{\frac{\kappa t}{\pi}} \exp\left(\frac{-x^2}{4\kappa t}\right) - x\,\mathrm{erfc}\left(\frac{x}{2\sqrt{\kappa t}}\right) \right\},$$

and in particular the surface temperature is given by

$$v(0, t) = \frac{w_0}{k}\sqrt{\frac{4\kappa t}{\pi}}. \tag{5.15}$$

For the heat-balance integral approximation we again assume a penetration depth $\rho(t)$ such that effectively heat conduction only occurs in $0 \le x \le \rho(t)$, and assume that both equations of (5.3) hold. For this problem, using $(5.14)_3$ the heat-balance equation (5.4) becomes

$$\theta(t) = \int_0^{\rho(t)} v(x,t)\,dx = \frac{\kappa w_0 t}{k}. \tag{5.16}$$

Now we derive a number of approximate temperature profiles using various polynomial expressions.

Quadratic temperature profile

On assuming that $v(x,t)$ may be approximated by (5.6) we may deduce from (5.3) and $(5.14)_3$ that

$$v(x,t) = \frac{w_0\rho(t)}{2k}\left(1 - \frac{x}{\rho(t)}\right)^2,$$

and that (5.16) yields

$$\rho(t) = \sqrt{6\kappa t}.$$

Thus the quadratic temperature profile predicts the surface temperature

$$v(0,t) = \frac{w_0}{k}\sqrt{\frac{3\kappa t}{2}},$$

so that the factor $\sqrt{4/\pi}$ in the exact expression (5.15) is approximated by $\sqrt{3/2}$.

Cubic temperature profile

Here we assume that $v(x,t)$ may be approximated by an expression of the form (5.10). In order to deduce an extra condition we differentiate $(5.14)_3$ with respect to time and use the governing partial differential equation $(5.14)_1$ to deduce

$$\frac{\partial^3 v}{\partial x^3}(0,t) = 0, \tag{5.17}$$

and either this equation of (5.12) provides the additional constraint. Clearly for this problem (5.12) is the appropriate equation to use since firstly it involves a lower order derivative, secondly (5.17) in any case reduces (5.10) to a quadratic expression and thirdly (5.12) results in a smoother temperature profile overall. From (5.3), (5.10) (5.12) and $(5.14)_3$ we may readily deduce

$$v(x,t) = \frac{w_0\rho(t)}{3k}\left(1 - \frac{x}{\rho(t)}\right)^3,$$

and from this result and (5.16) we have

$$\rho(t) = 2\sqrt{3\kappa t}.$$

Thus the cubic temperature profile predicts the surface temperature

$$v(0, t) = \frac{w_0}{k}\sqrt{\frac{4\kappa t}{3}},$$

which is an improvement on the quadratic result, since $\sqrt{4/3}$ represents a better approximation to $\sqrt{4/\pi}$ than does $\sqrt{3/2}$.

<u>Quartic temperature profile</u>

We now approximate the temperature by a quartic polynomial in x, namely

$$v(x,t) = a(t) + b(t)x + c(t)x^2 + d(t)x^3 + e(t)x^4, \tag{5.18}$$

where the five functions of time are determined from the five equations $(5.3)_{1,2}$, (5.12), $(5.14)_3$ and (5.17) which for ease of reference we restate here

$$\left. \begin{aligned} &\frac{\partial v}{\partial x}(0, t) = \frac{-w_0}{k}, \quad \frac{\partial^3 v}{\partial x^3}(0, t) = 0, \\[2mm] &v(\rho(t), t) - \frac{\partial v}{\partial x}(\rho(t), t) - \frac{\partial^2 v}{\partial x^2}(\rho(t), t) = 0. \end{aligned} \right\} \tag{5.19}$$

From these equations, after a straightforward calculation, we may deduce that the approximate temperature $v(x,t)$ is given by

$$v(x,t) = \frac{w_0\rho(t)}{8k}\left(3 + \frac{x}{\rho(t)}\right)\left(1 - \frac{x}{\rho(t)}\right)^3,$$

and from this expression and (5.16) we obtain

$$\rho(t) = \sqrt{10\kappa t},$$

for the penetration depth. Thus, for the quartic temperature profile we find that the surface is approximated by

$$v(0, t) = \frac{w_0}{k}\sqrt{\frac{45\kappa t}{32}},$$

which is an improvement on the quadratic profile, but not on the cubic profile.

<u>Quintic temperature profile</u>

Finally and for the purposes of illustration we approximate the temperature by a quintic expression of the form

$$v(x,t) = a(t) + b(t)x + c(t)x^2 + d(t)x^3 + e(t)x^4 + f(t)x^5. \qquad (5.20)$$

An additional constraint is obtained by differentiating the zero heat flux condition at $x = \rho(t)$, $(5.3)_2$, with respect to time, thus

$$\frac{\partial^2 v}{\partial x^2}(\rho(t), t)\frac{d\rho}{dt}(t) + \frac{\partial^2 v}{\partial x \partial t}(\rho(t), t) = 0,$$

which, on using (5.12) and the governing partial differential equation $(5.14)_1$ to eliminate the time partial derivative from the mixed second derivative, gives

$$\frac{\partial^3 v}{\partial x^3}(\rho(t), t) = 0.$$

Together this equation and the five equations (5.19) constitute the six conditions necessary for the determination of the six time dependent functions in the approximating quintic temperature profile (5.20). A long but relatively straightforward calculation gives the expression

$$v(x,t) = \frac{w_0 \rho(t)}{10k}\left(3 + \frac{2x}{\rho(t)}\right)\left(1 - \frac{x}{\rho(t)}\right)^4,$$

for the quintic approximation, and from (5.16) we obtain the expression

$$\rho(t) = \sqrt{15\kappa t},$$

for the penetration depth. Consequently we see that the quintic heat-balance approximation predicts a surface temperature

$$v(0,t) = \frac{w_0}{k}\sqrt{\frac{27\kappa t}{20}}.$$

This quintic approximation is superior to both the quadratic and quartic approximations, but is marginally inferior to the cubic approximation. This example again serves to emphasize that the accuracy of the heat-balance integral method is not necessarily improved by increasing the order of the approximating polynomial. The quadratic, cubic, quartic, but not the quintic, heat-balance solutions of problem (5.14) are compared graphically with the exact solution in Figure 5.2.

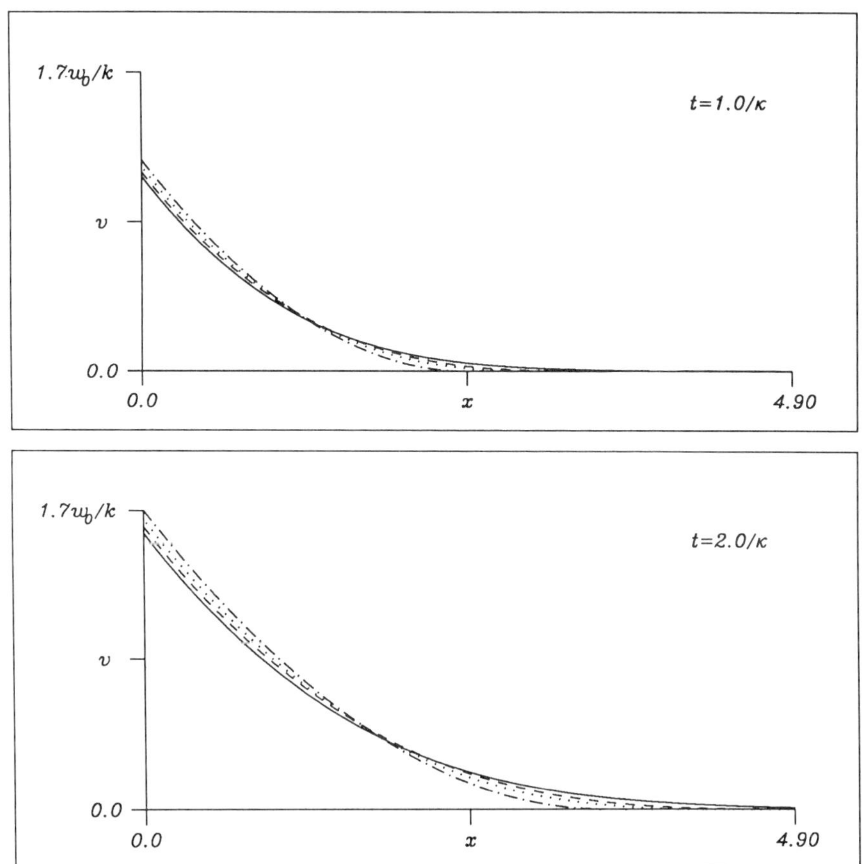

Figure 5.2. Comparison of heat-balance approximations for problem (5.14) with the exact solution (exact solution (—), the quadratic approximation ($- \cdot -$), the cubic approximation($- \, -$) and the quartic approximation ($\cdot \cdot \cdot$)).

5.4 Semi-infinite media with Newton cooling at $x = 0$ into a medium at constant temperature T_0

Again we consider the semi-infinite one dimensional region $0 \le x < \infty$ which is initially at zero temperature. At time zero and subsequently we suppose that Newton cooling takes place along the surface $x = 0$ into a medium at constant temperature T_0. We thus consider the special case of the problem given by equation (3.18), namely

$$
\left.
\begin{aligned}
\frac{\partial v}{\partial t} &= \kappa \frac{\partial^2 v}{\partial x^2}, \quad 0 < x < \infty, \\[2mm]
v(x,0) &= 0, \quad 0 \le x < \infty, \\[2mm]
-\frac{\partial v}{\partial x}(0,t) + \mu v(0,t) = \mu T_0, \quad t > 0, \qquad \frac{\partial v}{\partial x}(x,t) &\to 0 \text{ as } x \to \infty,
\end{aligned}
\right\}
\tag{5.21}
$$

the solution of which is given by (see Problem 3.8)

$$
v(x,t) = T_0\left\{ \mathrm{erfc}\left[\frac{x}{2\sqrt{\kappa t}}\right] - e^{\mu x + \kappa \mu^2 t}\, \mathrm{erfc}\left[\frac{x}{2\sqrt{\kappa t}} + \mu\sqrt{\kappa t}\right] \right\}.
$$

Substituting $x = 0$ into this expression we find that the surface temperature is given by given by

$$
v(0,t) = T_0\left[1 - e^{\kappa \mu^2 t}\, \mathrm{erfc}(\mu\sqrt{\kappa t})\right].
$$

Using the approximation $\mathrm{erf}(z) \approx 2z/\sqrt{\pi}$, valid for small z, we see that for small times ($t \ll 1$), the surface temperature is approximated by

$$
v(0,t) \approx \mu T_0 \sqrt{\frac{4\kappa t}{\pi}}, \quad t \ll 1,
\tag{5.22}
$$

and using the approximation $e^{z^2}\,\mathrm{erfc}(z) \approx 1/(z\sqrt{\pi})$, valid for large z, we find that for large times ($t \gg 1$) the surface temperature is approximately

$$
v(0,t) \approx T_0\left(1 - \frac{1}{\mu\sqrt{\pi \kappa t}}\right), \quad t \gg 1.
\tag{5.23}
$$

Again for the heat-balance integral approximation we assume a penetration depth $\rho(t)$ such that both equations of (5.3) still hold and we develop quadratic and cubic approximations for the temperature profile.

Quadratic temperature profile

From (5.3) and (5.6) we may readily deduce that the quadratic temperature takes the form

$$v(x,t) = a(t)\left(1 - \frac{x}{\rho(t)}\right)^2,$$

so that from the Newton cooling condition $(5.21)_3$ we find

$$a(t) = \frac{\mu T_0 \rho(t)}{[2 + \mu \rho(t)]}.$$

Now since $\theta(t) = a(t)\rho(t)/3$ we may deduce from these equations and the heat-balance equation (5.4) the first order ordinary differential equation

$$\frac{d}{dt}\left\{\frac{\rho(t)^2}{[2 + \mu\rho(t)]}\right\} = \frac{6\kappa}{[2 + \mu\rho(t)]},$$

for the penetration depth $\rho(t)$, which may be integrated to give the transcendental equation

$$(\mu\rho)^2 + 4\mu\rho - 8\log\left(1 + \frac{\mu\rho}{2}\right) = 12\kappa\mu^2 t, \tag{5.24}$$

where we have used $\rho(0) = 0$. For small times (and therefore small penetration depth) (5.24) gives $\rho(t) \approx \sqrt{6\kappa t}$ and therefore the surface temperature for small times is approximately

$$v(0,t) \approx \mu T_0 \sqrt{\frac{3\kappa t}{2}},$$

so that the factor $\sqrt{\pi/4}$ in the approximate expression (5.22) is approximated by $\sqrt{3/2}$. For large times (and therefore large penetration depth) we can neglect the logarithmic term in the transcendental equation (5.24) and obtain $\rho(t) \approx \left(-2 + 2\mu\sqrt{3\kappa t}\right)/\mu$ from which we have the approximation

$$v(0,t) \approx T_0\left(1 - \frac{1}{\mu\sqrt{3\kappa t}}\right).$$

Here we see that the factor of $\sqrt{\pi}$ in the approximate expression (5.23) is approximated by $\sqrt{3}$, indicating that the quadratic approximation is a good large time solution (see Figure 5.3).

Cubic temperature profile

If we employ a profile of the form (5.10) with the additional condition (5.12) then we readily deduce

$$v(x,t) = a(t)\left(1 - \frac{x}{\rho(t)}\right)^3,$$

where $a(t)$ is found from the Newton cooling condition $(5.21)_3$, thus

$$a(t) = \frac{\mu T_0 \rho(t)}{[3 + \mu\rho(t)]}.$$

In this case $\theta(t) = a(t)\rho(t)/4$ and from (5.4) we obtain

$$\frac{d}{dt}\left\{\frac{\rho(t)^2}{[3 + \mu\rho(t)]}\right\} = \frac{12\kappa}{[3 + \mu\rho(t)]},$$

which may be integrated to give the transcendental equation

$$(\mu\rho)^2 + 6\mu\rho - 18\log\left(1 + \frac{\mu\rho}{3}\right) = 24\kappa\mu^2 t, \tag{5.25}$$

for the penetration depth and where again we have used the initial condition $\rho(0) = 0$. For small times (5.25) gives $\rho(t) \approx \sqrt{12\kappa t}$ and therefore the surface temperature for small times becomes

$$v(0,t) \approx \mu T_0 \sqrt{\frac{4\kappa t}{3}},$$

which is evidently much closer to (5.22) than that predicted by the quadratic temperature. For large times we find that by neglecting the logarithmic term in (5.25) we obtain $\rho(t) \approx (-3 + \mu\sqrt{24\kappa t})/\mu$, and we have

$$v(0,t) \approx T_0\left(1 - \frac{1}{\mu\sqrt{\frac{8}{3}\kappa t}}\right),$$

so that the constant $\sqrt{\pi}$ in (5.23) is approximated by $\sqrt{8/3}$. Thus it is evident that the cubic approximation is inferior to the quadratic approximation as a large time solution (see Figure 5.3). For small times, when the surface temperature is small (see (5.22)), the problem essentially behaves as a prescribed flux problem and therefore, as in Section 5.3, we find the cubic approximation superior to the quadratic approximation. For large times, when the surface temperature is effectively T_0 (see (5.23)), the problem is essentially a prescribed temperature problem and we find, as in Section 5.2, that the quadratic approximation is superior to the cubic approximation.

5.5 One dimensional problems for finite media

In this section we consider the finite slab $0 \le x \le \ell$ initially at zero temperature with either prescribed surface temperatures, prescribed surface fluxes or Newton heat loss at the surfaces into media at prescribed temperatures. Thus we consider the three problems detailed in Sections 4.2, 4.3 and 4.4 and for each problem we only investigate a cubic approximation to the temperature profile. To accommodate both of the boundary conditions at $x = 0$ and at $x = \ell$, we take the temperature to be

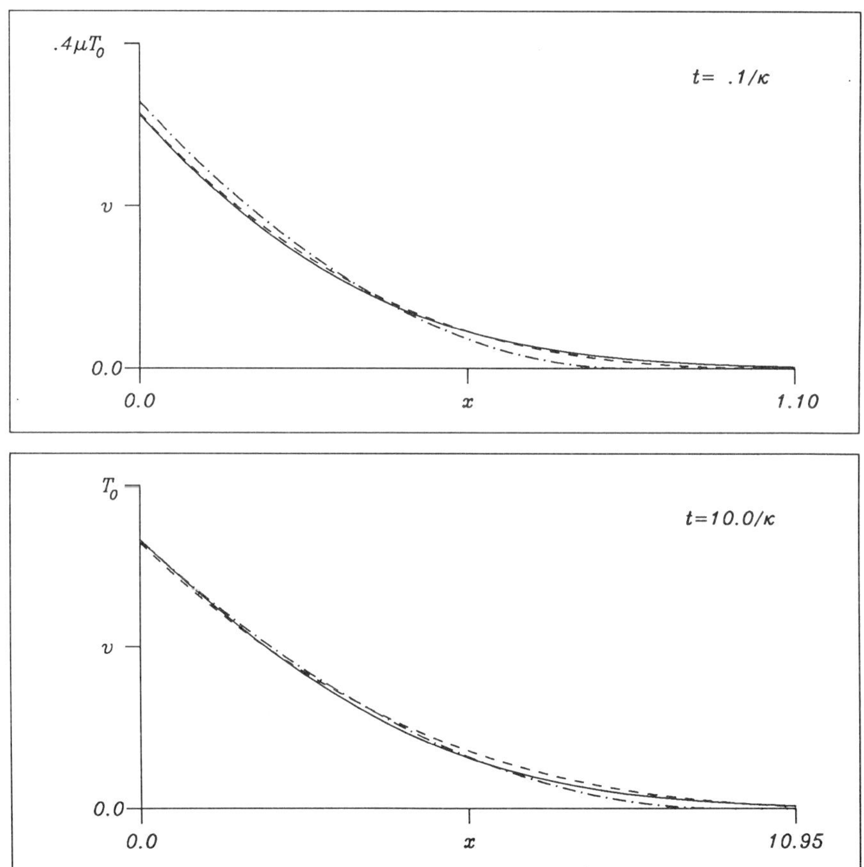

Figure 5.3. Comparison of the heat-balance approximations for problem (5.21) with the exact solution (exact solution (—), the quadratic approximation ($- \cdot -$) and the cubic approximation ($\cdot \cdot \cdot$)).

the sum of two independent components. Specifically, we introduce two independent heat-balance approximations, one of which satisfies the boundary condition at $x = 0$ and which has a penetration depth moving from $x = 0$ toward $x = \ell$ initially, and the other of which satisfies the boundary condition at $x = \ell$ and which has a penetration depth moving initially from $x = \ell$ toward $x = 0$. The approximate temperature is taken to be the sum of these two independent approximations. When one of the penetration depths equals the length of the body this approximation breaks down and another essentially large time approximation is used.

We first consider the problem given by equation (4.9), namely

$$\left.\begin{aligned} \frac{\partial v}{\partial t} &= \kappa \frac{\partial^2 v}{\partial x^2}, \quad 0 < x < \ell, \\[2mm] v(x,0) &= 0, \quad 0 \le x \le \ell, \\[2mm] v(0,t) &= g_1(t), \quad v(\ell,t) = g_2(t), \quad t > 0, \end{aligned}\right\} \tag{5.26}$$

where $g_1(t)$ and $g_2(t)$ are known functions of time t. We assume that

$$v(x,t) = v_1(x,t) + v_2(x,t),$$

where v_1 and v_2 are cubic heat-balance approximations with respective penetration depths $\rho_1(t)$ and $\rho_2(t)$ such that

$$v_1(0,t) = g_1(t), \quad v_2(\ell,t) = g_2(t),$$

and

$$\left.\begin{aligned} \rho_1(0) &= 0, \quad v_1(x,t) = \frac{\partial v_1}{\partial x}(x,t) = \frac{\partial^2 v_1}{\partial x^2}(x,t) = 0, \; x \ge \rho_1(t), \\[2mm] \rho_2(0) &= \ell, \quad v_2(x,t) = \frac{\partial v_2}{\partial x}(x,t) = \frac{\partial^2 v_2}{\partial x^2}(x,t) = 0, \; x \le \rho_2(t), \end{aligned}\right\} \tag{5.27}$$

as shown in Figure 5.4. Assuming a cubic approximation of the form (5.10) for both $v_1(x,t)$ and $v_2(x,t)$ and imposing the relevant boundary conditions and (5.27) we find that the two independent heat-balance components are given by

$$v_1(x,t) = g_1(t)\left(1 - \frac{x}{\rho_1(t)}\right)^3, \qquad v_2(x,t) = g_2(t)\left(\frac{x - \rho_2(t)}{\ell - \rho_2(t)}\right)^3, \tag{5.28}$$

so that the total heat-balance approximation $v(x,t)$ is given by

$$v(x,t) = H(\rho_1(t) - x)g_1(t)\left(1 - \frac{x}{\rho_1(t)}\right)^3 + H(x - \rho_2(t))g_2(t)\left(\frac{x - \rho_2(t)}{\ell - \rho_2(t)}\right)^3, \tag{5.29}$$

where $H(x)$ is the Heaviside unit step function. Applying the heat-balance integral method to (5.28) in the obvious manner, that is by requiring that each component

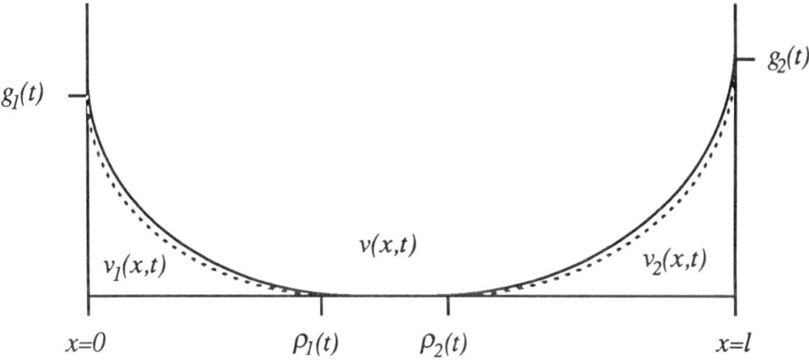

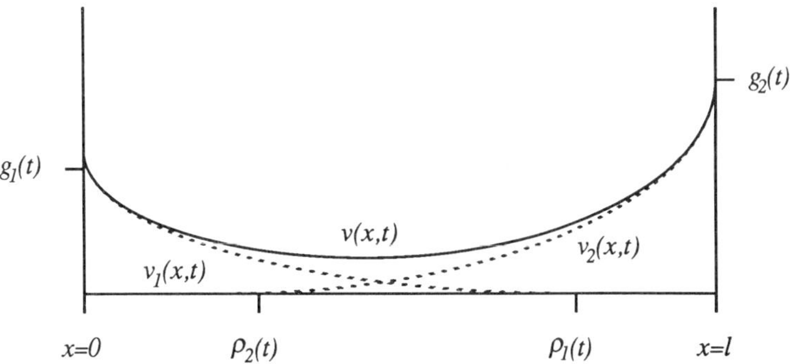

Figure 5.4. Construction of heat-balance approximations (—) for finite intervals by the addition of two independent components (− −).

$v_1(x, t)$ and $v_2(x, t)$ independently satisfies the heat-balance integral, namely

$$\frac{d}{dt} \int_0^{\rho_1(t)} v_1(x, t)\, dx = -\kappa \frac{\partial v_1}{\partial x}(0, t), \quad \frac{d}{dt} \int_{\rho_2(t)}^{\ell} v_2(x, t)\, dx = \kappa \frac{\partial v_2}{\partial x}(\ell, t), \quad (5.30)$$

we can easily deduce that

$$\rho_1(t) = \frac{2}{g_1(t)} \sqrt{6\kappa \int_0^t g_1(\tau)^2\, d\tau}, \qquad \rho_2(t) = \ell - \frac{2}{g_2(t)} \sqrt{6\kappa \int_0^t g_2(\tau)^2\, d\tau}. \quad (5.31)$$

However, once either $\rho_1(t) = \ell$ or $\rho_2(t) = 0$ this procedure breaks down as the boundary conditions on $v(x, t)$ are no longer satisfied. Assuming that t_c is the lesser of these two times, then for $t > t_c$ we still assume a cubic expression for the form (5.10) for $v(x, t)$ where the four functions of time are determined by the four conditions

$$v(0, t) = g_1(t), \quad v(\ell, t) = g_2(t),$$

$$\frac{\partial^2 v}{\partial x^2}(0, t) = \frac{g_1'(t)}{\kappa}, \quad \frac{\partial^2 v}{\partial x^2}(\ell, t) = \frac{g_2'(t)}{\kappa},$$

and where the final two equations are obtained by differentiating the prescribed temperature conditions on the surfaces and using the heat equation. A routine calculation gives

$$v(x, t) = g_1(t) + [g_2(t) - g_1(t)]\frac{x}{\ell} + \frac{g_1'(t)}{2\kappa}x(x - \ell) + \frac{\left[g_2'(t) - g_1'(t)\right]}{6\kappa\ell}x(x^2 - \ell^2). \quad (5.32)$$

In general the two approximate solutions (5.29) and (5.32) do not match at time t_c. The exact solutions of the following two examples are given in Examples 4.3 and 4.4 respectively.

<u>Example 5.1</u> For the finite region $0 \leq x \leq \ell$ with zero initial temperature and planes $x = 0$ and $x = \ell$ maintained at βt where β is a constant, deduce a heat-balance approximation utilizing a cubic temperature profile.

In this case, from (5.31) with $g_1(t) = \beta t$ and $g_2(t) = \beta t$, we deduce that the penetration depths $\rho_1(t)$ and $\rho_2(t)$ are given by

$$\rho_1(t) = \sqrt{8\kappa t}, \quad \rho_2(t) = \ell - \sqrt{8\kappa t},$$

so that $t_c = \ell^2/8\kappa$ and from (5.29) and (5.32) we have immediately (see Figure 5.5)

$$
\left.
\begin{aligned}
v(x,t) &= \beta t\left\{ H(\sqrt{8\kappa t} - x)\left(1 - \frac{x}{\sqrt{8\kappa t}}\right)^3 \right. \\
&\qquad \left. + H(x - \ell + \sqrt{8\kappa t})\left(1 - \frac{\ell - x}{\sqrt{8\kappa t}}\right)^3\right\} \qquad t \le t_c, \\
v(x,t) &= \beta t + \frac{\beta x(x - \ell)}{2\kappa}, \quad t > t_c.
\end{aligned}
\right\}
$$

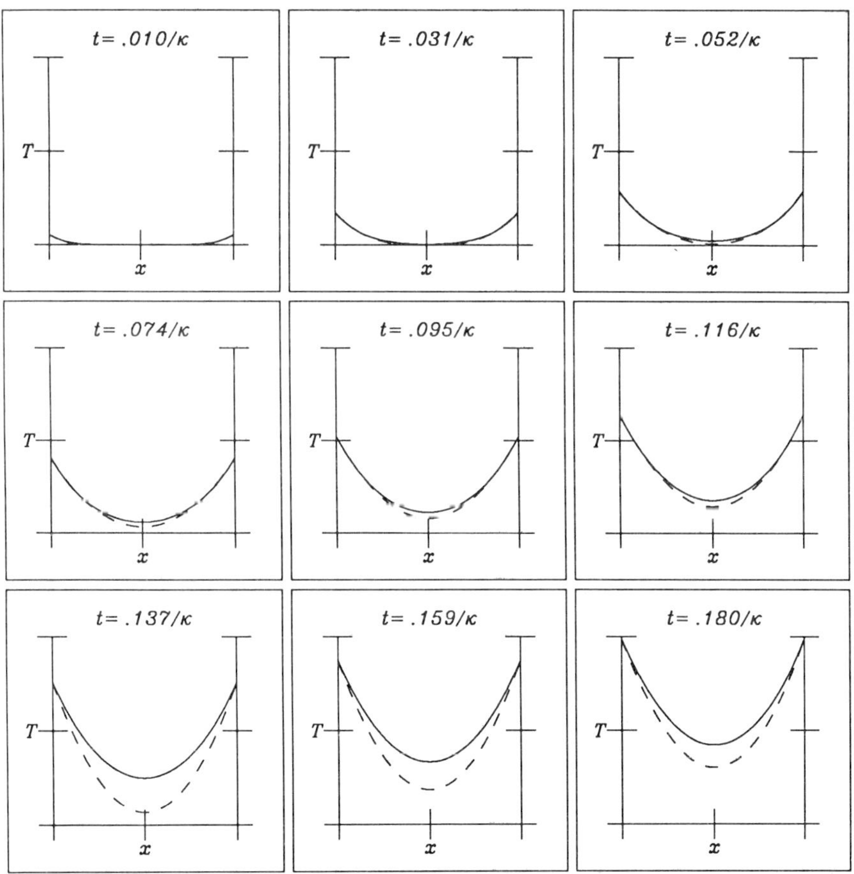

Figure 5.5. Comparison of the heat-balance approximation $(-\,-)$ and the exact solution $(-)$ for Example 5.1 with $\beta = 1.0$ and $\ell = 1.0$.

<u>Example 5.2</u> For the finite region $0 \leq x \leq \ell$ with zero initial temperature and planes $x = 0$ and $x = \ell$ maintained at constant temperatures v_{10} and v_{20} respectively, deduce a heat-balance approximation utilizing a cubic temperature profile.

For this problem (5.31) gives $\rho_1(t) = 2\sqrt{6\kappa t}$ and $\rho_2(t) = \ell - 2\sqrt{6\kappa t}$ so that $t_c = \ell^2/24\kappa$ and equations (5.29) and (5.32) give (see Figure 5.6)

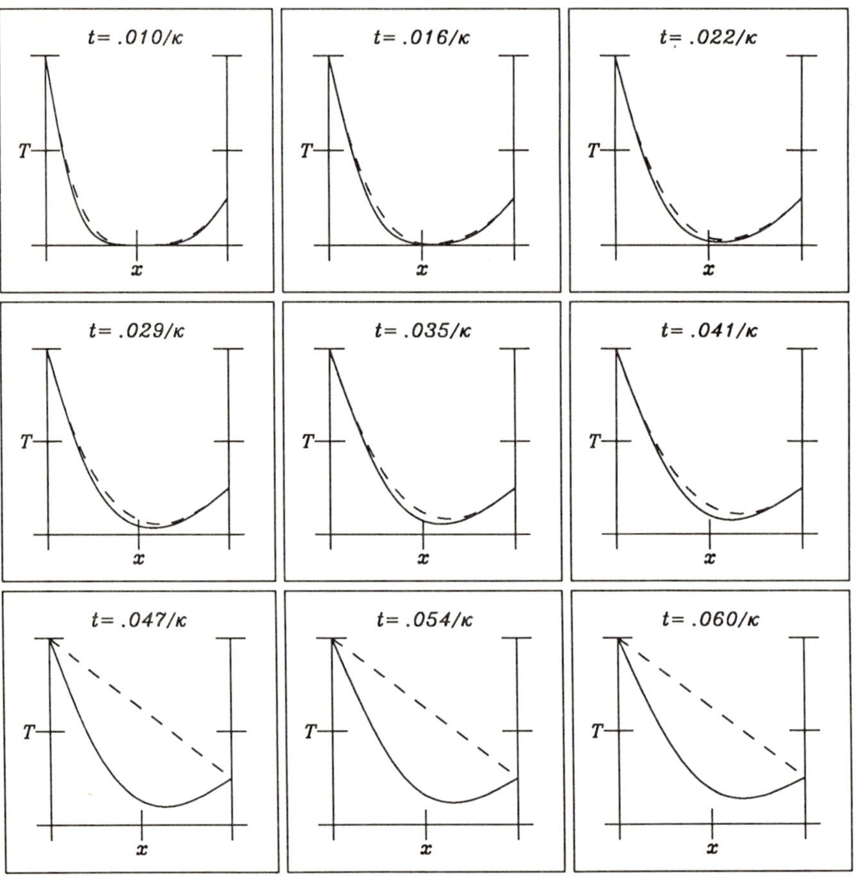

Figure 5.6. Comparison of the heat-balance approximation ($-\,-$) and the exact solution ($-\!-$) for Example 5.2 with $v_{10} = 1.0$, $v_{20} = 0.25$ and $\ell = 1.0$.

$$v(x,t) = v_{10}H(2\sqrt{6\kappa t} - x)\left(1 - \frac{x}{2\sqrt{6\kappa t}}\right)^3$$

$$+ v_{20}H(x - \ell + 2\sqrt{6\kappa t})\left(1 - \frac{\ell - x}{2\sqrt{6\kappa t}}\right)^3, \qquad t \le t_c,$$

$$v(x,t) = v_{10} + (v_{20} - v_{10})\frac{x}{\ell}, \qquad t > t_c.$$

We note that for both examples the given approximations for $t > t_c$ are simply the first terms of the exact solutions (see Examples 4.3 and 4.4), not involving the exponentially decaying terms in the infinite series. Thus the given heat-balance approximations for $t > t_c$ are essentially large time approximations.

We now consider the problem given by equation (4.20) for which the heat-balance method is not effective. Specifically, we consider the problem

$$\frac{\partial v}{\partial t} = \kappa\frac{\partial^2 v}{\partial x^2}, \qquad 0 < x < \ell,$$

$$v(x,0) = 0, \qquad 0 \le x \le \ell, \tag{5.33}$$

$$-k\frac{\partial v}{\partial x}(0,t) = h_1(t), \quad k\frac{\partial v}{\partial x}(\ell,t) = h_2(t), \qquad t > 0,$$

where $h_1(t)$ and $h_2(t)$ are prescribed functions of time. Proceeding as before, and assuming cubic approximations of the form (5.10) for v_1 and v_2 we have from (5.10), (5.27) and (5.33)$_3$

$$v_1(x,t) = \frac{h_1(t)\rho_1(t)}{3k}\left(1 - \frac{x}{\rho_1(t)}\right)^3, \qquad v_2(x,t) = \frac{h_2(t)}{3k}(\ell - \rho_2(t))\left(\frac{x - \rho_2(t)}{\ell - \rho_2(t)}\right)^3,$$

so that

$$v(x,t) = H(\rho_1(t) - x)\frac{h_1(t)\rho_1(t)}{3k}\left(1 - \frac{x}{\rho_1(t)}\right)^3$$

$$+ H(x - \rho_2(t))\frac{h_2(t)}{3k}(\ell - \rho_2(t))\left(\frac{x - \rho_2(t)}{\ell - \rho_2(t)}\right)^3. \tag{5.34}$$

The heat-balance equations (5.30) yield

$$\rho_1(t) = \sqrt{\frac{12\kappa}{h_1(t)}}\int_0^t h_1(\tau)\,d\tau, \qquad \rho_2(t) = \ell - \sqrt{\frac{12\kappa}{h_2(t)}}\int_0^t h_2(\tau)\,d\tau. \tag{5.35}$$

Still assuming a cubic expression for $t > t_c$, in principle the four arbitrary functions of time in (5.10) could be determined by the four conditions

$$
\frac{\partial v}{\partial x}(0,t) = \frac{-h_1(t)}{k}, \quad \frac{\partial v}{\partial x}(\ell,t) = \frac{h_2(t)}{k},
$$
$$
\frac{\partial^3 v}{\partial x^3}(0,t) = \frac{-h'_1(t)}{k\kappa}, \quad \frac{\partial^3 v}{\partial x^3}(\ell,t) = \frac{h'_2(t)}{k\kappa}. \tag{5.36}
$$

However, the first difficulty is that for a cubic expression we have

$$
\frac{\partial^3 v}{\partial x^3}(x,t) = 6d(t),
$$

which is independent of x and accordingly both of $(5.36)_3$ and $(5.36)_4$ cannot be accommodated. Assume for the time being that we take $(5.36)_3$ as the condition on the third derivative. Then in a routine manner we obtain

$$
v(x,t) = a(t) - \frac{h_1(t)}{k}x + \left[h_1(t) + h_2(t) + \frac{\ell^2}{2\kappa}h'_1(t)\right]\frac{x^2}{2k\ell} - \frac{h'_1(t)}{6k\kappa}x^3, \tag{5.37}
$$

and the second difficulty which arises is the determination of the function $a(t)$. These problems are due partially to the physical situation which this problem represents. That is, if we prescribe heat flux at both ends of the material, unless these are equal, a 'steady state' can never arise. The following related example is more amenable to the heat-balance procedure.

<u>Example 5.3</u> For the finite region $0 \le x \le \ell$ with zero initial temperature, heat flux $h_1(t)$ across $x = 0$ and $x = \ell$ maintained at zero temperature deduce a heat-balance approximation utilizing a cubic temperature profile.

Initially for this problem $(t \le t_c)$, equation $(5.35)_1$ still holds. However, since $v(\ell,t) = 0$ it is clear from (5.28) that $v_2(x,t)$ vanishes identically and that for all times $\rho_2(t) = \ell$. Thus the problem may be treated as a semi-infinite problem until such time as $\rho_1(t) = \ell$. For $t > t_c$ the four arbitrary functions of time in (5.10) are determined from

$$
\frac{\partial v}{\partial x}(0,t) = \frac{-h_1(t)}{k}, \quad v(\ell,t) = 0,
$$
$$
\frac{\partial^3 v}{\partial x^3}(0,t) = \frac{-h'_1(t)}{k\kappa}, \quad \frac{\partial^2 v}{\partial x^2}(\ell,t) = 0,
$$

from which we may deduce that

$$v(x,t) = \frac{(\ell - x)}{k}\left[h_1(t) + \frac{h_1'(t)}{6\kappa}(x^2 - 2x\ell - 2\ell^2)\right].$$

Finally in this section we consider the heat-balance approximation for the problem of the finite slab, initially at zero temperature and with Newton heat loss at the surfaces into media at temperatures $T_1(t)$ and $T_2(t)$. This problem is given by equation (4.29), namely

$$\left.\begin{array}{c} \dfrac{\partial v}{\partial t} = \kappa\dfrac{\partial^2 v}{\partial x^2}, \quad 0 < x < \ell, \\[2ex] v(x,0) = 0, \quad 0 \le x \le \ell, \\[2ex] -\dfrac{\partial v}{\partial x}(0,t) + \mu v(0,t) = \mu T_1(t), \quad \dfrac{\partial v}{\partial x}(\ell,t) + \mu v(\ell,t) = \mu T_2(t), \quad t > 0, \end{array}\right\} \quad (5.38)$$

and in this case the two difficulties noted above for the prescribed flux problem do not arise, essentially because the boundary conditions $(5.38)_3$ and $(5.38)_4$ contain the temperature explicitly and not just the derivatives of the temperature. For time $t \le t_c$ we have from (5.10), (5.27), $(5.38)_3$ and $(5.38)_4$

$$v_1(x,t) = \frac{\mu T_1(t)\rho_1(t)}{[3 + \mu\rho_1(t)]}\left(1 - \frac{x}{\rho_1(t)}\right)^3, \quad v_2(x,t) = \frac{\mu T_2(t)(\ell - \rho_2(t))}{[3 + \mu(\ell - \rho_2(t))]}\left(\frac{x - \rho_2(t)}{\ell - \rho_2(t)}\right)^3,$$

where the penetration depths are in principle obtained as solutions of the ordinary differential equations

$$\frac{d}{dt}\left\{\frac{T_1(t)\rho_1(t)^2}{[3 + \mu\rho_1(t)]}\right\} = \frac{12\kappa T_1(t)}{[3 + \mu\rho_1(t)]}, \quad \frac{d}{dt}\left\{\frac{T_2(t)(\ell - \rho_2(t))^2}{[3 + \mu(\ell - \rho_2(t))]}\right\} = \frac{12\kappa T_2(t)}{[3 + \mu(\ell - \rho_2(t))]},$$

subject to the initial conditions $\rho_1(0) = 0$ and $\rho_2(0) = \ell$. Assuming for particular $T_1(t)$ and $T_2(t)$ that this problem may be solved the temperature profile for $t > t_c$ is given by (5.10) where the four functions of time are determined by the four boundary conditions

$$\left.\begin{array}{c} -\dfrac{\partial v}{\partial x}(0,t) + \mu v(0,t) = \mu T_1(t), \quad \dfrac{\partial v}{\partial x}(\ell,t) + \mu v(\ell,t) = \mu T_2(t), \\[3ex] -\dfrac{\partial^3 v}{\partial x^3}(0,t) + \mu\dfrac{\partial^2 v}{\partial x^2}(0,t) = \dfrac{\mu}{\kappa}T_1'(t), \quad \dfrac{\partial^3 v}{\partial x^3}(\ell,t) + \mu\dfrac{\partial^2 v}{\partial x^2}(\ell,t) = \dfrac{\mu}{\kappa}T_2'(t). \end{array}\right\}$$

$$(5.39)$$

After a straightforward calculation we may deduce the following results for $a(t)$, $b(t)$, $c(t)$ and $d(t)$

$$a(t) = \frac{\phi(t)}{(2 + \mu\ell)} - \frac{\ell\left[3(2 + \mu\ell)\phi'(t) + \mu\ell(3 + \mu\ell)\psi'(t)\right]}{6\kappa\mu(2 + \mu\ell)^2},$$

$$b(t) = \frac{\mu\psi(t)}{(2 + \mu\ell)} - \frac{\ell\left[3(2 + \mu\ell)\phi'(t) + \mu\ell(3 + \mu\ell)\psi'(t)\right]}{6\kappa(2 + \mu\ell)^2},$$

$$c(t) = \frac{\phi'(t)}{2\kappa(2 + \mu\ell)},$$

$$d(t) = \frac{\mu\psi'(t)}{6\kappa(2 + \mu\ell)},$$

where the functions $\phi(t)$ and $\psi(t)$ are defined by

$$\phi(t) = T_2(t) + (1 + \mu\ell)T_1(t), \quad \psi(t) = T_2(t) - T_1(t).$$

Example 5.4 For the finite region $0 \leq x \leq \ell$ with zero initial temperature and Newton heat loss from the surfaces $x = 0$ and $x = \ell$ into media at constant temperature T_{10} and T_{20} respectively, deduce a heat balance approximation utilizing a cubic temperature profile.

From the above we may readily deduce that for $t \leq t_c$ we have

$$v_1(x, t) = \frac{\mu T_{10}\rho_1(t)}{[3 + \mu\rho_1(t)]}\left(1 - \frac{x}{\rho_1(t)}\right)^3, \quad v_2(x, t) = \frac{\mu T_{20}(\ell - \rho_2(t))}{[3 + \mu(\ell - \rho_2(t))]}\left(1 - \frac{\ell - x}{\ell - \rho_2(t)}\right)^3,$$

where $\rho_1(t)$ and $\rho_2(t)$ are determined as the roots of the respective transcendental equations

$$(\mu\rho_1)^2 + 6\mu\rho_1 - 18\log\left(1 + \frac{\mu\rho_1}{3}\right) = 24\mu^2\kappa t,$$

$$(\mu(\ell - \rho_2))^2 + 6\mu(\ell - \rho_2) - 18\log\left(1 + \frac{\mu(\ell - \rho_2)}{3}\right) = 24\mu^2\kappa t.$$

Further, for $t > t_c$ it is apparent from the above that in this case both $\phi(t)$ and $\psi(t)$ are constant and therefore there is no quadratic or cubic contribution to the temperature profile and further we have

$$v(x,t) = Ax + B,$$

where the constants A and B are given respectively by

$$A = \frac{\mu(T_{20} - T_{10})}{(2 + \mu\ell)}, \quad B = \frac{T_{20} + (1 + \mu\ell)T_{10}}{(2 + \mu\ell)}.$$

Again from Example 4.8 it is apparent that the heat-balance approximation for $t > t_c$ yields the first terms of the exact solution not involving the exponentially small terms in the infinite series. That is the heat-balance approximation is simply the steady or large time temperature distribution.

5.6 Radial flow in solid circular cylinders and spheres

In this section we consider the heat-balance approximation for infinite solid circular cylinders and solid spheres of radius a, initially at zero temperature and subject to a constant surface condition. We thus consider the problems

$$\left.\begin{aligned} \frac{\partial v}{\partial t} &= \kappa\left(\frac{\partial^2 v}{\partial r^2} + \frac{m}{r}\frac{\partial v}{\partial r}\right), \quad 0 < r < a, \\[2mm] v(r,0) &= 0, \quad 0 \le r \le a, \end{aligned}\right\} \tag{5.40}$$

subject to one of the following constant surface conditions for $t > 0$

$$\left.\begin{aligned} &\textbf{(i)} \quad v(a,t) = v_0, \\[2mm] &\textbf{(ii)} \quad k\frac{\partial v}{\partial r}(a,t) = w_0, \\[2mm] &\textbf{(iii)} \quad \frac{\partial v}{\partial r}(a,t) + \mu v(a,t) = \mu T_0, \end{aligned}\right\} \tag{5.41}$$

where as usual $\mu = h/k$ and $m = 1$ for cylinders and $m = 2$ for spheres. Again we assume a penetration radius $r = \rho(t)$ such that there is no heat conduction in the region $0 \le r < \rho(t)$, so that in particular we have

$$v(\rho(t),t) = \frac{\partial v}{\partial r}(\rho(t),t) = 0. \tag{5.42}$$

Based on general solutions for cylindrical geometry which are beyond the scope of this book, we assume for the cylinder that the solution of $(5.40)_1$ (with $m = 1$) subject to the condition $(5.42)_1$ has the general structure

$$v(r,t) = \left[r^2 - \rho(t)^2\right]F\left(r^2,\rho(t)^2\right) + G\left(r^2,\rho(t)^2\right)\log\left(\frac{r}{\rho(t)}\right),$$

where F and G denote functions which are assumed to be analytic and as such may be approximated in the usual manner by polynomial expressions. In the first instance these functions may be approximated simply by functions of time, thus

$$v(r,t) = a(t)\left[r^2 - \rho(t)^2\right] + b(t)\log\left(\frac{r}{\rho(t)}\right), \tag{5.43}$$

while higher order approximations may be generated as follows

$$v(r,t) = a(t)\left[r^2 - \rho(t)^2\right] + \left[b(t) + c(t)r^2\right]\log\left(\frac{r}{\rho(t)}\right), \tag{5.44}$$

$$v(r,t) = \left[a(t) + d(t)r^2\right]\left[r^2 - \rho(t)^2\right] + \left[b(t) + c(t)r^2\right]\log\left(\frac{r}{\rho(t)}\right), \tag{5.45}$$

and so on. We now utilize the approximation (5.43) to deduce heat-balance solutions of the problems (5.40) and (5.41) for the cylinder.

With the heat-balance integral $\theta(t)$ defined by

$$\theta(t) = \int_{\rho(t)}^{a} rv(r,t)\,dr, \tag{5.46}$$

(which is simply the integral of the temperature over the area into which heat has penetrated) we have from $(5.40)_1$, on multiplying by r and integrating r from $\rho(t)$ to a

$$\int_{\rho(t)}^{a} r\frac{\partial v}{\partial t}(r,t)\,dr = \kappa\left[r\frac{\partial v}{\partial r}(r,t)\right]_{\rho(t)}^{a},$$

so that on noting $(5.42)_2$ we have

$$\frac{d\theta}{dt}(t) = \kappa a\frac{\partial v}{\partial r}(a,t), \tag{5.47}$$

which is the basic heat-balance condition for the cylinder and simply expresses the fact that the rate at which the heat content of the cylinder is changing is proportional to the heat flux across its surface.

$(i)_c$ Cylinder with constant surface temperature v_0

From $(5.41)_1$, $(5.42)_2$ and (5.43) we obtain

$$v(r,t) = v_0 \frac{\left[(r^2 - \rho^2) - 2\rho^2 \log(r/\rho)\right]}{\left[(a^2 - \rho^2) - 2\rho^2 \log(a/\rho)\right]}, \tag{5.48}$$

and for convenience of notation we use ρ to denote the penetration radius $\rho(t)$. From (5.43) and (5.46) we can deduce

$$\theta(t) = \frac{a(t)}{4}\left(a^2 - \rho^2\right)^2 - \frac{b(t)}{4}\left[(a^2 - \rho^2) + 2a^2 \log(\rho/a)\right], \tag{5.49}$$

and from this equation and the known expression for $a(t)$ and $b(t)$ we obtain

$$\theta(t) = \frac{v_0}{4} \frac{\left[(a^4 - \rho^4) + 4a^2\rho^2 \log(\rho/a)\right]}{\left[(a^2 - \rho^2) + 2\rho^2 \log(\rho/a)\right]}.$$

Now from this equation, the heat-balance condition (5.47) and the expression

$$\frac{\partial v}{\partial r}(a,t) = \frac{2v_0(a^2 - \rho^2)}{a\left[(a^2 - \rho^2) + 2\rho^2 \log(\rho/a)\right]},$$

we may, on performing the time differentiation in (5.47), deduce

$$2\kappa \, dt = \frac{\left[(a^2 - \rho^2) + (a^2 + \rho^2)\log(\rho/a)\right]}{\left[(a^2 - \rho^2) + 2\rho^2 \log(\rho/a)\right]}\rho \, d\rho.$$

On noting that $\rho(0) = a$ and by making the substitution $\xi = (\rho/a)^2$ we have on integration

$$t = \frac{a^2}{8\kappa} \int_1^{(\rho/a)^2} \frac{[2(1 - \xi) + (1 + \xi)\log \xi]}{[(1 - \xi) + \xi \log \xi]} \, d\xi, \tag{5.50}$$

which does permit some simplification but to no real advantage. As noted in the introduction, non-planar heat-balance approximations, even for the simplest of problems, are frequently complicated and do not admit simple closed form expressions.

(ii)$_c$ <u>Cylinder with constant surface flux w_0</u>

In this case the functions $a(t)$ and $b(t)$ in (5.43) are determined from (5.41)$_2$ and (5.42)$_2$. We find that the temperature as given by (5.43) becomes

$$v(r,t) = \frac{a w_0}{2k(a^2 - \rho^2)}\left[(r^2 - \rho^2) - 2\rho^2 \log(r/\rho)\right],$$

and making use of (5.49) we can readily deduce

$$\theta(t) = \frac{a w_0\left[(a^4 - \rho^4) + 4a^2\rho^2 \log(\rho/a)\right]}{8k(a^2 - \rho^2)}.$$

However, in this case (5.47) integrates immediately to give

$$t = \frac{1}{8\kappa}\left\{(a^2 + \rho^2) + \frac{4a^2\rho^2}{(a^2 - \rho^2)}\log(\rho/a)\right\}, \tag{5.51}$$

as the determining transcendental equation for the penetration radius $\rho(t)$ and we observe that ρ is zero in time $a^2/8\kappa$.

(iii)$_c$ <u>with Newton cooling into media at constant temperature T_0</u>

Still employing the approximating temperature profile (5.43), we can deduce from (5.41)$_3$ and (5.42) that the temperature is given by

$$v(r,t) = \frac{\mu a T_0\left[(r^2 - \rho^2) - 2\rho^2 \log(r/\rho)\right]}{\left[(2 + \mu a)(a^2 - \rho^2) + 2\mu a\rho^2 \log(\rho/a)\right]},$$

and again making use of (5.49) we find

$$\theta(t) = \frac{\mu a T_0\left[(a^4 - \rho^4) + 4a^2\rho^2 \log(\rho/a)\right]}{4\left[(2 + \mu a)(a^2 - \rho^2) + 2\mu a\rho^2 \log(\rho/a)\right]}.$$

From these equations and (5.47) we may deduce, on performing the time differentiation in (5.47)

$$2\kappa\, dt =$$

$$\frac{\left\{(a^2 - \rho^2)\left[(2 + \mu a)(a^2 - \rho^2) + (a^2 + \rho^2)\right] + \left[\mu a(a^4 - \rho^4) + 4a^4\right]\log(\rho/a)\right\}}{(a^2 - \rho^2)\left[(2 + \mu a)(a^2 - \rho^2) + 2\mu a\rho^2 \log(\rho/a)\right]}\rho\, d\rho.$$

Again, on making the substitution $\xi = (\rho/a)^2$ we obtain

$$t = \frac{a^2}{8\kappa} \int_1^{(\rho/a)^2} \frac{\left\{2(1-\xi)[(2+\mu a)(1-\xi)+(1+\xi)] + \left[\mu a(1-\xi^2)+4\right]\log\xi\right\}}{(1-\xi)[(2+\mu a)(1-\xi)+\mu a\xi\log\xi]}\,d\xi,$$

$$(5.52)$$

and we can confirm from this result that the previously determined equations (5.50) and (5.51) emerge in the limits μ tending to infinity and zero, respectively.

For purely radial heat flow in solid spheres, as noted previously in Section 4.5, the transformation $v = w/r$ reduces (5.40) and (5.41) to

$$\left.\begin{aligned}
\frac{\partial w}{\partial t} &= \kappa\frac{\partial^2 w}{\partial r^2}, \quad 0 < r < a, \\[2mm]
w(r,0) &= 0, \quad 0 \le r \le a,
\end{aligned}\right\}$$

$$(5.53)$$

and one of the boundary conditions

$$\left.\begin{aligned}
&\text{(i)} \quad w(a,t) = a v_0, \\[2mm]
&\text{(ii)} \quad \frac{\partial w}{\partial r}(a,t) - \frac{w(a,t)}{a} = \frac{a w_0}{k}, \\[2mm]
&\text{(iii)} \quad \frac{\partial w}{\partial r}(a,t) + \left(\mu - \frac{1}{a}\right)w(a,t) = \mu a T_0,
\end{aligned}\right\}$$

$$(5.54)$$

and for times after the penetration radius has reached the origin (say $t \ge t_c$ where $\rho(t_c) = 0$) we require $w(0,t) = 0$ so that the temperature at the origin remains finite. Further, in place of (5.42) we have

$$w(\rho(t),t) = \frac{\partial w}{\partial r}(\rho(t),t) = 0,$$

$$(5.55)$$

and with $\theta(t)$ defined by

$$\theta(t) = \int_{\rho(t)}^a w(r,t)\,dr,$$

$$(5.56)$$

we find on integration of $(5.53)_1$ in the usual way

$$\frac{d\theta}{dt}(t) = \kappa\frac{\partial w}{\partial r}(a,t).$$

$$(5.57)$$

Exactly as described in Sections 5.2-5.5 we may now deduce approximate solutions of (5.53) and (5.54) by assuming a polynomial expression for $w(r,t)$. For example, using

$$w(r,t) = a(t) + b(t)r + c(t)r^2,$$

$$(5.58)$$

we can deduce the following:

(i)$_s$ <u>Sphere with constant surface temperature v_0</u>

From $(5.54)_1$, (5.55) and (5.58) we obtain

$$w(r,t) = a v_0 \left(\frac{r - \rho(t)}{a - \rho(t)}\right)^2,$$

and this equation together with (5.56) and (5.57) yields, in a routine manner

$$\rho(t) = a - 2\sqrt{3\kappa t},$$

where we have used the initial condition $\rho(0) = a$. We observe that in this case

$$t_c = a^2/12\kappa,$$

and for times $t > t_c$ the only simple polynomial expression satisfying appropriate conditions is

$$w(r,t) = v_0 r,$$

which corresponds to the large time solution for which the temperature throughout the sphere is constant at v_0.

(ii)$_s$ <u>Sphere with constant surface flux w_0</u>

In this case we find from $(5.54)_2$, (5.55) and (5.58)

$$w(r,t) = \frac{a^2 w_0}{k} \frac{[r - \rho(t)]^2}{\left[a^2 - \rho(t)^2\right]},$$

and this equation and (5.57) together give

$$\frac{\mathrm{d}}{\mathrm{d}t}\left\{\frac{[a - \rho(t)]^2}{[a + \rho(t)]}\right\} = \frac{6\kappa}{[a + \rho(t)]},$$

which on differentiating out and integrating, using the initial condition $\rho(0) = a$, ultimately yields the following transcendental equation for $\rho(t)$

$$t = \frac{1}{12\kappa}\left\{(\rho - a)(\rho + 3a) - 8a^2 \log \frac{(a + \rho)}{2a}\right\}.$$

(iii)$_s$ Sphere with Newton cooling into media at constant temperature T_0

From (5.54)$_3$, (5.55) and (5.58) we may deduce

$$w(r,t) = \frac{\mu a^2 T_0 [r - \rho(t)]^2}{[a - \rho(t)][a(1 + \mu a) + \rho(t)(1 - \mu a)]},$$

and (5.57) yields

$$\frac{d}{dt}\left\{\frac{[a - \rho(t)]^2}{[a(1 + \mu a) + \rho(t)(1 - \mu a)]}\right\} = \frac{6\kappa}{[a(1 + \mu a) + \rho(t)(1 - \mu a)]},$$

from which we can deduce, in the usual way

$$t = \frac{1}{12\kappa}\left\{(\rho - a)[(1 - \mu a)\rho + (3 + \mu a)a] \right.$$

$$\left. - \frac{4a^2(2 + \mu^2 a^2)}{(1 - \mu a)^2}\log\frac{[(1 - \mu a)\rho + (1 + \mu a)a]}{2a}\right\}.$$

Finally in this section, to emphasize the non-unique nature of the heat-balance method, we observe that alternatively, for the sphere we could define a heat-balance integral $\Theta(t)$ by

$$\Theta(t) = \int_{\rho(t)}^{a} r^2 v(r,t)\,dr,$$

(that is, a volume integral of the temperature $v(r,t)$), so that on multiplying (5.40)$_1$ (with $m = 2$) by r^2 we obtain

$$\frac{d\Theta}{dt}(t) = \kappa a^2 \frac{\partial v}{\partial r}(a,t).$$

This heat-balance condition, in general, leads to different approximate temperature profiles (given the same functional form of the approximate solution) since in terms of $w(r,t)$ this condition becomes

$$\frac{d}{dt}\int_{\rho(t)}^{a} r w(r,t)\,dr = \kappa\left(a\frac{\partial w}{\partial r}(a,t) - w(a,t)\right),$$

which is clearly distinct from (5.57). However, we do not pursue the matter further here.

PROBLEMS

1. For the exact solution of problem (5.1)

$$v(x,t) = v_0 \operatorname{erfc}\left(\frac{x}{2\sqrt{\kappa t}}\right),$$

calculate the exact values of the temperature and temperature gradient at the penetration depth $x = \rho(t)$ arising from the quadratic temperature approximation, namely

$$\rho(t) = 2\sqrt{3\kappa t}.$$

Also show that for the exact temperature $v(x,t)$

$$\int_0^{\rho(t)} v(x,t)\,dx = 2v_0\sqrt{\frac{\kappa t}{\pi}}\left\{1 - e^{-3} + \sqrt{3\pi}\,\operatorname{erfc}(\sqrt{3})\right\},$$

$$\int_{\rho(t)}^{\infty} v(x,t)\,dx = 2v_0\sqrt{\frac{\kappa t}{\pi}}\left\{e^{-3} - \sqrt{3\pi}\,\operatorname{erfc}(\sqrt{3})\right\}.$$

Show that the ratio of the quantity of heat in $x > \rho(t)$ to that in $0 \le x \le \rho(t)$ is

$$\frac{e^{-3} - \sqrt{3\pi}\,\operatorname{erfc}(\sqrt{3})}{1 - e^{-3} + \sqrt{3\pi}\,\operatorname{erfc}(\sqrt{3})} \approx 5.56 \times 10^{-5},$$

so that the total quantity of heat in $x > \rho(t)$ is an extremely small fraction of that in $0 \le x \le \rho(t)$ and hence the concept of penetration depth emerges.

2. For the semi-infinite region $0 \le x < \infty$, initially at zero temperature and surface $x = 0$ maintained at constant temperature v_0, show that the quartic approximation to the temperature profile is

$$v(x,t) = v_0\left(1 + \frac{x}{\rho(t)}\right)\left(1 - \frac{x}{\rho(t)}\right)^3,$$

and that

$$\theta(t) = \frac{3v_0}{10}\rho(t), \qquad \frac{\partial v}{\partial x}(0,t) = \frac{-2v_0}{\rho(t)},$$

where the penetration depth $\rho(t)$ is given by

$$\rho(t) = 2\sqrt{\frac{10\kappa t}{3}}.$$

$\left[\text{Observe that in this case the } \sqrt{\pi} \text{ in the exact expression (5.2) is approximated by } \sqrt{10/3}.\right]$

3. For the semi-infinite region $0 \le x < \infty$, initially at zero temperature and surface $x = 0$ maintained at temperature $g(t)$, show that the quadratic approximation to the temperature profile is

$$v(x,t) = g(t)\left(1 - \frac{x}{\rho(t)}\right)^2,$$

and that

$$\theta(t) = \frac{g(t)}{3}\rho(t), \qquad \frac{\partial v}{\partial x}(0,t) = \frac{-2g(t)}{\rho(t)},$$

where the penetration depth $\rho(t)$ is given by

$$\rho(t) = \frac{2}{g(t)}\sqrt{3\kappa \int_0^t g(\tau)^2 \, d\tau}.$$

In particular if $g(t) = \alpha\sqrt{t}$ for constant α, show that

$$\rho(t) = \sqrt{6\kappa t},$$

and that the approximation correctly predicts a constant value for $\frac{\partial v}{\partial x}$ at $x = 0$, except that the constant $\sqrt{\pi}$ in the correct expression is replaced by $\sqrt{8/3}$. $\Big[$ Hint, see Problem 3.2.$\Big]$

4. For the semi-infinite region $0 \le x < \infty$ initially at constant temperature u_0 and surface $x = 0$ maintained at constant temperature v_0, assuming that

$$T(\rho(t),t) = u_0, \qquad \frac{\partial T}{\partial x}(\rho(t),t) = 0,$$

show that the quadratic approximation to the temperature is

$$T(x,t) = u_0 + (v_0 - u_0)\left(1 - \frac{x}{\rho(t)}\right)^2,$$

and that

$$\theta(t) = (2u_0 + v_0)\frac{\rho(t)}{3}, \qquad \frac{\partial T}{\partial x}(0,t) = \frac{2(u_0 - v_0)}{\rho(t)},$$

where the penetration depth $\rho(t)$ is given by

$$\rho(t) = 2\sqrt{3\left(\frac{v_0 - u_0}{v_0 + 2u_0}\right)\kappa t}.$$

$\Big[$ Note that this solution is only meaningful if $(v_0 - u_0)/(v_0 + 2u_0) > 0$. $\Big]$

5.   Consider the semi-infinite region $0 \leq x < \infty$ initially at temperature $f(x)$ and surface $x = 0$ maintained at temperature $g(t)$.

(i)   Assuming that

$$T(\rho(t), t) = f(\rho(t)), \quad \frac{\partial T}{\partial x}(\rho(t), t) = f'(\rho(t)),$$

show that the quadratic approximation to the temperature profile is

$$T(x, t) = f(\rho) + (x - \rho)f'(\rho) + [g(t) + \rho f'(\rho) - f(\rho)]\left(1 - \frac{x}{\rho}\right)^2.$$

(ii)   Show that

$$\frac{\partial^2 T}{\partial x^2}(0, t) = \frac{g'(t)}{\kappa}, \quad \frac{\partial^2 T}{\partial x^2}(\rho(t), t) = 0,$$

and hence show that the quartic approximation to the temperature is

$$T(x, t) = f(\rho) + (x - \rho)f'(\rho) + [g(t) + \rho f'(\rho) - f(\rho)]\left(1 + \frac{x}{\rho}\right)\left(1 - \frac{x}{\rho}\right)^3$$

$$- \frac{\rho g'(t)}{6\kappa}x\left(1 - \frac{x}{\rho}\right)^3.$$

$\left[\right.$Note, that primes here denote differentiation with respect to the indicated

argument.$\left.\right]$

6.   Consider problem (5.14) and let $v_n(x, t)$ be an nth degree polynomial approximation to the temperature

$$v_n(x, t) = a_0(t) + a_1(t)x + \cdots + a_n(t)x^n.$$

Instead of choosing boundary conditions involving the lowest order derivatives, choose
boundary conditions to make $v_n(x, t)$ as smooth as possible at $x = \rho(t)$ so that $v_n(x, t)$
satisfies the boundary condition

$$\frac{\partial v_n}{\partial x}(0, t) = \frac{-w_0}{k},$$

as well as the conditions

$$v_n(\rho(t), t) = \frac{\partial v_n}{\partial x}(\rho(t), t) = \cdots = \frac{\partial^{n-1} v_n}{\partial x^{n-1}}(\rho(t).t) = 0,$$

(i) Show that the resulting heat-balance approximation for problem (5.14) is given by

$$v_n(x,t) = \frac{w_0}{nk}\rho(t)\left(1 - \frac{x}{\rho(t)}\right)^n,$$

where the penetration depth $\rho(t)$ is given by

$$\rho(t) = \sqrt{n(n+1)\kappa t}.$$

(ii) In particular, observe that

$$v_n(0,t) = \frac{w_0}{k}\sqrt{\frac{(n+1)}{n}\kappa t},$$

and show that $v_n(0,t)$ is closest to the exact surface temperature (see Example 3.3) $(w_0/k)\sqrt{4\kappa t/\pi}$ when $n = 5$.

(iii) Investigate the effect of replacing the condition

$$\frac{\partial^{n-1}v_n}{\partial x^{n-1}}(\rho(t),t) = 0,$$

by the condition

$$\frac{\partial^3 v_n}{\partial x^3}(0,t) = 0,$$

for $n > 5$.

7. For the semi-infinite region $0 \le x < \infty$ initially at zero temperature and a flux $h(t)$ maintained across the surface $x = 0$, show that the quadratic approximation to the temperature profile is

$$v(x,t) = \frac{h(t)\rho(t)}{2k}\left(1 - \frac{x}{\rho(t)}\right)^2,$$

where the penetration depth $\rho(t)$ is given by

$$\rho(t) = \sqrt{\frac{6\kappa}{h(t)}\int_0^t h(\tau)\,d\tau}.$$

8. For the semi-infinite region $0 \le x < \infty$ initially at constant temperature u_0 and a constant flux w_0 maintained across $x = 0$, use the conditions

$$T(\rho(t),t) = u_0, \qquad \frac{\partial T}{\partial x}(\rho(t),t) = 0,$$

to show that the quadratic approximation to the temperature is

$$T(x,t) = u_0 + \frac{w_0\rho(t)}{2k}\left(1 - \frac{x}{\rho(t)}\right)^2,$$

and that the heat-balance integral is

$$\theta(t) = u_0\rho(t) + \frac{w_0\rho(t)^2}{6k},$$

where the penetration depth is given by

$$\rho(t) = \sqrt{6\kappa t}.$$

$$\left[\text{Hint, observe that here}\right.$$

$$\int_0^{\rho(t)} \frac{\partial T}{\partial t}(x,t)\,dx = \frac{d}{dt}\{\theta(t) - u_0\rho(t)\} = \frac{\kappa w_0}{k}. \quad \left.\right]$$

9. For the semi-infinite region $0 \le x < \infty$ initially at constant temperature u_0 and the surface $x = 0$ having Newton cooling into a medium at constant temperature T_0, use the conditions

$$T(\rho(t),t) = u_0, \quad \frac{\partial T}{\partial x}(\rho(t),t) = \frac{\partial^2 T}{\partial x^2}(\rho(t),t) = 0,$$

to show that the cubic approximation to the temperature is

$$T(x,t) = u_0 + \frac{\mu(T_0 - u_0)\rho(t)}{[3 + \mu\rho(t)]}\left(1 - \frac{x}{\rho(t)}\right)^3,$$

where the penetration depth $\rho(t)$ is obtained from the transcendental equation

$$(\mu\rho)^2 + 6\mu\rho - 18\log\left(1 + \frac{\mu\rho}{3}\right) = 24\mu^2\kappa t,$$

which is independent of the constant initial temperature u_0.

10. For the infinite solid circular cylinder initially at zero temperature and with its surface held at constant temperature v_0, show that on assuming the profile

$$v(r,t) = a(t)(r^2 - \rho^2) + \left[b(t) + c(t)r^2\right]\log\frac{r}{\rho},$$

the conditions

$$v(a,t) = v_0, \quad \frac{\partial v}{\partial t}(a,t) = \frac{\partial v}{\partial r}(\rho(t),t) = 0,$$

yield

$$a(t) = \left[\log \frac{\rho}{a} - 1\right]c(t), \quad b(t) = \rho^2\left[1 - 2\log \frac{\rho}{a}\right]c(t),$$

where $c(t)$ is given by

$$c(t) = \frac{v_0}{\left\{(\rho^2 - a^2) - 2\rho^2 \log \frac{\rho}{a} + 2\rho^2\left[\log \frac{\rho}{a}\right]^2\right\}},$$

and where ρ denotes the penetration radius $\rho(t)$. Further show that

$$\frac{d\theta}{dt} = \frac{\kappa v_0\left[(\rho^2 - a^2) - 2\rho^2 \log \frac{\rho}{a}\right]}{\left\{(\rho^2 - a^2) - 2\rho^2 \log \frac{\rho}{a} + 2\rho^2\left[\log \frac{\rho}{a}\right]^2\right\}},$$

where $\theta(t)$ is given by

$$\theta(t) = \frac{v_0\left\{(\rho^2 - a^2)(\rho^2 + 5a^2) - 4\rho^2(\rho^2 + 2a^2)\log \frac{\rho}{a} + 16a^2\rho^2\left[\log \frac{\rho}{a}\right]^2\right\}}{16\left\{(\rho^2 - a^2) - 2\rho^2 \log \frac{\rho}{a} + 2\rho^2\left[\log \frac{\rho}{a}\right]^2\right\}}.$$

Hence, deduce an expression for the penetration radius $\rho(t)$.

11. For the solid sphere initially at zero temperature and with its surface held at constant temperature v_0, show that on assuming the cubic profile

$$w(r,t) = a(t) + b(t)r + c(t)r^2 + d(t)r^3,$$

the conditions

$$w(a,t) = av_0, \quad w(\rho(t),t) = \frac{\partial w}{\partial r}(\rho(t),t) = \frac{\partial^2 w}{\partial r^2}(\rho(t),t) = 0,$$

yield

$$w(r,t) = av_0\left(\frac{r - \rho(t)}{a - \rho(t)}\right)^3,$$

where the penetration radius $\rho(t)$ is given by

$$\rho(t) = a - 2\sqrt{6\kappa t}.$$

12. For the solid sphere initially at zero temperature with constant flux w_0 at its surface, show that the alternative heat-balance condition

$$\frac{\mathrm{d}}{\mathrm{d}t} \int_{\rho(t)}^{a} r^2 v(r,t)\,\mathrm{d}r = \kappa a^2 \frac{\partial v}{\partial r}(a,t),$$

together with the assumed profile

$$v(r,t) = \frac{1}{r}\left\{a(t) + b(t)r + c(t)r^2\right\},$$

yields

$$v(r,t) = \frac{a^2 w_0 [r - \rho(t)]^2}{kr\left[a^2 - \rho(t)^2\right]},$$

where the penetration radius $\rho(t)$ is determined from

$$t = \frac{(a - \rho)^2(3a + \rho)}{12\kappa(a + \rho)}.$$

Chapter Six
Numerical solutions

6.1 Introduction

Numerical solution methods for the heat equation are necessary for most geometries and boundary conditions encountered in practice since simple exact solutions cannot always be obtained in terms of standard analytical functions. Even if an exact solution can be obtained, this still has to be evaluated numerically and this process may be almost as arduous as the full numerical solution of the problem. Further, for many problems numerical solutions are frequently more convenient, efficient and accurate than approximate analytical solutions. This is not to say that numerical solutions are always preferable to exact or approximate analytical solutions, for the latter usually offer greater insights into the solution, its overall behaviour and the parameter dependence of the problem. Nevertheless, for the purposes of obtaining numerical results for a given problem a numerical solution procedure is often preferable. In this chapter we examine three numerical methods for the heat equation, namely *finite difference* techniques, *finite element* techniques and the *boundary integral* method. We shall only consider the solution of one dimensional problems, the extension of these methods to two and three dimensions being a problem of scale rather than kind. Sample FORTRAN subroutines, used in the examples, are given in the final section of this chapter.

Finite difference methods are particularly well suited to one dimensional problems, or higher dimensional problems involving geometries with linear boundaries, such as rectangles or 'L' shapes. In addition finite difference methods are conceptually simple, arising as they do from a simple space-time discretization of the governing partial differential equation and its boundary conditions. Finite difference schemes, however, lose much of their attractiveness when applied to problems in two and three dimensions involving irregular geometries. For such situations finite element or boundary integral techniques are more effective. Finite element methods are based on a weak or variational formulation of the heat conduction problem, while boundary integral methods are based on the use of Green's functions as discussed in Section 2.6. We note that in one dimension many of the advantages of these latter methods over the finite difference method are obscured and therefore we urge the reader not to form the opinion that these methods introduce unnecessary complications. Unfortunately in this book it is not possible to treat a general two dimensional problem by all three methods and compare the results.

In numerical work it is convenient to introduce at the outset the non-dimensional variables (so named because they have no physical dimensions) $\bar{x}$, $\bar{t}$ and $\overline{T}$ defined by

$$\bar{x} = x/\ell, \quad \bar{t} = \kappa t/\ell^2, \quad \overline{T}(\bar{x},\bar{t}) = T(x,t)/T_0, \tag{6.1}$$

where x, t, T and κ denote physical (dimensional) length, time, temperature and thermal conductivity and where ℓ and T_0 denote a characteristic length (such as the width of a finite block) and a characteristic temperature (such as the fixed or maximum temperature at a surface). In terms of the non-dimensional variables (6.1) we have

$$\frac{\partial T}{\partial t} = \frac{\kappa T_0}{\ell^2}\frac{\partial \overline{T}}{\partial \bar{t}}, \quad \frac{\partial T}{\partial x} = \frac{T_0}{\ell}\frac{\partial \overline{T}}{\partial \bar{x}}, \quad \frac{\partial^2 T}{\partial x^2} = \frac{T_0}{\ell^2}\frac{\partial^2 \overline{T}}{\partial \bar{x}^2}. \tag{6.2}$$

Thus, the heat equation (posed on the interval $[0, \ell]$) becomes

$$\frac{\partial \overline{T}}{\partial \bar{t}} = \frac{\partial^2 \overline{T}}{\partial \bar{x}^2}, \quad 0 < \bar{x} < 1. \tag{6.3}$$

Similarly we use (6.1) and (6.2) to transform the dimensional boundary and initial conditions to their non-dimensional forms (see Problem 1). This does not change the functional form of either the boundary or initial conditions.

It is of course possible to apply similar non-dimensionalizations to problems involving two and three spatial dimensions and to employ non-dimensional variables in analytic work. We have chosen not to adopt the latter course in previous chapters because the use of non-dimensional variables tends to obscure the parameter dependence of exact and approximate analytical solutions. Since numerical solutions do not make such dependencies readily apparent anyway, there is scant justification for using dimensional variables, as this simply adds two superfluous parameters to the problem. In particular, entire families of dimensional solutions for different physical parameter values can be obtained from a single non-dimensional solution by rescaling (that is, by inverting (6.1)), which is clearly computationally more efficient than calculating each individually. Thus we consider the numerical solution of (6.3), subject to suitable non-dimensionalized boundary conditions.

To discuss the accuracy, convergence and general behaviour of numerical solutions we need to introduce the *order notation*. If $f(x)$ and $g(x)$ are two functions such that

$$\lim_{x \to a} \frac{f(x)}{g(x)} = A,$$

where A is some finite real number, we say that $f(x)$ has the same order of magnitude as $g(x)$ as $x \to a$ and we write

$$f(x) = O(g(x)) \text{ as } x \to a.$$

This concept is particularly useful when discussing the nature of the singularities and zeros of a function. For example we have

$$x + \log x = O(\log x) \text{ as } x \to 0,$$

since

$$\lim_{x \to 0} \frac{x + \log x}{\log x} = 1,$$

but

$$x + \log x = O(x) \text{ as } x \to \infty,$$

because

$$\lim_{x \to \infty} \frac{x + \log x}{x} = 1.$$

These two results show that for small x, $x + \log x$ behaves as $\log x$, while for large x it behaves like x. In these two examples the order relationship is symmetric, that is we also have

$$\log x = O(x + \log x) \text{ as } x \to 0, \quad \text{and} \quad x = O(x + \log x) \text{ as } x \to \infty.$$

However, in general the order relationship is not symmetric since for example we have

$$x^2 = O(x) \text{ as } x \to 0, \quad \text{but} \quad x \neq O(x^2) \text{ as } x \to 0,$$

because $\lim_{x \to 0} x^2/x = 0$ but $\lim_{x \to 0} x/x^2$ does not exist. Normally and throughout this book the order notation is used to indicate the essential behaviour of a function near one of its singularities or zeros.

Frequently in our work we have some small quantity such as Δx or Δt and we will say that an approximation is accurate to order $O(\Delta x)$ or $O((\Delta t)^2)$ or some similar expression, and by this we mean that the difference between the exact and approximate solution tends to zero at least as rapidly as Δx or $(\Delta t)^2$ respectively. For example if $f(x)$ is a continuously differentiable function then the approximation

$$f(x + \Delta x) \approx f(x),$$

is accurate to order $O(\Delta x)$, since by Taylor's theorem we have

$$f(x + \Delta x) = f(x) + \Delta x \frac{\mathrm{d}f}{\mathrm{d}x}(x + \theta \Delta x),$$

where $0 \le \theta \le 1$, and therefore

$$\lim_{\Delta x \to 0} \frac{f(x + \Delta x) - f(x)}{\Delta x} = \frac{\mathrm{d}f}{\mathrm{d}x}(x).$$

Accordingly we write this approximation in the form

$$f(x + \Delta x) = f(x) + O(\Delta x).$$

If $f(x)$ is twice continuously differentiable, then similarly

$$f(x + \Delta x) = f(x) + \Delta x \frac{\mathrm{d}f}{\mathrm{d}x}(x) + O\big((\Delta x)^2\big).$$

Although an approximation which is accurate to order $O\big((\Delta x)^2\big)$ will be more accurate than an approximation to order $O(\Delta x)$ for all sufficiently small Δx, this says nothing about how small Δx must become before the higher order approximation becomes more accurate than the lower order.

6.2 Explicit finite difference methods

Finite difference procedures are based on the idea of representing the problem in terms of discretized (non-dimensional) temperature. The interval $[0, 1]$ on which the problem is posed is partitioned into N subintervals. For our purposes these subintervals are of equal length $\Delta \bar{x} = 1/N$. A time step $\Delta \bar{t}$ is chosen and we then deal only with the temperature at exact positive integer multiples of $\Delta \bar{x}$ and $\Delta \bar{t}$, that is at $\bar{x} = i\Delta \bar{x}$ and $\bar{t} = j\Delta \bar{t}$ for $0 \le i \le N$ and $j \ge 0$, and we denote the exact temperature there by

$\overline{T}_{ij}$, that is (see Figure 6.1)

$$\overline{T}_{ij} = \overline{T}(i\Delta\overline{x}, j\Delta\overline{t}), \quad 0 \le i \le N, \; j \ge 0.$$

The central feature of finite difference methods is the approximation of the partial derivatives in the governing heat equation (6.3) and boundary conditions by differences of nodal temperature values $\overline{T}_{ij}$. We do this (assuming sufficient differentiability) by employing Taylor series expansions from which we may deduce

$$\left. \begin{aligned} \frac{\partial \overline{T}}{\partial \overline{t}}(i\Delta\overline{x}, j\Delta\overline{t}) &= \frac{1}{\Delta\overline{t}}\left(\overline{T}_{ij+1} - \overline{T}_{ij}\right) - \frac{\partial^2 \overline{T}}{\partial \overline{t}^2}\left(i\Delta\overline{x}, (j + \xi_{ij})\Delta\overline{t}\right)\frac{\Delta\overline{t}}{2}, \\[2ex] \frac{\partial \overline{T}}{\partial \overline{x}}(i\Delta\overline{x}, j\Delta\overline{t}) &= \frac{1}{2\Delta\overline{x}}\left(\overline{T}_{i+1j} - \overline{T}_{i-1j}\right) - \frac{\partial^3 \overline{T}}{\partial \overline{x}^3}\left((i + \eta_{ij})\Delta\overline{x}, j\Delta\overline{t}\right)\frac{(\Delta\overline{x})^2}{6}, \\[2ex] \frac{\partial^2 \overline{T}}{\partial \overline{x}^2}(i\Delta\overline{x}, j\Delta\overline{t}) &= \frac{1}{(\Delta\overline{x})^2}\left(\overline{T}_{i+1j} + \overline{T}_{i-1j} - 2\overline{T}_{ij}\right) - \frac{\partial^4 \overline{T}}{\partial \overline{x}^4}\left((i + \zeta_{ij})\Delta\overline{x}, j\Delta\overline{t}\right)\frac{(\Delta\overline{x})^2}{12}, \end{aligned} \right\}$$

$$(6.4)$$

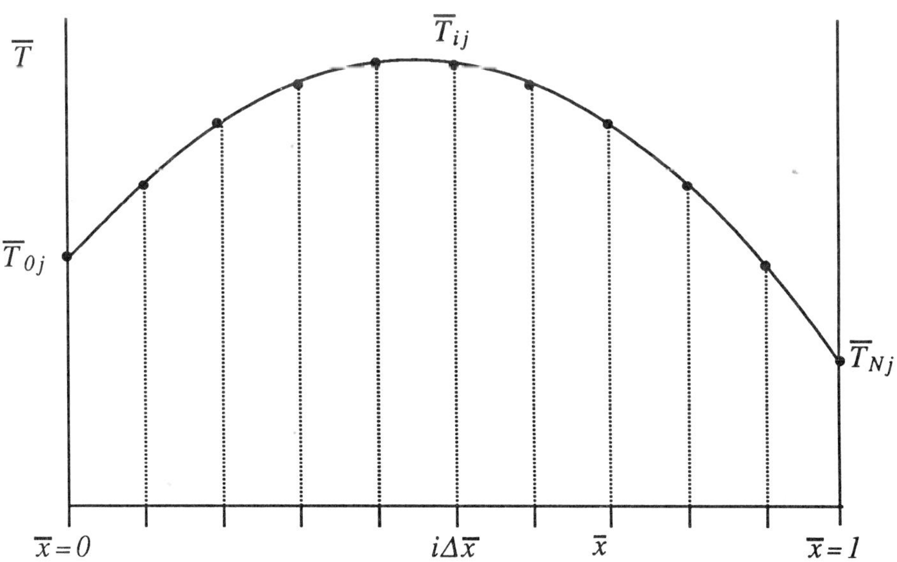

Figure 6.1. Spatial discretization of the temperature $\overline{T}$ on the interval $[0, 1]$.

where ξ_{ij}, η_{ij} and ζ_{ij} are constants such that $0 \le \xi_{ij}, \eta_{ij} \le 1$, $-1 \le \zeta_{ij} \le 1$. If we neglect terms of order $O(\Delta \bar{t})$ and order $O\left((\Delta \bar{x})^2\right)$ in these expressions we obtain the finite difference approximations

$$
\left.
\begin{aligned}
\frac{\partial \overline{T}}{\partial \bar{t}}(i\Delta\bar{x}, j\Delta\bar{t}) &\approx \frac{1}{\Delta\bar{t}}\left(\overline{T}_{ij+1} - \overline{T}_{ij}\right), \\[2mm]
\frac{\partial \overline{T}}{\partial \bar{x}}(i\Delta\bar{x}, j\Delta\bar{t}) &\approx \frac{1}{2\Delta\bar{x}}\left(\overline{T}_{i+1j} - \overline{T}_{i-1j}\right), \\[2mm]
\frac{\partial^2 \overline{T}}{\partial \bar{x}^2}(i\Delta\bar{x}, j\Delta\bar{t}) &\approx \frac{1}{(\Delta\bar{x})^2}\left(\overline{T}_{i+1j} + \overline{T}_{i-1j} - 2\overline{T}_{ij}\right).
\end{aligned}
\right\}
\tag{6.5}
$$

Although the *central difference* approximation $(6.5)_2$ has a higher order of accuracy than the *forward difference* approximation $(6.5)_1$, it is inappropriate to use a central difference approximation of the form $(6.5)_2$ for the time partial derivative since this choice leads to an unstable scheme. We use the approximations (6.5) to replace the heat equation by a system of difference equations which can be solved using a computer.

By way of example we first consider the explicit finite difference solution of problem (4.6), which in non-dimensional form becomes

$$
\left.
\begin{aligned}
\frac{\partial \overline{T}}{\partial \bar{t}} &= \frac{\partial^2 \overline{T}}{\partial \bar{x}^2}, \quad 0 < \bar{x} < 1, \\[2mm]
\overline{T}(\bar{x}, 0) &= f(\bar{x}), \quad 0 \le \bar{x} \le 1, \\[2mm]
\overline{T}(0, \bar{t}) &= g_1(\bar{t}), \quad \overline{T}(1, \bar{t}) = g_2(\bar{t}).
\end{aligned}
\right\}
\tag{6.6}
$$

Using the forward difference approximation $(6.5)_1$ for $\frac{\partial \overline{T}}{\partial \bar{t}}$ and $(6.5)_3$ for $\frac{\partial^2 \overline{T}}{\partial \bar{x}^2}$ we find that, to order $O\left((\Delta\bar{x})^2 + \Delta\bar{t}\right)$, the heat equation becomes

$$
\frac{1}{\Delta\bar{t}}\left(\tilde{T}_{ij+1} - \tilde{T}_{ij}\right) = \frac{1}{(\Delta\bar{x})^2}\left(\tilde{T}_{i+1j} + \tilde{T}_{i-1j} - 2\tilde{T}_{ij}\right), \quad 1 \le i \le N-1, \quad j \ge 0,
$$

which after rearrangement gives the difference equations

$$
\tilde{T}_{ij+1} = \tilde{T}_{ij} + \delta\left(\tilde{T}_{i+1j} + \tilde{T}_{i-1j} - 2\tilde{T}_{ij}\right), \quad 1 \le i \le N-1, \quad j \ge 0, \tag{6.7}
$$

where

$$
\delta = \frac{\Delta\bar{t}}{(\Delta\bar{x})^2}, \tag{6.8}
$$

and we have used the notation $\tilde{T}_{ij}$ to distinguish the solution of the finite difference equation (6.7) from the exact value of the temperature $\overline{T}_{ij}$. We remind the reader of the probabalistic interpretation of (6.7) as outlined in Section 1.4.

A numerical scheme for solving the heat equation based on (6.7) is known as an *explicit* method, since we have explicit formulae for $\tilde{T}_{ij+1}$ in terms of $\tilde{T}_{i+1j}$, $\tilde{T}_{ij}$ and $\tilde{T}_{i-1j}$ (see Figure 6.2). We discretize the initial condition $(6.6)_2$ to find the initial approximation $\tilde{T}_{i0}$, thus

$$\tilde{T}_{i0} = f(i\Delta\overline{x}), \quad 0 \le i \le N, \tag{6.9}$$

and the boundary conditions $(6.6)_3$ are also discretized to provide expressions for $\tilde{T}_{0j}$ and $\tilde{T}_{Nj}$

$$\tilde{T}_{0j} = g_1(j\Delta\overline{t}), \quad \tilde{T}_{Nj} = g_2(j\Delta\overline{t}), \quad j \ge 1. \tag{6.10}$$

Once $\tilde{T}_{ij}$ is known for $0 \le i \le N$, we find $\tilde{T}_{ij+1}$ for $1 \le i \le N - 1$ using (6.7) and for $i = 0$ and $i = N$ using (6.10). Since we know $\tilde{T}_{i0}$ for $0 \le i \le N$ from (6.9), we can find any $\tilde{T}_{ij}$ by applying (6.7) and (6.10) j times.

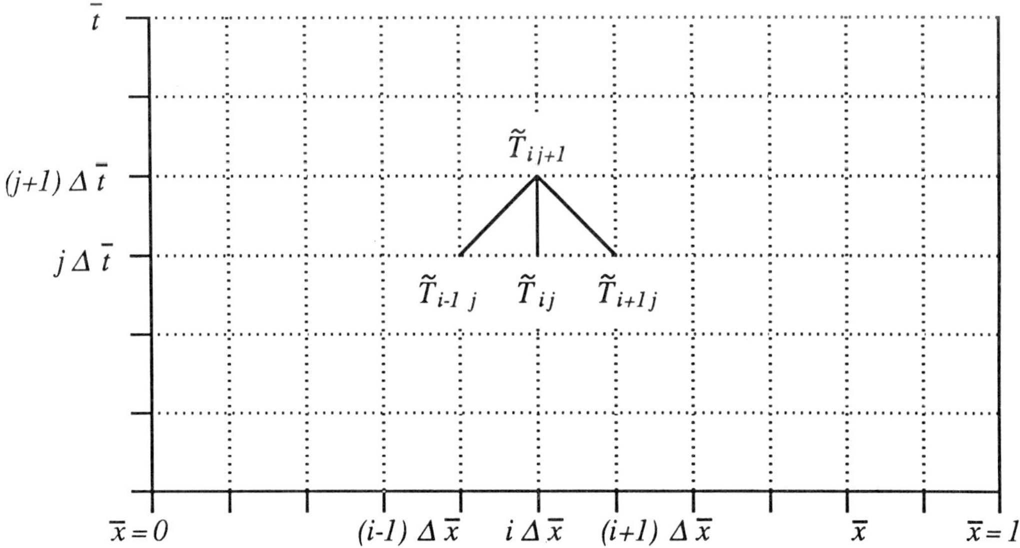

Figure 6.2. Explicit finite difference spatial and temporal discretization.

<u>Example 6.1</u> For the finite region $0 \le \bar{x} \le 1$ with constant initial temperature $\bar{u}_0 = 1$ and prescribed zero temperature at $\bar{x} = 0$ and $\bar{x} = 1$ describe an algorithm to solve the non-dimensional heat equation using an explicit finite difference algorithm.

To find the approximate temperature at time $\bar{t} + \Delta\bar{t}$ we only require the approximate temperature at time $\bar{t}$, so we can implement the algorithm using two arrays or vectors to hold the approximate nodal temperature values, one to contain the current temperature and one to receive the updated temperature. Our basic algorithm consists of setting the temperature initially to 1 and then repeatedly applying (6.7) and (6.10) with $g_1(t) = g_2(t) = 0$, until we have stepped through to the desired time. A sample FORTRAN subroutine which performs this process is given in Figure 6.7 and discussed in Section 6.6.

In Table 6.1 we compare numerical solutions of the problem, with various values of N and δ (and therefore of $\Delta\bar{x}$ and $\Delta\bar{t}$) with the exact solution (see Example 4.1 with $\ell = \kappa = 1$) at a fixed time $\bar{t} = 0.1$. Note that since the solutions are symmetric about $\bar{x} = 0.5$ values are only listed for $0 \le \bar{x} \le 0.5$. For values of $\delta \le 0.5$ there is good agreement between the exact and numerical solutions, but for $\delta = 1.0$ the numerical solution is clearly meaningless. To explain this phenomena we must examine the *convergence* and *stability* of the explicit finite difference scheme (6.7). The question of convergence is basically the following. Does the solution of the finite difference scheme (6.7), $\tilde{T}_{ij}$, approach the exact value of the temperature $\overline{T}_{ij}$ as $\Delta\bar{x}$ and $\Delta\bar{t}$ approach zero? This question deals with behaviour of the difference between the exact temperature $\overline{T}_{ij}$ and the solution $\tilde{T}_{ij}$ of the finite difference scheme (6.7). The stability problem arises

		$\delta = 0.25$		$\delta = 0.5$		$\delta = 1.0$	
$\bar{x}$	$\overline{T}(\bar{x}, \bar{t})$	$N = 10$	$N = 20$	$N = 10$	$N = 20$	$N = 10$	$N = 20$
0.0	0.0000	0.0000	0.0000	0.0000	0.0000	0.0000	0.000E00
0.1	0.1467	0.1485	0.1471	0.1466	0.1485	-233.00	1.133E17
0.2	0.2790	0.2825	0.2799	0.2933	0.2825	378.00	1.184E17
0.3	0.3839	0.3887	0.3851	0.3838	0.3887	-408.00	2.026E17
0.4	0.4513	0.4569	0.4527	0.4745	0.4569	375.00	1.898E17
0.5	0.4745	0.4804	0.4760	0.4745	0.4804	-351.00	1.803E17

Table 6.1. Comparison of the exact solution with explicit finite difference solutions of Example 6.1 for three values of δ and two values of N at time $\bar{t} = 0.1$.

because in practice we use finite precision computer arithmetic to solve the difference equation (6.7), which introduces various errors into the *numerical* solution of (6.7). The system (6.7) is said to be *stable* if these errors are not magnified to the point where the numerical solution of (6.7) becomes meaningless.

To answer the convergence question we start by defining the local truncation error E_{ij} as the difference between the exact temperature $\overline{T}_{ij}$ and the finite difference approximation $\tilde{T}_{ij}$ (assuming perfectly accurate calculations are used to obtain $\tilde{T}_{ij}$)

$$E_{ij} = \overline{T}_{ij} - \tilde{T}_{ij}. \tag{6.11}$$

We assume that we can evaluate the prescribed boundary temperature perfectly accurately so that $E_{0j} = E_{Nj} = 0$. Using the fact that the exact solution satisfies the heat equation, we find from (6.4) and (6.7)

$$E_{ij+1} = (1 - 2\delta)E_{ij} + \delta\left(E_{i+1\,j} + E_{i-1\,j}\right)$$

$$+ \Delta \overline{t}\left[\frac{\partial^2 \overline{T}}{\partial \overline{t}^2}\left(i\Delta\overline{x},(j+\xi_{ij})\Delta\overline{t}\right)\frac{\Delta\overline{t}}{2} - \frac{\partial^4 \overline{T}}{\partial \overline{x}^4}\left((i+\zeta_{ij})\Delta\overline{x},j\Delta\overline{t}\right)\frac{(\Delta\overline{x})^2}{12}\right], \quad 1 \le i \le N-1,$$

$$\tag{6.12}$$

where $0 \le \xi_{ij} \le 1$ and $-1 \le \zeta_{ij} \le 1$. If $\overline{T}(\overline{x},\overline{t})$ is sufficiently differentiable for $0 \le \overline{t} \le t_0$, where t_0 is some fixed finite time, we can find constants M_1 and M_2 such that

$$\left|\frac{1}{2}\frac{\partial^2 \overline{T}}{\partial \overline{t}^2}(\overline{x},\overline{t})\right| \le M_1, \qquad \left|\frac{1}{12}\frac{\partial^4 \overline{T}}{\partial \overline{x}^4}(\overline{x},\overline{t})\right| \le M_2, \quad 0 \le \overline{x} \le 1, \quad 0 \le \overline{t} \le t_0.$$

Thus from (6.12) we have

$$|E_{ij+1}| \le |(1-2\delta)E_{ij}| + |\delta E_{i+1\,j}| + |\delta E_{i-1\,j}| + \Delta\overline{t}\left(M_1\Delta\overline{t} + M_2(\Delta\overline{x})^2\right),$$

and taking $\hat{E}_j = \sup_{0<i<N}|E_{ij}|$ we have

$$\hat{E}_{j+1} \le \hat{E}_j + \Delta\overline{t}\left(M_1\Delta\overline{t} + M_2(\Delta\overline{x})^2\right), \tag{6.13}$$

provided $\delta > 0$ and $(1-2\delta) \ge 0$, that is if $0 < \delta \le 0.5$. Arguing inductively on (6.13) we have

$$\hat{E}_j \le \hat{E}_0 + j\Delta\overline{t}\left[M_1\Delta\overline{t} + M_2(\Delta\overline{x})^2\right], \quad j \ge 0,$$

and assuming that we evaluate (6.9) perfectly accurately, so that $\hat{E}_0 = 0$, it is clear that as $\Delta\overline{x} \to 0$ and $\Delta\overline{t} \to 0$ the maximum local truncation error tends to zero and the finite difference solution converges to the actual solution, provided the ratio δ remains positive but less than or equal to 0.5.

To treat the stability question we introduce an $(N - 1)$ dimensional vector $\underset{\sim}{T}^j$ of internal (unknown) numerical temperature values for time $j\Delta t$ and an $(N - 1)$ vector $\underset{\sim}{U}^j$ containing prescribed boundary temperature values at time $j\Delta t$

$$\underset{\sim}{T}^j = \begin{pmatrix} \tilde{T}_{1j} \\ \tilde{T}_{2j} \\ \vdots \\ \tilde{T}_{N-2j} \\ \tilde{T}_{N-1j} \end{pmatrix}, \quad \underset{\sim}{U}^j = \begin{pmatrix} \tilde{T}_{0j} \\ 0 \\ \vdots \\ 0 \\ \tilde{T}_{Nj} \end{pmatrix}, \tag{6.14}$$

so that we can rewrite the explicit finite difference method (6.7) in the matrix form

$$\underset{\sim}{T}^{j+1} = \mathbf{A}\underset{\sim}{T}^j + \delta\,\underset{\sim}{U}^j, \tag{6.15}$$

where $\mathbf{A}$ is an $(N - 1) \times (N - 1)$ symmetric tridiagonal matrix given by

$$\mathbf{A} = \begin{pmatrix} 1-2\delta & \delta & 0 & \cdots & & 0 \\ \delta & 1-2\delta & \delta & & & 0 \\ 0 & \delta & 1-2\delta & & & 0 \\ \vdots & & & & & \vdots \\ 0 & & & & 1-2\delta & \delta \\ 0 & \cdots & & 0 & \delta & 1-2\delta \end{pmatrix}. \tag{6.16}$$

In order that (6.15) be stable, we require that any error present in $\underset{\sim}{T}^j$ not be amplified in calculating $\underset{\sim}{T}^{j+1}$. To be specific, let $\underset{\sim}{E}^j$ denote the $(N - 1)$ dimensional vector of errors present at time step j. Then we have

$$\mathbf{A}(\underset{\sim}{T}^j + \underset{\sim}{E}^j) = \mathbf{A}\underset{\sim}{T}^j + \mathbf{A}\underset{\sim}{E}^j,$$

so that the error propagated to $\underset{\sim}{T}^{j+1}$ from the previous time step is $\mathbf{A}\underset{\sim}{E}^j$. If this is not to exceed the error $\underset{\sim}{E}^j$ then we require

$$|\mathbf{A}\underset{\sim}{E}^j| \le |\underset{\sim}{E}^j|, \tag{6.17}$$

where, as usual, we take the norm of a vector to be the Euclidean norm, namely $|\underset{\sim}{x}| = \sqrt{\Sigma_i x_i^2}$. It is well known that if (6.17) is to hold for all possible error vectors, the spectral radius of $\mathbf{A}$ must be less than equal to 1, that is

$$\rho(\mathbf{A}) \le 1, \tag{6.18}$$

where the spectral radius of a matrix is the largest of the absolute values of its eigenvalues. It can be shown (see Problem 2) that the eigenvalues of $\mathbf{A}$, as defined by (6.16), are

$$\lambda_i = 1 - 4\delta \sin^2\left(\frac{i\pi}{2N}\right), \quad 1 \le i \le N - 1.$$

Thus in order to satisfy (6.18), independent of N, we must take $0 \le \delta \le 0.5$.

Both the convergence and stability analyses show that the explicit finite difference method is effective when $0 < \delta \le 0.5$. This places very severe restrictions on the size of the time steps. For example, taking a space step of $\Delta \bar{x} = 0.1$ ($N = 10$) restricts the time step to $\Delta \bar{t} \le 0.005$, so that 2000 iterations are necessary to find the temperature at $\bar{t} = 1$. For smaller space steps the situation is worse, for the time step must vary as the square of the space step to maintain convergence and stability. This can lead to extensive computation in order to obtain temperature values even for quite moderate times. We address this problem in the following section.

We now consider the numerical implementation of boundary conditions involving temperature gradients. Specifically, let us suppose that a temperature gradient appears in the boundary condition at $\bar{x} = 0$. We assume that the heat equation holds at the point $\bar{x} = 0$, so that in explicit finite difference form we have

$$\tilde{T}_{0j+1} = \tilde{T}_{0j} + \delta\left(\tilde{T}_{1j} + \tilde{T}_{-1j} - 2\tilde{T}_{0j}\right). \tag{6.19}$$

Here, $\tilde{T}_{-1j}$ is a fictitious temperature introduced at the point $-\Delta \bar{x}$, which is outside the region of interest. However, the same fictitious temperature also occurs in the central difference approximation for the temperature gradient at $\bar{x} = 0$, namely

$$\frac{\partial \tilde{T}}{\partial \bar{x}}(0, \bar{t}) \approx \frac{1}{2\Delta \bar{x}}\left(\tilde{T}_{1j} - \tilde{T}_{-1j}\right), \tag{6.20}$$

so that we can incorporate this approximation into the boundary condition at $\bar{x} = 0$ and then solve for $\tilde{T}_{-1j}$ in terms of $\tilde{T}_{0j}$, $\tilde{T}_{1j}$ and other known quantities. This allows us to eliminate $\tilde{T}_{-1j}$ from (6.19) completely. This method is generally preferable to the use of forward or backward difference approximations for the temperature gradient, as it has a higher order of accuracy.

For example, for a prescribed temperature gradient

$$\frac{\partial \overline{T}}{\partial \overline{x}}(0, \overline{t}) = h_1(\overline{t}),$$

at $\overline{x} = 0$ we have the finite difference approximation

$$\frac{1}{2\Delta \overline{x}}\left(\tilde{T}_{1j} - \tilde{T}_{-1j}\right) = h_1(j\Delta \overline{t}),$$

and on solving for $\tilde{T}_{-1j}$ and substituting into (6.19) we have

$$\tilde{T}_{0j+1} = \tilde{T}_{0j} + 2\delta\left(\tilde{T}_{1j} - \tilde{T}_{0j} - \Delta \overline{x} h_1(j\Delta \overline{t})\right).$$

Similarly, for a non-dimensional Newton cooling condition imposed at $\overline{x} = 1$

$$\frac{\partial \overline{T}}{\partial \overline{x}}(1, \overline{t}) + \overline{T}(1, \overline{t}) = \overline{T}_2(\overline{t}),$$

we assume the heat equation holds at $\overline{x} = 1$ so that

$$\tilde{T}_{Nj+1} = \tilde{T}_{Nj} + \delta\left(\tilde{T}_{N+1j} + \tilde{T}_{N-1j} - 2\tilde{T}_{Nj}\right). \tag{6.21}$$

The boundary condition in this case becomes

$$\frac{1}{2\Delta \overline{x}}\left(\tilde{T}_{N+1j} - \tilde{T}_{N-1j}\right) + \tilde{T}_{Nj} = \overline{T}_2(j\Delta \overline{t}),$$

and on solving for the fictitious temperature $\tilde{T}_{N+1j}$ we reduce (6.21) to

$$\tilde{T}_{Nj+1} = \tilde{T}_{Nj} + 2\delta\left(T_{N-1j} - (1 + \Delta \overline{x})\tilde{T}_{Nj} + \Delta \overline{x}\overline{T}_2(j\Delta \overline{t})\right).$$

<u>Example 6.2</u> For the finite region $0 \leq \overline{x} \leq 1$ with initial temperature $\overline{u}_0 = 1$, Newton heat loss into a medium at zero temperature at $\overline{x} = 0$

$$-\frac{\partial \overline{T}}{\partial \overline{x}}(0, \overline{t}) + \overline{T}(0, \overline{t}) = 0,$$

and at $\overline{x} = 1$ genuine non-linear radiation into a medium at zero temperature

$$\frac{\partial \overline{T}}{\partial \overline{x}}(1, \overline{t}) + \left(\overline{T}(1, \overline{t})\right)^4 = 0,$$

devise an explicit finite difference scheme to solve the non-dimensional heat equation.

On discretizing these boundary conditions, using central difference approximations for the temperature gradients, and substituting into the discretized heat equation at $\bar{x} = 0$ and $\bar{x} = 1$ we have

$$\left. \begin{aligned} \tilde{T}_{0j+1} &= \tilde{T}_{0j} + 2\delta\left(\tilde{T}_{1j} - (1 + \Delta\bar{x})\tilde{T}_{0j}\right), \\ \tilde{T}_{Nj+1} &= \tilde{T}_{Nj} + 2\delta\left(\tilde{T}_{N-1j} - \tilde{T}_{Nj} - \Delta\bar{x}\tilde{T}_{Nj}^4\right). \end{aligned} \right\} \tag{6.22}$$

Since (6.7) remains valid for $1 \le i \le N - 1$, we have only to modify the scheme described in Example 6.1 to incorporate the expressions (6.22) for the approximate temperatures $\tilde{T}_{0j+1}$ and $\tilde{T}_{Nj+1}$ at $\bar{x} = 0$ and $\bar{x} = 1$ respectively (see Section 6.6). In Table 6.2 we present values of the temperature calculated using this scheme. We note that, because of the non-linear radiation boundary condition at $\bar{x} = 1$, there is no known exact solution to this problem.

The convergence and stability analyses of explicit finite difference schemes incorporating linear derivative boundary conditions are essentially the same as those already

	$\delta = 0.25$			$\delta = 0.5$		
	$N = 20$	$N = 40$	$N = 80$	$N = 20$	$N = 40$	$N = 80$
			$\bar{t} = 0.1$			
$\bar{x} = 0.0$	0.7189	0.7188	0.7190	0.7212	0.7198	0.7191
0.2	0.8372	0.8372	0.8372	0.8392	0.8377	0.8373
0.4	0.9022	0.9022	0.9021	0.9043	0.9027	0.9023
0.6	0.9155	0.9154	0.9154	0.9183	0.9162	0.9156
0.8	0.8826	0.8826	0.8826	0.8870	0.8838	0.8829
1.0	0.8115	0.8116	0.8117	0.8184	0.8134	0.8121
			$\bar{t} = 0.2$			
$\bar{x} = 0.0$	0.6162	0.6163	0.6163	0.6200	0.6173	0.6166
0.2	0.7209	0.7209	0.7209	0.7242	0.7218	0.7212
0.4	0.7854	0.7854	0.7855	0.7891	0.7863	0.7857
0.6	0.8088	0.8088	0.8088	0.8135	0.8100	0.8091
0.8	0.7936	0.7937	0.7937	0.8003	0.7954	0.7951
1.0	0.7457	0.7458	0.7458	0.7550	0.7481	0.7464

Table 6.2. Comparison of the explicit finite difference solutions of Example 6.2 for two values of δ and three values of N at times $\bar{t} = 0.1$ and $\bar{t} = 0.2$.

given for fixed temperature boundary conditions. However, the end points at which derivative boundary conditions apply must now be included. For the stability analysis this leads to $N \times N$ or $(N + 1) \times (N + 1)$ asymmetric tridiagonal matrices, for which it is difficult to obtain exact analytic expressions for the eigenvalues (see Problem 3) although it is possible to bound the eigenvalues using standard techniques, such as the Gershgorin circle theorem for for those tridiagonal matrices for which eigenvalues cannot be obtained from simple analytic formulae. The eigenvalues can be significantly perturbed from the eigenvalues of (6.16) by the introduction of derivative boundary conditions and under certain circumstances it is necessary to reduce δ below 0.5 to obtain stability. It is usually obvious from the temperature values whether or not the scheme has become unstable. The convergence and stability analyses of explicit schemes incorporating non-linear boundary conditions, such as the radiation boundary condition, are much more complex and we do not consider such matters at all here.

6.3 Implicit finite difference methods

Implicit finite difference schemes are used to overcome the stability limitations of explicit finite difference methods. We consider two such methods, first the fully implicit scheme and then the Crank-Nicolson procedure. The fully implicit method is similar to the explicit finite difference scheme except that the time partial derivative is approximated by a backward difference approximation (6.23) instead of a forward difference approximation. Thus on using

$$\frac{\partial \overline{T}}{\partial \overline{t}}(i\Delta\overline{x}, j\Delta\overline{t}) \approx \frac{1}{\Delta\overline{t}}\left(\overline{T}_{ij} - \overline{T}_{ij-1}\right) + O\left(\Delta\overline{t}\right), \quad j \geq 1, \tag{6.23}$$

we find to order $O\left((\Delta\overline{x})^2 + \Delta\overline{t}\right)$ that the heat equation is approximated by the difference scheme

$$\frac{1}{\Delta\overline{t}}\left(\tilde{T}_{ij} - \tilde{T}_{ij-1}\right) = \frac{1}{(\Delta\overline{x})^2}\left(\tilde{T}_{i+1j} + \tilde{T}_{i-1j} - 2\tilde{T}_{ij}\right), \quad 1 \leq i \leq N - 1, \quad j \geq 1,$$

which after a simple rearrangement becomes

$$(1 + 2\delta)\tilde{T}_{ij} - \delta\left(\tilde{T}_{i+1j} + \tilde{T}_{i-1j}\right) = \tilde{T}_{ij-1}, \quad 1 \leq i \leq N - 1, \quad j \geq 1, \tag{6.24}$$

where δ is defined by (6.8). We note that now the terms $\tilde{T}_{i+1\,j}$, $\tilde{T}_{ij}$ and $\tilde{T}_{i-1\,j}$ are all implicitly defined in terms of $\tilde{T}_{i\,j-1}$ (see Figure 6.3).

For the present we assume prescribed temperature boundary conditions, so that $\tilde{T}_{0j}$ and $\tilde{T}_{Nj}$ are given by (6.10). Using the notation (6.14) for the $(N-1)$ dimensional vectors $\underset{\sim}{T}^j$ and $\underset{\sim}{U}^j$ we write the difference equation (6.24) for $\tilde{T}_{ij}$ in the form

$$\mathbf{B}\underset{\sim}{T}^j = \underset{\sim}{T}^{j-1} + \delta\,\underset{\sim}{U}^j, \quad j \geq 1, \tag{6.25}$$

where the $(N-1) \times (N-1)$ symmetric tridiagonal matrix $\mathbf{B}$ is given by

$$\mathbf{B} = \begin{pmatrix} 1+2\delta & -\delta & 0 & \cdots & & 0 \\ -\delta & 1+2\delta & -\delta & & & 0 \\ 0 & -\delta & 1+2\delta & & & 0 \\ \cdot & & & & & \cdot \\ \cdot & & & & & \cdot \\ \cdot & & & & & \cdot \\ 0 & 0 & & 1+2\delta & -\delta \\ 0 & 0 & \cdots & & -\delta & 1+2\delta \end{pmatrix}. \tag{6.26}$$

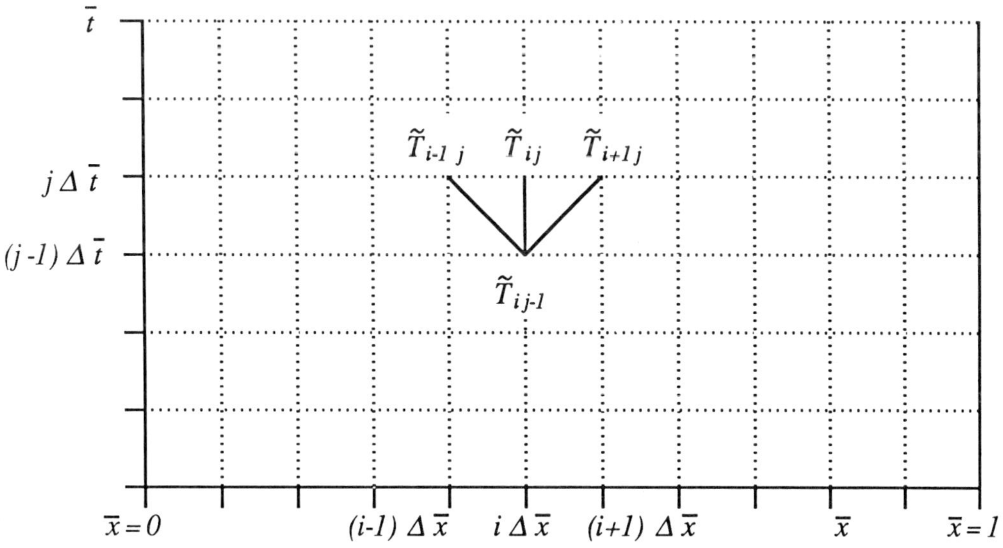

Figure 6.3. Fully implicit finite difference spatial and temporal discretization.

Since **B** is a diagonally dominant matrix for any $\delta > 0$ (that is $|B_{ii}| > \Sigma_{j \neq i}|B_{ij}|$ for every row of the matrix **B**), **B** is invertible and we can write

$$\underline{T}^j = \mathbf{B}^{-1}\big(\underline{T}^{j-1} + \delta\,\underline{U}^j\big), \quad j \geq 1. \tag{6.27}$$

We can find $\underline{T}^0$ and $\underline{U}^j$ from (6.9) and (6.10) respectively. Thus we can find $\underline{T}^j$ for any $j \geq 1$ by starting with $\underline{T}^0$ and applying (6.27) j times.

To prove the convergence of the fully implicit scheme we argue along the same lines as for the explicit scheme. With the local truncation error E_{ij} defined by (6.11) we have from (6.23) and (6.24)

$$(1 + 2\delta)E_{ij} = E_{i\,j-1} + \delta\big(E_{i+1\,j} + E_{i-1\,j}\big)$$

$$+\Delta\bar{t}\left[\frac{\partial^2\overline{T}}{\partial\bar{t}^2}(i\Delta\bar{x},(j-1+\xi_{ij})\Delta\bar{t})\frac{\Delta\bar{t}}{2} - \frac{\partial^4\overline{T}}{\partial\bar{x}^4}\big((i+\zeta_{ij})\Delta\bar{x},j\Delta\bar{t}\big)\frac{(\Delta x)^2}{12}\right], \quad \begin{array}{l} 1 \leq i \leq N-1, \\ \\ \end{array}$$

$$\tag{6.28}$$

while $E_{0j} = E_{Nj} = 0$. Assuming $\overline{T}(\bar{x},\bar{t})$ sufficiently differentiable, we can find positive constants M_1 and M_2 such that

$$|(1 + 2\delta)E_{ij}| \leq |E_{i\,j-1}| + |\delta E_{i+1\,j}| + |\delta E_{i-1\,j}| + \Delta\bar{t}\big(M_1\Delta\bar{t} + M_2(\Delta\bar{x})^2\big),$$

and letting $\hat{E}_j = \sup\limits_{0<i<N}|E_{ij}|$ we see that for any $\delta > 0$

$$\hat{E}_j \leq \hat{E}_{j-1} + \Delta\bar{t}\Big[M_1\Delta\bar{t} + M_2(\Delta x)^2\Big].$$

Hence, inductively

$$\hat{E}_j \leq j\Delta\bar{t}\Big[M_1\Delta\bar{t} + M_2(\Delta\bar{x})^2\Big],$$

so that the local truncation error vanishes as $\Delta\bar{t}$ and $\Delta\bar{x}$ tend to zero, for any constant value of $\delta > 0$.

We also treat the stability problem for the fully implicit method in the same way as for the explicit scheme. Specifically, we require from (6.27) that for any error vector $\underline{E}^j$ we should have

$$|\mathbf{B}^{-1}\underline{E}^j| \leq |\underline{E}^j|,$$

so that the spectral radius of $\mathbf{B}^{-1}$ satisfies $\rho\big(\mathbf{B}^{-1}\big) \leq 1$. We can show (see Problem 4) that the eigenvalues of **B**, as given by (6.26), are

$$\lambda_i = 1 + 4\delta\,\sin^2\Big(\frac{i\pi}{2N}\Big), \quad 1 \leq i \leq N.$$

Since the eigenvalues of $\mathbf{B}^{-1}$ are the reciprocals of the eigenvalues of $\mathbf{B}$, it follows that for any $\delta > 0$, $\rho(\mathbf{B}^{-1}) < 1$, so the fully implicit finite difference method is stable for all $\delta > 0$.

<u>Example 6.3</u> For the finite region $0 \le \bar{x} \le 1$ with constant initial temperature $\bar{u}_0 = 1$ and prescribed zero temperature at $\bar{x} = 0$ and $\bar{x} = 1$, describe an algorithm to solve the non-dimensional heat equation using a fully implicit finite difference algorithm.

As for the explicit algorithm we will store the discretized temperature as an array. We also require arrays to store the diagonal and off-diagonal elements of $\mathbf{B}$, since storing $\mathbf{B}$ as a full matrix is wasteful, as most of its entries are zero. Since $\mathbf{B}$ is diagonally dominant, tridiagonal, positive definite and symmetric it is usually more efficient to solve (6.25) using a method adapted for such matrices rather than to invert $\mathbf{B}$ and evaluate (6.27) using full matrix multiplication (since $\mathbf{B}^{-1}$ is not tridiagonal). Efficient routines to solve systems of equations such as (6.25) may be found in any of the standard mathematical libraries or packages. For those without access to such facilities a Cholesky LU factorization method is more than adequate. Since $\underset{\sim}{U}^j$ is identically zero in this case, we have only to repeatedly solve $\mathbf{B}\underset{\sim}{T}^j = \underset{\sim}{T}^{j-1}$ until we have stepped through to the desired time. A FORTRAN subroutine which performs this process is shown in Figure 6.8 and discussed in Section 6.6.

In Table 6.3 we compare the implicit finite difference solution of this problem, for a variety of N and δ with the exact solution at time $\bar{t} = 0.1$ obtained from Example 4.1. Since the solutions are symmetric about $\bar{x} = 0.5$ values are only quoted

		$\delta = 0.50$		$\delta = 1.00$		$\delta = 10.0$	
$\bar{x}$	$\bar{T}(\bar{x}, \bar{t})$	$N = 10$	$N = 20$	$N = 10$	$N = 20$	$N = 10$	$N = 20$
0.0	0.0000	0.0000	0.0000	0.0000	0.0000	0.0000	0.0000
0.1	0.1467	0.1504	0.1476	0.1541	0.1485	0.2438	0.1662
0.2	0.2790	0.2858	0.2807	0.2926	0.2824	0.4120	0.3135
0.3	0.3839	0.3930	0.3862	0.4018	0.3886	0.5214	0.4273
0.4	0.4513	0.4616	0.4540	0.4714	0.4566	0.5830	0.4986
0.5	0.4745	0.4852	0.4773	0.4952	0.4801	0.6028	0.5228

Figure 6.3. Comparison of the exact solution with fully implicit finite difference solutions of Example 6.3 for three values of δ and two values of N at time $\bar{t} = 0.1$.

for $0 \leq \bar{x} \leq 0.5$. As a result of the convergence and stability of the fully implicit scheme, even with $\delta = 10.0$ the solution is reasonable. Although each individual time step of the fully implicit method is more computationally expensive than a time step for the explicit method, the ability to take larger time steps (that is, $\delta > 0.5$) means that overall the fully implicit scheme is more computationally efficient than the explicit method. However, since the fully implicit scheme is accurate only to order $O\big((\Delta \bar{x})^2 + \Delta \bar{t}\big)$ large values of δ can lead to inaccurate approximations (see Table 6.3). This is the motivation to introduce the Crank-Nicolson scheme which is stable and convergent for all $\delta > 0$ and is accurate to order $O\big((\Delta \bar{x})^2 + (\Delta \bar{t})^2\big)$.

In the Crank-Nicolson scheme we approximate the partial derivative $\frac{\partial^2 \bar{T}}{\partial \bar{x}^2}$ by the average of its finite difference representations at times $j\Delta \bar{t}$ and $(j + 1)\Delta \bar{t}$. Thus the non-dimensional heat equation (6.6) is approximated by

$$
\frac{1}{\Delta \bar{t}}\big(\tilde{T}_{i\,j+1} - \tilde{T}_{ij}\big)
$$
$$
= \frac{1}{2(\Delta \bar{x})^2}\Big(\big(\tilde{T}_{i+1\,j+1} + \tilde{T}_{i-1\,j+1} - 2\tilde{T}_{i\,j+1}\big) + \big(\tilde{T}_{i+1\,j} + \tilde{T}_{i-1\,j} - 2\tilde{T}_{ij}\big)\Big), \tag{6.29}
$$

which is just the average of the fully implicit and explicit representations of the heat equation. We can also regard (6.29) as a finite difference approximation about the points $(i\Delta \bar{x}, (j + \frac{1}{2})\Delta \bar{t})$, where $\tilde{T}_{i\,j+\frac{1}{2}} = (\tilde{T}_{ij} + \tilde{T}_{i\,j+1})/2$. Viewed from this perspective the approximation for the time partial derivative is a central difference approximation arising from

$$
\frac{\partial \bar{T}}{\partial \bar{t}}(i\Delta \bar{x}, (j + \tfrac{1}{2})\Delta \bar{t}) = \frac{1}{2(\Delta \bar{t}/2)}\big(\bar{T}_{i\,j+1} - \bar{T}_{ij}\big) - \frac{\partial^3 \bar{T}}{\partial \bar{t}^3}(i\Delta \bar{x}, (j + \xi_{ij})\Delta \bar{t})\frac{(\Delta \bar{t}/2)^2}{6},
$$

where $0 \leq \xi_{ij} \leq 1$, and is accurate to order $O\big((\Delta \bar{t})^2\big)$. In this way we see that (6.29) represents the heat equation accurately to order $O\big((\Delta \bar{x})^2 + (\Delta \bar{t})^2\big)$. A simple rearrangement of the terms in (6.29) leads to the difference equation

$$
(1 + \delta)\tilde{T}_{i\,j+1} - \frac{\delta}{2}\big(\tilde{T}_{i+1\,j+1} + \tilde{T}_{i-1\,j+1}\big)
$$
$$
= (1 - \delta)\tilde{T}_{ij} + \frac{\delta}{2}\big(\tilde{T}_{i+1\,j} + \tilde{T}_{i-1\,j}\big), \qquad 1 \leq i \leq N - 1, \tag{6.30}
$$

where δ is given by (6.8).

Again assuming prescribed temperature boundary conditions, so that $\tilde{T}_{0j}$ and $\tilde{T}_{Nj}$ are defined by (6.10), and with the vectors $\underset{\sim}{T}^j$ and $\underset{\sim}{U}^j$ defined by (6.14) we can write the Crank-Nicolson finite difference scheme in the form

$$\mathbf{C}\underset{\sim}{T}^{j+1} = \mathbf{D}\underset{\sim}{T}^j + \frac{\delta}{2}\left(\underset{\sim}{U}^j + \underset{\sim}{U}^{j+1}\right). \tag{6.31}$$

Here $\mathbf{C}$ and $\mathbf{D}$ are the $(N-1) \times (N-1)$ symmetric tridiagonal matrices

$$\mathbf{C} = \begin{bmatrix} 1+\delta & -\delta/2 & 0 & \cdots & & 0 \\ -\delta/2 & 1+\delta & -\delta/2 & & & 0 \\ 0 & -\delta/2 & 1+\delta & & & 0 \\ \cdot & & & & & \cdot \\ \cdot & & & & & \cdot \\ \cdot & & & & & \cdot \\ 0 & 0 & \cdots & & 1+\delta & -\delta/2 \\ 0 & 0 & \cdots & & -\delta/2 & 1+\delta \end{bmatrix}, \quad \mathbf{D} = \begin{bmatrix} 1-\delta & \delta/2 & 0 & \cdots & & 0 \\ \delta/2 & 1-\delta & \delta/2 & & & 0 \\ 0 & \delta/2 & 1-\delta & & & 0 \\ \cdot & & & & & \cdot \\ \cdot & & & & & \cdot \\ \cdot & & & & & \cdot \\ 0 & 0 & \cdots & & 1-\delta & \delta/2 \\ 0 & 0 & \cdots & & \delta/2 & 1-\delta \end{bmatrix}.$$

$$\tag{6.32}$$

Since $\mathbf{C}$ is diagonally dominant, for any $\delta > 0$, it is invertible and we may write (6.31) as

$$\underset{\sim}{T}^{j+1} = \mathbf{C}^{-1}\left(\mathbf{D}\underset{\sim}{T}^j + \frac{\delta}{2}\left(\underset{\sim}{U}^{j+1} + \underset{\sim}{U}^j\right)\right). \tag{6.33}$$

As for the fully implicit scheme, all the quantities on the right hand side of (6.33) are either known or calculable. The vectors $\underset{\sim}{T}^0$, $\underset{\sim}{U}^j$ and $\underset{\sim}{U}^{j+1}$ are known from the initial condition (6.9) and the boundary conditions (6.10). Thus, we find $\underset{\sim}{T}^j$ by starting with $\underset{\sim}{T}^0$ and applying (6.33) j times. The Crank-Nicolson scheme is convergent and stable for all $\delta > 0$. We leave the details to the reader as an exercise (see Problem 5). Because the Crank-Nicolson scheme is accurate to order $O\left((\Delta\bar{x})^2 + (\Delta\bar{t})^2\right)$ we can in theory take larger time steps than for the fully implicit scheme without too much loss of accuracy.

<u>Example 6.4</u> For the finite region $0 \le \bar{x} \le 1$ with constant initial temperature $\bar{u}_0 = 1$ and prescribed zero temperature at $\bar{x} = 0$ and $\bar{x} = 1$, describe an algorithm to solve the non-dimensional heat equation using a Crank-Nicolson method.

In this case $\underset{\sim}{U}^{j}$ is zero for all $j \geq 0$, so that (6.31) becomes $\mathbf{C}\underset{\sim}{T}^{j+1} = \mathbf{D}\underset{\sim}{T}^{j}$ and (6.33) becomes $\underset{\sim}{T}^{j+1} = \mathbf{C}^{-1}\mathbf{D}\underset{\sim}{T}^{j}$. We need arrays to hold the new and old temperature vectors and arrays to hold the diagonal and off-diagonal elements of $\mathbf{C}$ and $\mathbf{D}$, as again it is inefficient to store $\mathbf{C}$ and $\mathbf{D}$ as full matrices since most of their elements are zeros. As for the previous example, it is more efficient to solve (6.31) than to find $\mathbf{C}^{-1}\mathbf{D}$ and solve (6.33) using full matrix multiplication. Once the initial temperature vector $\underset{\sim}{T}^{0}$ is established (all entries are 1), we have only to apply and solve the equation $\mathbf{C}\underset{\sim}{T}^{j+1} = \mathbf{D}\underset{\sim}{T}^{j}$ for $\underset{\sim}{T}^{j+1}$ repeatedly until the desired time is reached. As noted in Example 6.3, there are many routines which can solve systems involving positive definite symmetric tridiagonal matrices efficiently. A FORTRAN subroutine which carries out this process is shown in Figure 6.9 and discussed in Section 6.6.

In Table 6.4 we compare Crank-Nicolson finite difference solutions of this problem for various values of N and δ with the exact solution, for time $\bar{t} = 0.1$. As the solutions are symmetric about $\bar{x} = 0.5$ values are only quoted for $0 \leq \bar{x} \leq 0.5$. For the values of N and δ given in this table, the fully implicit solutions (see Table 6.3) are marginally superior to the Crank-Nicolson solutions. This emphasizes that even though the Crank-Nicolson method is accurate to order $O\!\left((\Delta\bar{x})^{2} + (\Delta\bar{t})^{2}\right)$, and the fully implicit method is only accurate to $O\!\left((\Delta\bar{x})^{2} + \Delta\bar{t}\right)$, in practice it may be necessary to take N to be quite large (that is $\Delta\bar{x}$ and $\Delta\bar{t}$ to be relatively small) before this theoretical property manifests itself.

		$\delta = 0.50$		$\delta = 1.00$		$\delta = 10.0$	
$\bar{x}$	$\overline{T}(\bar{x},\bar{t})$	$N = 10$	$N = 20$	$N = 10$	$N = 20$	$N = 10$	$N = 20$
0.0	0.0000	0.0000	0.0000	0.0000	0.0000	0.0000	0.0000
0.1	0.1467	0.1504	0.1476	0.1541	0.1485	0.3475	0.1562
0.2	0.2790	0.2859	0.2807	0.2931	0.2825	0.5646	0.3241
0.3	0.3839	0.3934	0.3863	0.4033	0.3887	0.6945	0.4384
0.4	0.4513	0.4624	0.4541	0.4740	0.4569	0.7633	0.5104
0.5	0.4745	0.4862	0.4774	0.4984	0.4803	0.7848	0.5359

Table 6.4. Comparison of the exact solution with Crank-Nicolson finite difference solutions of Example 6.4 for three values of δ and two values of N at time $\bar{t} = 0.1$.

To treat problems with derivative boundary conditions using either of the implicit methods discussed we use the same strategy as for explicit methods. That is, we assume the difference equation for the approximate temperature holds at the appropriate end point, introducing fictitious temperatures at points outside the interval $[0, 1]$. We then use the finite difference approximations for the boundary conditions to eliminate these fictitious temperatures from the problem. The stability and convergence analysis for such problems is similar to that already given, but is complicated by the asymmetry introduced by the difference relations for $i = 0$ and $i = N$. It is worth noting that the treatment of non-linear boundary conditions by implicit schemes requires the solution of a non-linear system of algebraic equations and the additional computational overhead involved negates, to a large degree, the advantages of implicit methods over explicit ones.

<u>Example 6.5</u> Devise both a fully implicit and a Crank-Nicolson finite difference method to solve the heat flow problem for the finite region $0 \leq \bar{x} \leq 1$ with initial temperature $\bar{u}_0 = 1$, no heat flow across $\bar{x} = 0$

$$\frac{\partial \overline{T}}{\partial \bar{x}}(0, \bar{t}) = 0,$$

and Newton heat loss into a medium at zero temperature at $\bar{x} = 1$,

$$\frac{\partial \overline{T}}{\partial \bar{x}}(1, \bar{t}) + \overline{T}(1, \bar{t}) = 0.$$

Using central difference approximations for the temperature gradients, the condition $\frac{\partial \overline{T}}{\partial \bar{x}}(0, \bar{t}) = 0$ becomes $\left(\tilde{T}_{1j} - \tilde{T}_{-1j}\right)/(2\Delta \bar{x}) = 0$ and on solving for the fictitious temperature $\tilde{T}_{-1j}$ we have

$$\tilde{T}_{-1j} = \tilde{T}_{1j}. \tag{6.34}$$

The boundary condition at $\bar{x} = 1$ is approximated by $\left(\tilde{T}_{N+1j} - \tilde{T}_{N-1j}\right)/(2\Delta \bar{x}) + \tilde{T}_{Nj} = 0$, and on solving for the fictitious temperature $\tilde{T}_{N+1j}$ we have

$$\tilde{T}_{N+1j} = \tilde{T}_{N-1j} - 2\Delta \bar{x} \tilde{T}_{Nj}. \tag{6.35}$$

For the fully implicit scheme, (6.24) remains valid, but instead of (6.10) we now have

$$\left.\begin{aligned}
(1 + 2\delta)\tilde{T}_{0j} - \delta\left(\tilde{T}_{1j} + \tilde{T}_{-1j}\right) &= \tilde{T}_{0j-1}, \\
(1 + 2\delta)\tilde{T}_{Nj} - \delta\left(\tilde{T}_{N+1j} + \tilde{T}_{N-1j}\right) &= \tilde{T}_{Nj-1},
\end{aligned}\right\} \quad j \geq 1,$$

so that on using (6.34) and (6.35) to eliminate $\tilde{T}_{-1j}$ and $\tilde{T}_{N+1j}$ these become

$$\left.\begin{array}{l}(1 + 2\delta)\tilde{T}_{0j} - 2\delta\tilde{T}_{1j} = \tilde{T}_{0j-1}, \\[2mm] (1 + 2\delta(1 + \Delta\bar{x}))\tilde{T}_{Nj} - 2\delta\tilde{T}_{N-1j} = \tilde{T}_{Nj-1},\end{array}\right\} \quad j \geq 1,$$

and altogether we have the linear system

$$\mathbf{B}\underline{T}^{j} = \underline{T}^{j-1}, \tag{6.36}$$

where the asymmetric $(N + 1) \times (N + 1)$ tridiagonal matrix $\mathbf{B}$ and the $(N + 1)$ dimensional vector $\underline{T}^{j}$ are given by

$$\mathbf{B} = \begin{pmatrix} 1+2\delta & -2\delta & 0 & \cdots & & 0 \\ -\delta & 1+2\delta & -\delta & & & 0 \\ 0 & -\delta & 1+2\delta & & & 0 \\ \cdot & & & & & \cdot \\ \cdot & & & & & \cdot \\ \cdot & & & & & \cdot \\ 0 & & -\delta & 1+2\delta & -\delta \\ 0 & \cdots & & 0 & -2\delta & 1+2\delta(1+\Delta\bar{x}) \end{pmatrix}, \quad \underline{T}^{j} = \begin{pmatrix} \tilde{T}_{0j} \\ \tilde{T}_{1j} \\ \vdots \\ \vdots \\ \tilde{T}_{N-1j} \\ \tilde{T}_{Nj} \end{pmatrix}. \tag{6.37}$$

Since $\tilde{T}_{i0} = 1$ for $i = 0, 1, \ldots, N$, we know $\underline{T}^{0}$, and can find $\underline{T}^{j}$ by starting from $\underline{T}^{0}$ and applying and solving (6.36) j times. A FORTRAN subroutine which performs this process is given in Figure 6.10 and discussed in Section 6.6

For the Crank-Nicolson scheme, (6.30) still applies, but assuming that (6.30) also holds at $i = 0$ and at $i = N$ and applying (6.34) and (6.35) to eliminate the fictitious temperatures $\tilde{T}_{-1j}$ and $\tilde{T}_{N+1j}$ we have instead of (6.10)

$$\left.\begin{array}{l}(1 + \delta)\tilde{T}_{0j+1} - \delta\tilde{T}_{1j+1} = (1 - \delta)\tilde{T}_{0j} + \delta\tilde{T}_{1j}, \\[2mm] (1 + \delta(1 + \Delta\bar{x}))\tilde{T}_{Nj+1} - \delta\tilde{T}_{N-1j+1} = (1 - \delta(1 + \Delta\bar{x}))\tilde{T}_{Nj} + \delta\tilde{T}_{N-1j},\end{array}\right\} \quad j \geq 1,$$

for $\tilde{T}_{0j}$ and $\tilde{T}_{Nj}$. In total we have the linear system

$$\mathbf{C}\underline{T}^{j+1} = \mathbf{D}\underline{T}^{j}, \quad j = 0, 1, \ldots \tag{6.38}$$

where the $(N + 1) \times (N + 1)$ matrices $\mathbf{C}$ and $\mathbf{D}$ are given by

$$\mathbf{C} = \begin{bmatrix} 1+\delta & -\delta & 0 & \cdots & & 0 \\ -\delta/2 & 1+\delta & -\delta/2 & & & 0 \\ 0 & -\delta/2 & 1+\delta & & & 0 \\ \cdot & & & & & \cdot \\ \cdot & & & & & \cdot \\ \cdot & & & & & \cdot \\ 0 & & -\delta/2 & 1+\delta & & -\delta/2 \\ 0 & \cdots & 0 & -\delta & & 1+\delta(1+\Delta\overline{x}) \end{bmatrix}, \quad \mathbf{D} = \begin{bmatrix} 1-\delta & \delta & 0 & \cdots & & 0 \\ \delta/2 & 1-\delta & \delta/2 & & & 0 \\ 0 & \delta/2 & 1-\delta & & & 0 \\ \cdot & & & & & \cdot \\ \cdot & & & & & \cdot \\ \cdot & & & & & \cdot \\ 0 & & \delta/2 & 1-\delta & & \delta/2 \\ 0 & \cdots & 0 & \delta & & 1-\delta(1+\Delta\overline{x}) \end{bmatrix}.$$

and $\underset{\sim}{T}^{j}$ is as given by $(6.37)_2$. Since $\underset{\sim}{T}^{0}$ is known, we may find any $\underset{\sim}{T}^{j}$ by starting with $\underset{\sim}{T}^{0}$ and applying (6.38) j times. A FORTRAN subroutine which does this is given in Figure 6.11.

Even though the matrices $\mathbf{B}$ and $\mathbf{C}$ are not symmetric, they are still tridiagonal, positive definite matrices and it is still more efficient to solve (6.36) and (6.38) as they stand using specialized routines than to perform matrix inversions to find $\mathbf{B}^{-1}$ or $\mathbf{C}^{-1}$ and then solve (6.36) and (6.38) by matrix multiplication. In Table 6.5 we compare fully implicit finite difference and Crank-Nicolson solutions of Example 6.5

		Crank-Nicolson		Fully implicit	
$\overline{x}$	$\overline{T}(\overline{x},\overline{t})$	$\delta = 1.0$	$\delta = 10.0$	$\delta = 1.0$	$\delta = 10.0$
0.0	0.9931	0.9931	0.9931	0.9925	0.9883
0.1	0.9918	0.9918	0.9920	0.9912	0.9871
0.2	0.9878	0.9878	0.9880	0.9873	0.9834
0.3	0.9804	0.9804	0.9807	0.9799	0.9767
0.4	0.9684	0.9684	0.9687	0.9682	0.9661
0.5	0.9505	0.9506	0.9507	0.9506	0.9503
0.6	0.9252	0.9253	0.9253	0.9256	0.9277
0.7	0.8908	0.8910	0.8914	0.8916	0.8964
0.8	0.8462	0.8464	0.8462	0.8473	0.8545
0.9	0.7905	0.7907	0.7861	0.7917	0.8004
1.0	0.7236	0.7237	0.7382	0.7247	0.7334

Table 6.5. Comparison of fully implicit and Crank-Nicolson finite difference solutions (with $N = 20$ and two values of δ) with the exact solution of Example 6.5 at time $\overline{t} = 0.1$.

for $N = 20$ and $\delta = 1.0$ and $\delta = 10.0$ with the exact solution (see Problem 4.7 with $\mu = \kappa = \ell = 1$) at time $\bar{t} = 0.1$. Inspection of this table shows that for this problem the Crank-Nicolson scheme is superior to the fully implicit method, particularly for the larger value $\delta = 10.0$.

Finally it should be pointed out that in our derivation and analysis of the various finite difference schemes in this and the previous section we have assumed that the exact temperature $\overline{T}(\overline{x}, \overline{t})$ has a well behaved Taylor series expansion about any of the points $(i\Delta\overline{x}, j\Delta\overline{t})$. When this is not the case, for example either when inconsistent boundary and initial conditions or when discontinuous boundary or initial data are prescribed, the various finite difference approximations $\tilde{T}_{ij}$ may not converge to the actual temperature $\overline{T}_{ij}$ at all. Fortunately the discrepancies tend to vanish after a small number of time steps as the singularities in the actual temperature are smoothed out. This is not always the case, but we must refer the reader to a specialist text on the numerical treatment of singularities for more detailed information.

6.4 Finite element method

The finite element method is a systematic means of generating approximate solutions to a *weak* formulation of the heat equation. Originally finite elements were developed to solve time independent problems governed by elliptic partial differential equations, and this ancestory is reflected in the finite element method we present in this section, which is in fact a hybrid finite element and finite difference method where only the spatial component of the problem is treated by the finite element method. To set up the weak form of the heat equation we need to introduce a space of *test functions*, and throughout this section (and unlike Chapter 1) a test function will mean a function $\phi(\overline{x})$ such that $\phi(\overline{x})$ and its derivative $\phi'(\overline{x})$ exist on the interval $[0, 1]$ and square integrable in the sense that

$$\int_0^1 \left[(\phi(\overline{x}))^2 + \left(\phi'(\overline{x})\right)^2 \right] d\overline{x},$$

exists and is finite. The linear space of all test functions is denoted $H[0, 1]$. If the heat equation (6.3) is multiplied by an arbitrary test function $\phi(\overline{x}) \in H[0, 1]$ and the product integrated from $\overline{x} = 0$ to $\overline{x} = 1$ we have

$$\int_0^1 \phi(\overline{x}) \left(\frac{\partial \overline{T}}{\partial \overline{t}} - \frac{\partial^2 \overline{T}}{\partial \overline{x}^2} \right) d\overline{x} = 0,$$

which on integrating by parts becomes

$$\int_0^1 \left(\phi(\overline{x})\frac{\partial \overline{T}}{\partial t} + \frac{d\phi}{d\overline{x}}\frac{\partial \overline{T}}{\partial \overline{x}} \right) d\overline{x} = \phi(1)\frac{\partial \overline{T}}{\partial \overline{x}}(1,\overline{t}) - \phi(0)\frac{\partial \overline{T}}{\partial \overline{x}}(0,\overline{t}). \qquad (6.39)$$

We take equation (6.39) as our basic weak form of the heat equation, although we do not as yet give a precise definition because such a definition depends on the nature of the boundary conditions applied to $\overline{T}$. We do note however that (6.39) implies only that $\frac{\partial \overline{T}}{\partial \overline{x}}$ and $\frac{\partial \overline{T}}{\partial \overline{t}}$ exist as square integrable functions of $\overline{x}$ whereas the heat equation (6.3) implies the stronger condition that $\frac{\partial^2 \overline{T}}{\partial \overline{x}^2}$ exists.

We construct approximate solutions of (6.39) by employing Galerkin's method, which is as follows. We introduce a finite linearly independent set of N *trial basis functions* $\{\psi_1(\overline{x}), \psi_2(\overline{x}), \ldots, \psi_N(\overline{x})\}$ and assume an approximate solution $\tilde{T}(\overline{x},\overline{t})$ which is a linear combination of the trial basis functions

$$\tilde{T}(\overline{x},\overline{t}) = \sum_{i=1}^{N} b_i(\overline{t})\psi_i(\overline{x}), \qquad (6.40)$$

where each $\psi_i(\overline{x}) \in H[0,1]$ $(1 \le i \le N)$ and the functions $b_i(\overline{t})$ $(1 \le i \le N)$ are as yet undetermined differentiable functions of time. Further, we choose a finite linearly independent set of N *test basis functions* $\{\phi_1(\overline{x}), \phi_2(\overline{x}), \ldots, \phi_N(\overline{x})\}$, where each $\phi_i(\overline{x}) \in H[0,1]$ $(1 \le i \le N)$, from which we construct our test functions $\phi(\overline{x})$. With these assumptions problem (6.39) is reduced to finding a function $\tilde{T}(\overline{x},\overline{t})$ of the form (6.40) such that

$$\int_0^1 \left(\phi(\overline{x})\frac{\partial \tilde{T}}{\partial t} + \frac{d\phi}{d\overline{x}}\frac{\partial \tilde{T}}{\partial \overline{x}} \right) d\overline{x} = \phi(1)\frac{\partial \tilde{T}}{\partial \overline{x}}(1,\overline{t}) - \phi(0)\frac{\partial \tilde{T}}{\partial \overline{x}}(0,\overline{t}), \qquad (6.41)$$

for every test function $\phi(\overline{x})$ which is a linear combination of the N test basis functions

$$\phi(\overline{x}) = \sum_{i=1}^{N} a_i\phi_i(\overline{x}), \qquad (6.42)$$

where the coefficients a_i $(1 \le i \le N)$ are arbitrary real numbers. Problem (6.41) eventually reduces to a system of N linear first order ordinary differential equations for the unknown functions $b_i(\overline{t})$ $(1 \le i \le N)$, which can then be solved using a finite difference method. The initial conditions, that is $b_i(0)$ $(1 \le i \le N)$, are obtained from the initial temperature. Although in Galerkin's method there is no particular reason why the subspaces of $H[0,1]$ spanned by the trial basis functions $\psi_i(\overline{x})$ $(1 \le i \le N)$ and the test basis functions $\phi_i(\overline{x})$ $(1 \le i \le N)$ should coincide, it is nevertheless convenient to assume that the trial basis functions and test basis functions are equal, that is

$$\psi_i(\overline{x}) = \phi_i(\overline{x}), \quad 1 \le i \le N,$$

and we will assume this in what follows. It is customary when discussing finite elements to refer to the identical test and trial basis functions as *shape functions*.

The finite element method uses a systematic technique for selecting the shape functions $\phi_i(\overline{x})$ ($1 \le i \le N$). The interval $[0, 1]$ is divided into $N - 1$ subintervals or elements $\Omega_1, \Omega_2, \ldots, \Omega_{N-1}$ of length $h_1, h_2, \ldots, h_{N-1}$ respectively. The endpoints of the ith element Ω_i are denoted $\overline{x}_i$ and $\overline{x}_{i+1}$

$$\Omega_i = [\overline{x}_i, \overline{x}_{i+1}], \quad h_i = \overline{x}_{i+1} - \overline{x}_i, \quad 1 \le i \le N - 1,$$

so that $\overline{x}_1 = 0$ and $\overline{x}_N = 1$, and we refer to the endpoints as nodes (see Figure 6.4). The collection of elements and nodes is referred to as the finite element mesh for the problem. In more sophisticated treatments each element can contain more than two nodes. We then define our shape functions $\phi_i(\overline{x})$ ($1 \le i \le N$) by

$$\phi_1(\overline{x}) = \begin{cases} 1 - \overline{x}/h_1, & \overline{x} \in \Omega_1, \\ 0, & \overline{x} \notin \Omega_1, \end{cases}$$

$$\phi_i(\overline{x}) = \begin{cases} (\overline{x} - \overline{x}_{i-1})/h_{i-1}, & \overline{x} \in \Omega_{i-1}, \\ 1 - (\overline{x} - \overline{x}_i)/h_i, & \overline{x} \in \Omega_i, \\ 0, & \overline{x} \notin \Omega_i \cup \Omega_{i-1}, \end{cases} \qquad 2 \le i \le N - 1, \qquad (6.43)$$

$$\phi_N(\overline{x}) = \begin{cases} 1 - (\overline{x} - 1)/h_{N-1}, & \overline{x} \in \Omega_{N-1}, \\ 0, & \overline{x} \notin \Omega_{N-1}, \end{cases}$$

as shown in Figure 6.5. These shape functions $\phi_i(\overline{x})$ ($1 \le i \le N$) are chosen so that they satisfy the following important properties:

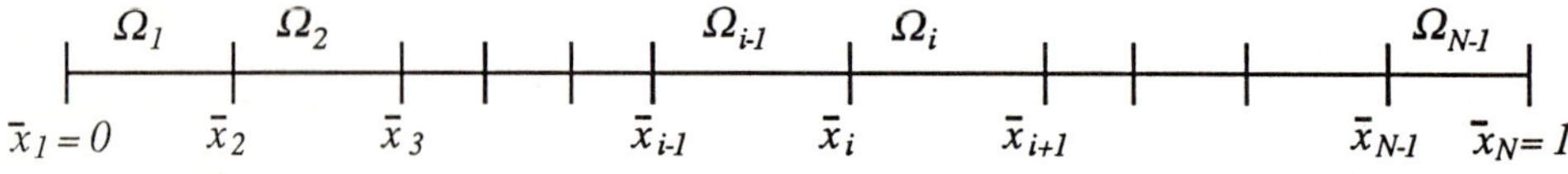

Figure 6.4. Finite element mesh for spatial interval [0,1].

(i) Each shape function $\phi_i(\overline{x})$ $(1 \le i \le N)$ is in $H[0, 1]$, that is it and its piecewise constant derivative are both square integrable over $[0, 1]$.

(ii) Each shape function $\phi_i(\overline{x})$ $(1 \le i \le N)$ is defined piecewise over the elements and in this case is a linear function on each element Ω_j $(1 \le j \le N - 1)$.

(iii) $\phi_i(\overline{x}_j) = \delta_{ij}$ $(1 \le i, j \le N)$ so that the value of the test function $\phi(\overline{x}) = \sum_{i=1}^{N} a_i \phi_i(\overline{x})$ is $\phi(\overline{x}_j) = a_j$ at the node $\overline{x}_j$.

(iv) $\phi_i(\overline{x})\phi_j(\overline{x}) = 0$ if $|i - j| \ge 2$, that is if i and j differ by more than one.

It is possible to introduce more sophisticated functions satisfying these requirements as shape functions such as element-wise quadratic or cubic functions. This approach requires additional nodes to be added to each element, and we do not pursue the matter.

Let us now consider a finite element treatment of problem (6.6). Here we have prescribed boundary temperatures at $\overline{x} = 0$ and $\overline{x} = 1$ which are also known in finite element parlance as *essential* boundary conditions. Since neither $\frac{\partial \overline{T}}{\partial \overline{x}}(0, \overline{t})$ nor $\frac{\partial \overline{T}}{\partial \overline{x}}(1, \overline{t})$ are known a priori, we eliminate them from equation (6.39) by insisting that only test functions $\phi(\overline{x}) \in H[0, 1]$ which also vanish at $\overline{x} = 0$ and $\overline{x} = 1$ be admitted. That is, the weak form of problem (6.6) is to find a function $\overline{T}(\overline{x}, \overline{t})$ which satisfies the initial

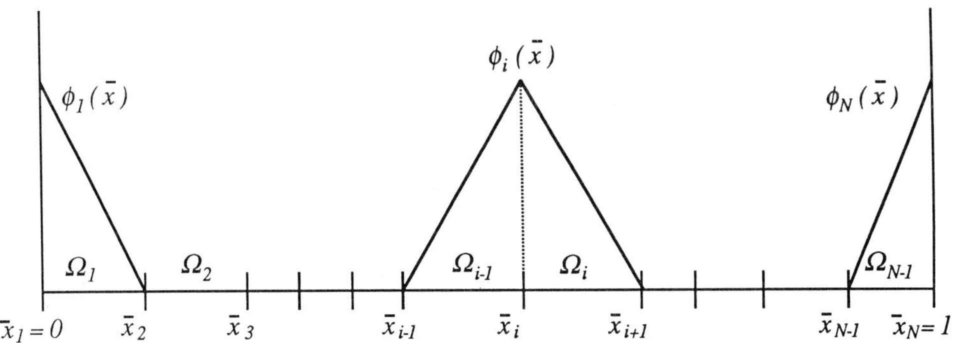

Figure 6.5. Finite element shape functions defined on ×the interval [0,1].

and boundary conditions $(6.6)_2$ and $(6.6)_3$ and is such that

$$\int_0^1 \left(\phi(\overline{x}) \frac{\partial \overline{T}}{\partial \overline{t}} + \frac{\mathrm{d}\phi}{\mathrm{d}\overline{x}} \frac{\partial \overline{T}}{\partial \overline{x}} \right) \mathrm{d}\overline{x} = 0, \tag{6.44}$$

for every $\overline{t} > 0$ and all test functions $\phi(\overline{x}) \in H[0,1]$ which satisfy $\phi(0) = \phi(1) = 0$. We implement a finite element solution of this problem as follows. We choose a finite element mesh with elements $\Omega_1, \Omega_2, \ldots \Omega_{N-1}$ and nodes $0 = \overline{x}_1, \overline{x}_2, \ldots, \overline{x}_N = 1$ and shape functions $\phi_1(\overline{x}), \phi_2(\overline{x}), \ldots, \phi_N(\overline{x})$. The lengths of the elements Ω_i are denoted by h_i $(1 \le i \le N - 1)$ and we let h be the largest of these lengths. The approximate solution is denoted $\overline{T}_h(\overline{x}, \overline{t})$, the idea being that as h tends to zero (that is the length of the largest element tends to zero) $\overline{T}_h(\overline{x}, \overline{t})$ tends to the actual weak solution $\overline{T}(\overline{x}, \overline{t})$ of problem (6.44), $(6.6)_{2,3}$ in the sense that

$$\lim_{h \to 0} \int_0^1 \left[\left(\overline{T}(\overline{x}, \overline{t}) - \overline{T}_h(\overline{x}, \overline{t}) \right)^2 + \left(\frac{\partial \overline{T}}{\partial \overline{x}}(\overline{x}, \overline{t}) - \frac{\partial \overline{T}_h}{\partial \overline{x}}(\overline{x}, \overline{t}) \right)^2 \right] \mathrm{d}\overline{x} = 0,$$

for all $\overline{t} > 0$. For a proof of this assertion we refer the interested reader to one of the many specialist texts on the finite element method.

Thus with the finite element approximation given by

$$\overline{T}_h(\overline{x}, \overline{t}) = \sum_{i=1}^{N} b_i(\overline{t}) \phi_i(\overline{x}),$$

the approximate form of (6.44) becomes to find $b_i(\overline{t})$ $(1 \le i \le N)$ such that

$$\int_0^1 \left(\phi(\overline{x}) \frac{\partial \overline{T}_h}{\partial \overline{t}} + \frac{\mathrm{d}\phi}{\mathrm{d}\overline{x}} \frac{\partial \overline{T}_h}{\partial \overline{x}} \right) \mathrm{d}\overline{x} = 0, \tag{6.45}$$

for all test functions $\phi(\overline{x})$ of the form (6.42) which vanish at $\overline{x} = 0$ and $\overline{x} = 1$. In particular if the test functions are to vanish at $\overline{x} = 0$ and $\overline{x} = 1$, then in view of condition **(iii)** stated above we must have $a_1 = a_N = 0$, so that the test functions actually have the form

$$\phi(\overline{x}) = \sum_{i=2}^{N-1} a_i \phi_i(\overline{x}), \tag{6.46}$$

where the coefficients a_i $(2 \le i \le N - 1)$ are arbitrary real numbers. Similarly since $\overline{T}_h(\overline{x}, \overline{t})$ must satisfy the boundary conditions

$$\overline{T}_h(0, \overline{t}) = g_1(\overline{t}), \quad \overline{T}_h(1, \overline{t}) = g_2(\overline{t}),$$

derived from $(6.6)_3$, we must also have $b_1(\bar{t}) = g_1(\bar{t})$ and $b_N(\bar{t}) = g_2(\bar{t})$, so that

$$\bar{T}_h(\bar{x}, \bar{t}) = g_1(\bar{t})\phi_1(\bar{x}) + \sum_{i=2}^{N-1} b_i(\bar{t})\phi_i(\bar{x}) + g_2(\bar{t})\phi_N(\bar{x}), \tag{6.47}$$

leaving only $N - 2$ unknowns $b_2(\bar{t}), b_3(\bar{t}), \ldots, b_{N-1}(\bar{t})$ to be found.

With the test functions defined by (6.46) and the approximate temperature given by (6.47) problem (6.45) becomes

$$\sum_{i=2}^{N-1}\sum_{j=2}^{N-1} a_i \left(\frac{db_j}{d\bar{t}} \int_0^1 \phi_i(\bar{x})\phi_j(\bar{x})\,d\bar{x} + b_j(\bar{t}) \int_0^1 \phi_i'(\bar{x})\phi_j'(\bar{x})\,d\bar{x} \right)$$

$$= -\sum_{i=2}^{N-2} a_i \left(\int_0^1 \phi_i(\bar{x}) \left[\frac{dg_1}{d\bar{t}}\phi_1(\bar{x}) + \frac{dg_2}{d\bar{t}}\phi_N(\bar{x}) \right] d\bar{x} \right.$$

$$\left. + \int_0^1 \phi_i'(\bar{x}) \left[g_1(\bar{t})\phi_1'(\bar{x}) + g_2(\bar{t})\phi_N'(\bar{x}) \right] d\bar{x} \right),$$

for all possible values of a_i ($2 \le i \le N - 1$). This can only happen if the $b_i(\bar{t})$ ($2 \le i \le N - 1$) satisfy the system of $N - 2$ first order linear ordinary differential equations

$$\sum_{j=2}^{N-1} \left(C_{ij}\frac{db_j}{d\bar{t}} + K_{ij}b_j(\bar{t}) \right) =$$

$$\qquad\qquad\qquad\qquad\qquad\qquad 2 \le i \le N - 1, \tag{6.48}$$

$$-\left(C_{i1}\frac{dg_1}{d\bar{t}} + C_{iN}\frac{dg_2}{d\bar{t}} + K_{i1}g_1(\bar{t}) + K_{iN}g_2(\bar{t}) \right),$$

where C_{ij} and K_{ij} represent the integrals

$$C_{ij} = C_{ji} = \int_0^1 \phi_i(\bar{x})\phi_j(\bar{x})\,d\bar{x}, \quad K_{ij} = K_{ji} = \int_0^1 \phi_i'(\bar{x})\phi_j'(\bar{x})\,d\bar{x}, \quad 1 \le i,j \le N. \tag{6.49}$$

We leave it to the reader to show that with the shape functions given in (6.43) we have

$$C_{11} = \tfrac{1}{3}h_1, \quad C_{NN} = \tfrac{1}{3}h_{N-1}, \quad C_{ii} = \tfrac{1}{3}(h_{i-1} + h_i),\ 2 \le i \le N - 1,$$

$$K_{11} = \frac{1}{h_1}, \quad K_{NN} = \frac{1}{h_{N-1}}, \quad K_{ii} = \frac{1}{h_{i-1}} + \frac{1}{h_i},\ 2 \le i \le N - 1, \tag{6.50}$$

$$C_{i\,i+1} = \tfrac{1}{6}h_i, \quad K_{i\,i+1} = \frac{-1}{h_i}, \quad 1 \le i \le N - 1,$$

where h_i denotes the length of element Ω_i ($1 \le i \le N - 1$), while if i and j differ by more than one, then

$$C_{ij} = 0, \quad K_{ij} = 0, \quad |i - j| > 1. \tag{6.51}$$

We can write (6.48) in matrix form

$$\mathbf{C}\frac{\mathrm{d}}{\mathrm{d}\bar{t}}\underline{b}(\bar{t}) + \mathbf{K}\underline{b}(\bar{t}) = \frac{\mathrm{d}}{\mathrm{d}\bar{t}}\underline{c}(\bar{t}) + \underline{d}(\bar{t}), \tag{6.52}$$

where the $(N - 2) \times (N - 2)$ tridiagonal positive definite symmetric matrices $\mathbf{C}$ and $\mathbf{K}$ (known for historical reasons as the *capacitance* and *stiffness* matrices respectively) are given by

$$\mathbf{C} = \begin{bmatrix} C_{22} & \cdots & C_{2(N-1)} \\ \vdots & & \vdots \\ C_{(N-1)2} & \cdots & C_{(N-1)(N-1)} \end{bmatrix}, \quad \mathbf{K} = \begin{bmatrix} K_{22} & \cdots & K_{2(N-1)} \\ \vdots & & \vdots \\ K_{(N-1)2} & \cdots & K_{(N-1)(N-1)} \end{bmatrix}, \quad \tag{6.53}$$

and the vectors $\underline{b}(\bar{t})$, $\underline{c}(\bar{t})$ and $\underline{d}(\bar{t})$ are given by

$$\underline{b}(\bar{t}) = \begin{bmatrix} b_2(\bar{t}) \\ b_3(\bar{t}) \\ \vdots \\ b_{N-2}(\bar{t}) \\ b_{N-1}(\bar{t}) \end{bmatrix}, \quad \underline{c}(\bar{t}) = - \begin{bmatrix} C_{21}g_1(\bar{t}) \\ 0 \\ \vdots \\ 0 \\ C_{(N-1)N}g_2(\bar{t}) \end{bmatrix}, \quad \underline{d}(\bar{t}) = - \begin{bmatrix} K_{21}g_1(\bar{t}) \\ 0 \\ \vdots \\ 0 \\ K_{(N-1)N}g_2(\bar{t}) \end{bmatrix}.$$
$$\tag{6.54}$$

As well we have the initial conditions

$$b_i(0) = f(\bar{x}_i), \quad 2 \le i \le N - 1, \tag{6.55}$$

which we derive from the initial condition $(6.6)_2$. As an alternative, we can determine the initial values of $b_i(\bar{t})$ ($1 \le i \le N$) by choosing $b_i(0)$ so as to minimize the integral

$$\int_0^1 \left[\left(\bar{T}_h(\bar{x}, 0) - f(\bar{x})\right)^2 + \left(\frac{\partial \bar{T}_h}{\partial \bar{x}}(\bar{x}, 0) - f'(\bar{x})\right)^2 \right] \mathrm{d}\bar{x},$$

but while this procedure is more in keeping with the spirit of the finite element method, it is also more complicated than (6.55) and does not in any event improve the order of accuracy of the approximate solution.

We solve (6.52) using a finite difference method. With time step $\Delta \bar{t}$, the finite difference approximation $\underline{b}^j$ to the vector $\underline{b}$ and the vectors $\underline{c}^j$ and $\underline{d}^j$ are defined by

$$\underline{b}^j \approx \underline{b}(j\Delta\bar{t}), \quad \underline{c}^j = \underline{c}(j\Delta\bar{t}), \quad \underline{d}^j = \underline{d}(j\Delta\bar{t}),$$

while the time derivatives in (6.52) are approximated by

$$\frac{\mathrm{d}}{\mathrm{d}t}\underline{b} \approx \frac{1}{\Delta t}\big(\underline{b}^{j+1} - \underline{b}^j\big), \quad \frac{\mathrm{d}}{\mathrm{d}t}\underline{c} \approx \frac{1}{\Delta t}\big(\underline{c}^{j+1} - \underline{c}^j\big).$$

The non-derivative terms in (6.52) are approximated by weighted averages of their values at times $\bar{t}$ and $\bar{t} + \Delta\bar{t}$, so that altogether equation (6.52) is approximated by

$$\frac{1}{\Delta t}\mathbf{C}\big(\underline{b}^{j+1} - \underline{b}^j\big) + \mathbf{K}\big(\theta\,\underline{b}^{j+1} + (1-\theta)\underline{b}^j\big) = \frac{1}{\Delta t}\big(\underline{c}^{j+1} - \underline{c}^j\big) + \theta\,\underline{d}^{j+1} + (1-\theta)\underline{d}^j,$$

where $0 \le \theta \le 1$ is the weight factor. In particular the cases $\theta = 0$, $\theta = \frac{1}{2}$ and $\theta = 1$ correspond respectively to the explicit, Crank-Nicolson and fully implicit finite difference methods for solving (6.52) and we do not have to treat each case individually (see Problem 6). A simple rearrangement of this equation gives

$$\mathbf{D}\underline{b}^{j+1} = \mathbf{E}\underline{b}^j + \underline{c}^{j+1} - \underline{c}^j + \Delta\bar{t}\big(\theta\,\underline{d}^{j+1} + (1-\theta)\underline{d}^j\big), \tag{6.56}$$

where the symmetric tridiagonal positive definite matrices $\mathbf{D}$ and $\mathbf{E}$ are given by

$$\mathbf{D} = \mathbf{C} + \theta\Delta\bar{t}\,\mathbf{K}, \quad \mathbf{E} = \mathbf{C} - (1-\theta)\Delta\bar{t}\,\mathbf{K}. \tag{6.57}$$

Note that for all three special cases (and indeed for any $0 \le \theta \le 1$) a non-trivial linear system has to be solved in order to determine $\underline{b}^{j+1}$ so that the explicit method ($\theta = 0$) is no simpler than the two implicit methods $\theta = \frac{1}{2}$ or $\theta = 1$, and the term explicit is more conventional than appropriate. Further the explicit method is particularly unattractive in this situation because the system of difference equations (6.56) can be shown to suffer from a stability problem for $\theta < \frac{1}{2}$ (see Problem 8), which places an upper bound on the ratio of the time step $\Delta\bar{t}$ to the squares of the element lengths, and limits the size of the time step severely. For $\theta \ge \frac{1}{2}$ system (6.56) is stable for all time steps $\Delta\bar{t}$ and the only restriction on the time step for the Crank-Nicolson and fully implicit methods are those relating to accuracy. The Crank-Nicolson method ($\theta = \frac{1}{2}$) for solving (6.52) is accurate to order $O\big((\Delta\bar{t})^2\big)$ whereas the fully implicit ($\theta = 1$) and explicit ($\theta = 0$) schemes are accurate to order $O(\overline{\Delta t})$. Whatever value of $0 \leq \theta \leq 1$ is chosen, $\mathbf{E}$ is an invertible matrix, and we can therefore determine $\underline{b}^{j+1}$ given $\underline{b}^j$ and the temperatures $g_1(\overline{t})$ and $g_2(\overline{t})$ at the endpoints. Thus, starting with $\underline{b}^0$, determined by (6.55), we can successively generate $\underline{b}^1$, $\underline{b}^2$ and so forth up to $\underline{b}^j$. The components $b_i^j \approx b_i(j\overline{\Delta t})$ of $\underline{b}^j$ can then be substituted into (6.47) to determine $\tilde{T}_h(\overline{x}, j\overline{\Delta t})$, the finite difference approximation to $\overline{T}_h(\overline{x}, \overline{t})$.

<u>Example 6.6</u> Describe an algorithm which uses the finite element method outlined in this section to solve the heat flow problem for the finite region $0 \leq \overline{x} \leq 1$ with initial temperature $\overline{u}_0 = 1$ and zero boundary temperature at $\overline{x} = 0$ and $\overline{x} = 1$.

To implement the finite element method discussed in this section we require some means of representing and storing the finite element mesh. Since the mesh is fully determined by the N nodes in our case, we can represent and store the mesh as an array of the nodes $\overline{x}_i$ ($1 \leq i \leq N$). Using these values we can determine the length of each element, namely $h_i = \overline{x}_{i+1} - \overline{x}_i$, $1 \leq i \leq N - 1$ and hence using (6.50) and (6.51) we can generate the capacitance and stiffness matrices $\mathbf{C}$ and $\mathbf{K}$. Once the time step $\overline{\Delta t}$ and θ are prescribed we can determine the matrices $\mathbf{D}$ and $\mathbf{E}$ from $\mathbf{C}$ and $\mathbf{K}$, or more efficiently, we can bypass the capacitance and stiffness matrices totally and find $\mathbf{D}$ and $\mathbf{E}$ directly from $\overline{\Delta t}$, θ and h_i using (6.50), (6.51), (6.53) and (6.57). Since the boundary temperatures are zero ($g_1(\overline{t}) = g_2(\overline{t}) = 0$) the vectors $\underline{c}^j$ and $\underline{d}^j$ are identically zero and (6.56) becomes

$$\mathbf{D}\underline{b}^{j+1} = \mathbf{E}\underline{b}^j, \tag{6.58}$$

or, since $\mathbf{D}$ is invertible

$$\underline{b}^{j+1} = \mathbf{D}^{-1}\mathbf{E}\underline{b}^j. \tag{6.59}$$

Generally it is more efficient to solve the system of equations (6.58) than to generate the matrix $\mathbf{D}^{-1}\mathbf{E}$ and evaluate (6.59), due to the special tridiagonal symmetric structure of these matrices. We determine $\underline{b}^0$ from the initial condition $\overline{T}(\overline{x}, \overline{t}) = 1$ so that the ($N - 2$) components of $\underline{b}^0$ are all unity. Solving (6.58) or applying (6.59) j times gives $\underline{b}^j$ and we determine $\tilde{T}_h(\overline{x}, j\overline{\Delta t})$ as

$$\tilde{T}_h(\bar{x}, j\Delta\bar{t}) = \sum_{i=2}^{N-1} b_i^j \phi_i(\bar{x}),$$

where the coefficients b_i^j ($2 \le i \le N - 1$) are the $(i - 1)$st components of $\underline{b}^j$. A FORTRAN subroutine which carries out this process is given in Figure 6.12 and discussed in Section 6.6. In Table 6.6 we compare the exact solution of this problem with finite element solutions for $\theta = 0, \frac{1}{2}, 1$ and two different time steps $\Delta\bar{t} = 0.00025$ and $\Delta\bar{t} = 0.00100$, generated by this subroutine using a mesh with 20 elements and 21 equally spaced nodes ($h = 0.05$). Since both the problem and the finite element mesh are symmetric about $\bar{x} = 0.5$, so are the exact and numerical solutions, and therefore they are only tabulated for $0 \le \bar{x} \le 0.5$. Notice that the finite element solution for $\theta = 0$ and $\Delta\bar{t} = 0.00100$ is meaningless due to the stability restrictions mentioned previously.

We now turn to the finite element method for heat flow problems involving *natural* boundary conditions, that is boundary conditions involving temperature gradients such as prescribed temperature gradients or Newton cooling conditions. We consider a general natural boundary condition problem, $(6.6)_{1,2}$ subject to the generalized Newton cooling boundary conditions

$$-\frac{\partial\overline{T}}{\partial\bar{x}}(0,\bar{t}) + \mu_0\overline{T}(0,\bar{t}) = \overline{T}_0(\bar{t}), \quad \frac{\partial\overline{T}}{\partial\bar{x}}(1,\bar{t}) + \mu_1\overline{T}(1,\bar{t}) = \overline{T}_1(\bar{t}), \tag{6.60}$$

which reduce to prescribed temperature gradient boundary conditions if μ_0 and μ_1 vanish. We then solve for the unknown boundary temperature gradients in (6.39) in

		$\Delta\bar{t} = 0.00025$			$\Delta\bar{t} = 0.00100$		
$\bar{x}$	$\overline{T}(\bar{x},\bar{t})$	$\theta = 1.0$	$\theta = 0.5$	$\theta = 0.0$	$\theta = 1.0$	$\theta = 0.5$	$\theta = 0.0$
0.0	0.0000	0.0000	0.0000	0.0000	0.0000	0.0000	0.000E00
0.1	0.1468	0.1463	0.1461	0.1459	0.1468	0.1461	-1.309E38
0.2	0.2790	0.2782	0.2778	0.2775	0.2792	0.2778	0.000E00
0.3	0.3839	0.3828	0.3827	0.3819	0.3842	0.3824	0.000E00
0.4	0.4513	0.4500	0.4494	0.4489	0.4516	0.4494	0.000E00
0.5	0.4745	0.4731	0.4726	0.4720	0.4748	0.4726	0.000E00

Table 6.6. Comparison of the exact solution with finite element solutions of Example 6.6 with 20 elements of equal length for three values of θ and two values of $\Delta\bar{t}$ at time $\bar{t} = 0.1$.

terms of $\overline{T}_0(\overline{t})$, $\overline{T}(0,\overline{t})$, $\overline{T}_1(\overline{t})$ and $\overline{T}(1,\overline{t})$, so that the weak formulation of the problem $(6.6)_{1,2}$ and (6.60) is to find a function $\overline{T}(\overline{x},\overline{t})$ which satisfies $(6.6)_2$ at time $\overline{t}=0$ and is such that

$$\int_0^1 \left(\phi(\overline{x}) \frac{\partial \overline{T}}{\partial \overline{t}} + \frac{d\phi}{d\overline{x}} \frac{\partial \overline{T}}{\partial \overline{x}} \right) d\overline{x} = \phi(1)\left[\overline{T}_1(\overline{t}) - \mu_1 \overline{T}(1,\overline{t}) \right] + \phi(0)\left[\overline{T}_0(\overline{t}) - \mu_0 \overline{T}(0,\overline{t}) \right],$$

$$(6.61)$$

for all times $\overline{t}>0$ and all test functions $\phi(\overline{x}) \in H[0,1]$. We note that no boundary conditions are explicitly imposed on either $\overline{T}(\overline{x},\overline{t})$ or the test functions $\phi(\overline{x})$. The boundary conditions (6.60) are known as *natural* because they can be incorporated into the weak formulation (6.61), rather than explicitly imposed on the solution. We develop a finite element solution of this problem as follows.

As for the previous problem, we choose a finite element mesh with elements $\Omega_1, \Omega_2, \ldots, \Omega_{N-1}$, nodes $\overline{x}_1, \overline{x}_2, \ldots, \overline{x}_N$ and shape functions $\phi_1(\overline{x}), \phi_2(\overline{x}), \ldots, \phi_N(\overline{x})$. We assume an approximate solution of the form

$$\overline{T}_h(\overline{x},\overline{t}) = \sum_{i=1}^{N} b_i(\overline{t})\phi_i(\overline{x}),$$

$$(6.62)$$

where h denotes the length of the largest element, and the finite element form of (6.61) is then to find $\overline{T}_h(\overline{x},\overline{t})$ such that

$$\int_0^1 \left(\phi(\overline{x}) \frac{\partial \overline{T}_h}{\partial \overline{t}} + \frac{d\phi}{d\overline{x}} \frac{\partial \overline{T}_h}{\partial \overline{x}} \right) d\overline{x} =$$

$$\phi(1)\left[\overline{T}_1(\overline{t}) - \mu_1 \overline{T}_h(1,\overline{t}) \right] + \phi(0)\left[\overline{T}_0(\overline{t}) - \mu_0 \overline{T}_h(0,\overline{t}) \right],$$

$$(6.63)$$

for all time $\overline{t}>0$ and all test functions of the form (6.42). As before (6.63) reduces to a system of first order linear ordinary differential equations for the functions $b_i(\overline{t})$ $(1 \leq i \leq N)$, namely

$$\sum_{j=1}^{N} \left(C_{ij} \frac{db_j}{d\overline{t}} + K_{ij}b_j(\overline{t}) \right) = \delta_{iN}\left[\overline{T}_1(\overline{t}) - \mu_1 \delta_{jN} b_j(\overline{t}) \right] + \delta_{i1}\left[\overline{T}_0(\overline{t}) - \mu_0 \delta_{j1} b_j(\overline{t}) \right],$$

where C_{ij} and K_{ij} are given by (6.49) and δ_{ij} is the Kronecker delta. Note that since $b_1(\overline{t})$ and $b_N(\overline{t})$ are unknown, and there are no constraints on the boundary values of the test functions $\phi(\overline{x})$, we have, in this case, a system of N equations for the N unknown functions $b_i(\overline{t})$, which we can write in matrix form as

$$\mathbf{C}\frac{d}{d\overline{t}}\underline{b} + \mathbf{K}\underline{b}(\overline{t}) = \underline{c}(\overline{t}),$$

$$(6.64)$$

where the $N \times N$ positive definite symmetric tridiagonal capacitance and stiffness matrices are given by

$$\mathbf{C} = \begin{pmatrix} C_{11} & \cdots & C_{1N} \\ \vdots & & \vdots \\ C_{N1} & \cdots & C_{NN} \end{pmatrix}, \quad \mathbf{K} = \begin{pmatrix} K_{11}+\mu_0 & K_{12} & \cdots & & K_{1N} \\ K_{21} & K_{22} & \cdots & & K_{2N} \\ \vdots & & & & \vdots \\ K_{N-11} & \cdots & K_{N-1N-1} & K_{N-1N} \\ K_{N1} & \cdots & K_{NN-1} & K_{NN}+\mu_1 \end{pmatrix}, \quad (6.65)$$

and the N dimensional vectors $\underline{b}(\bar{t})$ and $\underline{c}(\bar{t})$ are given by

$$\underline{b}(\bar{t}) = \begin{pmatrix} b_1(\bar{t}) \\ b_2(\bar{t}) \\ \vdots \\ b_{N-1}(\bar{t}) \\ b_N(\bar{t}) \end{pmatrix}, \quad \underline{c}(\bar{t}) = \begin{pmatrix} \bar{T}_0(\bar{t}) \\ 0 \\ \vdots \\ 0 \\ \bar{T}_1(\bar{t}) \end{pmatrix}. \quad (6.66)$$

As with the previous problem, we solve (6.64) using finite differences, and in exactly the same way as before we obtain the system of equations

$$\mathbf{D}\underline{b}^{j+1} = \mathbf{E}\underline{b}^{j} + \theta\underline{c}^{j+1} + (1-\theta)\underline{c}^{j}, \quad (6.67)$$

where the matrices $\mathbf{D}$ and $\mathbf{E}$ are given in terms of the matrices $\mathbf{C}$ and $\mathbf{K}$ given by (6.65) and $\Delta\bar{t}$ and θ by (6.57). Further, the vectors $\underline{b}^{j}$ and $\underline{c}^{j}$ are given by

$$\underline{b}^{j} \approx \underline{b}(j\Delta\bar{t}), \quad \underline{c}^{j} = \underline{c}(j\Delta\bar{t}). \quad (6.68)$$

Values of θ in the range $\frac{1}{2} \le \theta \le 1$ are preferred since for this range (6.67) is stable for all time steps $\Delta\bar{t}$. We determine the initial vector $\underline{b}^{0}$ from the initial conditions $(6.6)_2$, so that the components b_i^{0} $(1 \le i \le N)$ of $\underline{b}^{0}$ are given by

$$b_i^{0} = f(\bar{x}_i), \quad 1 \le i \le N. \quad (6.69)$$

With $\underline{b}^{0}$ so determined we can find $\underline{b}^{j}$ for $j \ge 1$ by applying and solving (6.67) j times, and hence $\tilde{T}_h(\bar{x}, j\Delta\bar{t})$ for any $j \ge 0$.

Example 6.7 Describe a finite element method to solve the heat flow problem for the finite region $0 \le \bar{x} \le 1$ with initial temperature $\bar{u}_0 = 1$, no heat flow across $\bar{x} = 0$

$$\frac{\partial \bar{T}}{\partial \bar{x}}(0, \bar{t}) = 0,$$

and Newton heat loss into a medium at zero temperature at $\bar{x} = 1$,

$$\frac{\partial \bar{T}}{\partial \bar{x}}(1, \bar{t}) + \bar{T}(1, \bar{t}) = 0.$$

This is simply problem (6.63) with $\mu_0 = 0$, $\mu_1 = 1$, $\bar{T}_0(\bar{t}) = 0$, $\bar{T}_1(\bar{t}) = 0$ and initial condition

$$\bar{T}(\bar{x}, 0) = 1.$$

As before, to implement the finite element method we represent and store the mesh as an array of the nodes $\bar{x}_i$ $(1 \le i \le N)$. Using these values we determine the length of each element, namely $h_i = \bar{x}_{i+1} - \bar{x}_i$, $1 \le i \le N - 1$ and hence using (6.50) and (6.51) and the values of $\mu_0 = 0$ and $\mu_1 = 1$, we can generate the capacitance and stiffness matrices $\mathbf{C}$ and $\mathbf{K}$ defined by (6.65). With the time step $\Delta \bar{t}$ and θ prescribed we determine the matrices $\mathbf{D}$ and $\mathbf{E}$. Since both $\bar{T}_0(\bar{t}) = \bar{T}_1(\bar{t}) = 0$ the vector $\underset{\sim}{c}^j$ is identically zero and (6.67) becomes

		$\Delta \bar{t} = 0.00025$			$\Delta \bar{t} = 0.00100$		
$\bar{x}$	$\bar{T}(\bar{x}, \bar{t})$	$\theta = 1.0$	$\theta = 0.5$	$\theta = 0.0$	$\theta = 1.0$	$\theta = 0.5$	$\theta = 0.0$
0.0	0.9931	0.9932	0.9933	0.9933	0.9930	0.9933	0.000E00
0.1	0.9918	0.9919	0.9920	0.9920	0.9918	0.9920	1.517E38
0.2	0.9878	0.9879	0.9880	0.9880	0.9878	0.9880	1.443E38
0.3	0.9804	0.9805	0.9805	0.9806	0.9804	0.9805	0.000E00
0.4	0.9684	0.9685	0.9685	0.9686	0.9685	0.9685	0.000E00
0.5	0.9505	0.9507	0.9507	0.9507	0.9507	0.9507	0.000E00
0.6	0.9252	0.9253	0.9253	0.9253	0.9254	0.9253	1.665E38
0.7	0.8908	0.8910	0.8909	0.8909	0.8912	0.8909	0.000E00
0.8	0.8462	0.8464	0.8463	0.8462	0.8466	0.8463	0.000E00
0.9	0.7905	0.7907	0.7906	0.7905	0.7910	0.7906	1.667E38
1.0	0.7236	0.7237	0.7236	0.7235	0.7240	0.7236	1.112E38

Table 6.7. Comparison of finite element solutions with 20 elements of equal length for three values of θ and two values of $\Delta \bar{t}$ with the exact solution of Example 6.7 at time $\bar{t} = 0.1$.

$$\mathbf{D}\underset{\sim}{b}^{j+1} = \mathbf{E}\underset{\sim}{b}^{j}. \tag{6.70}$$

Again, it is generally more computationally efficient to solve this matrix equation as it stands, rather than to invert $\mathbf{D}$ and determine $\underset{\sim}{b}^{j+1}$ by matrix multiplication using

$$\underset{\sim}{b}^{j+1} = \mathbf{D}^{-1}\mathbf{E}\underset{\sim}{b}^{j}.$$

From the initial condition we know that the N components of $\underset{\sim}{b}^{0}$ are all unity, and thus we can determine $\underset{\sim}{b}^{j}$ by starting with $\underset{\sim}{b}^{0}$ and solving (6.70) j times. A FORTRAN subroutine which carries out this task is given in Figure 6.13 and discussed in Section 6.6. In Table 6.7 we compare the exact solution of this problem with finite element solutions for $\theta = 0, \frac{1}{2}, 1$ and two different time steps $\Delta \bar{t} = 0.00025$ and $\Delta \bar{t} = 0.00100$, generated using a mesh with 20 elements and 21 equally spaced nodes ($h = 0.05$). Notice that the finite element solution for $\theta = 0$ and $\Delta \bar{t} = 0.00100$ is meaningless, again due to the stability restrictions for $\theta < \frac{1}{2}$.

Finally in this section we discuss in general terms some of the advantages and disadvantages of the finite element method. The main advantage of the finite element approach to the numerical solution of the heat equation is that in two and three spatial dimensions the elements (which become two and three dimensional polygons, respectively) can be adapted to essentially any spatial geometry, whereas a finite difference grid cannot. Even for the problems considered in this section, it is apparent that the size of the elements do not have to be uniform, so that for instance the entire theory presented here remains applicable if we choose to refine our mesh (that is, make the elements smaller) in some particular region if, for instance, we suspect there are large local variations in the temperature or temperature gradient in that region. In two and three spatial dimensions this advantage of the finite element method over the finite difference method becomes more pronounced and more important.

There is however one quite serious limitation of the finite element method as outlined in this section. Because the spatial convergence of the method is not pointwise but convergence in the mean, there can arise local oscillations in the approximate temperature about points where the initial temperature is discontinuous or at end points where the initial and boundary data are inconsistent. Provided the finite element method is stable, these oscillations are eventually damped out and vanish, but if the scheme is only just stable these oscillations may remain for a large number of time steps. If the scheme is unstable, these oscillations accelerate the rate at which the numerical solution collapses. In Figure 6.6 we show a finite element solution to Example 6.6 for $\theta = 0.5$ and draw the readers' attention to the behaviour of the solution near the end points $\bar{x} = 0$ and $\bar{x} = 1.0$ where the initial and boundary data are inconsistent. For the fully implicit finite element numerical solution $\theta = 1.0$ these oscillations do not arise, since the first time step is essentially 'unaware' of the inconsistencies in initial and boundary data due to the forward time difference approximation, and therefore the oscillations do not occur at the first time step.

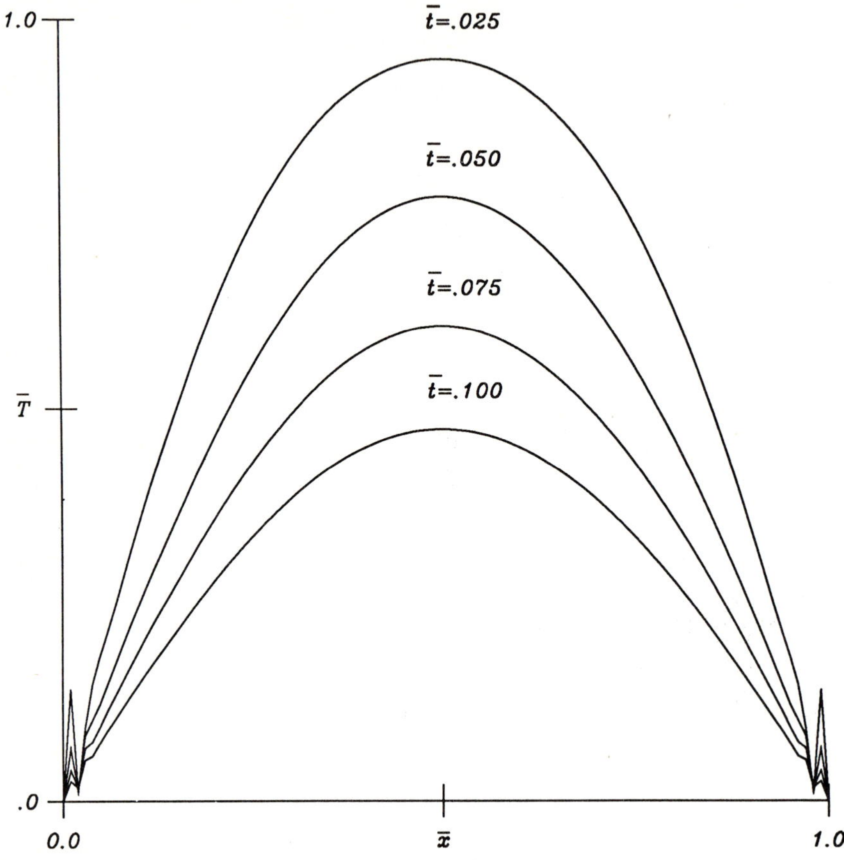

Figure 6.6. Finite element solution to Example 6.6 for $\theta = 0.5$ and $t = 0.1$.

6.5 Boundary integral method

The boundary integral method for obtaining numerical solutions of heat conduction problems is based on the integral representation (2.67) of the temperature $\overline{T}(\overline{x},\overline{t})$ in terms of its initial and boundary values and a Green's function (see Section 2.6). For the moment we consider the heat equation on $[0, 1]$ with only the initial temperature given

$$\left.\begin{array}{c} \dfrac{\partial \overline{T}}{\partial \overline{t}} = \dfrac{\partial^2 \overline{T}}{\partial \overline{x}^2}, \quad 0 < \overline{x} < 1, \\[2mm] \overline{T}(\overline{x},0) = f(\overline{x}), \quad 0 \le \overline{x} \le 1, \end{array}\right\} \tag{6.71}$$

and we use the notation

$$u_1(\overline{t}) = \overline{T}(0,\overline{t}), \quad u_2(\overline{t}) = \overline{T}(1,\overline{t}), \quad v_1(\overline{t}) = \frac{\partial \overline{T}}{\partial \overline{x}}(0,\overline{t}), \quad v_2(\overline{t}) = \frac{\partial \overline{T}}{\partial \overline{x}}(1,\overline{t}). \tag{6.72}$$

We emphasize that, for the while, none of $u_1(\overline{t})$, $u_2(\overline{t})$, $v_1(\overline{t})$ and $v_2(\overline{t})$ are prescribed functions but merely stand for the boundary values of the temperature and temperature gradient. As in Section 2.6, but now in non-dimensional variables, we find that for $0 \le \overline{x} \le 1$ we have

$$\lambda \overline{T}(\overline{x},\overline{t}) = \int_0^1 f(\xi)\overline{G}(\overline{x},\xi,\overline{t},0)\,d\xi$$

$$+ \int_0^{\overline{t}} u_1(\tau)\frac{\partial \overline{G}}{\partial \xi}(\overline{x},0,\overline{t},\tau)\,d\tau - \int_0^{\overline{t}} u_2(\tau)\frac{\partial \overline{G}}{\partial \xi}(\overline{x},1,\overline{t},\tau)\,d\tau \quad 0 \le \overline{x} \le 1,$$

$$- \int_0^{\overline{t}} v_1(\tau)\overline{G}(\overline{x},0,\overline{t},\tau)\,d\tau + \int_0^{\overline{t}} v_2(\tau)\overline{G}(\overline{x},1,\overline{t},\tau)\,d\tau, \tag{6.73}$$

where $\overline{G}(\overline{x},\xi,\overline{t},\tau)$ is a non-dimensional Green's function which we will take to be

$$\overline{G}(\overline{x},\xi,\overline{t},\tau) = \frac{H(\overline{t}-\tau)}{2\sqrt{\pi(\overline{t}-\tau)}} \exp \frac{-(\overline{x}-\xi)^2}{4(\overline{t}-\tau)}, \tag{6.74}$$

and where λ is given by (see Problem 10)

$$\lambda = \begin{cases} 1, & 0 < \overline{x} < 1, \\ \dfrac{1}{2}, & \overline{x} = 0 \text{ or } \overline{x} = 1. \end{cases} \tag{6.75}$$

In particular substituting $\bar{x} = 0$ and $\bar{x} = 1$ gives the integral equations

$$\frac{1}{2}u_1(\bar{t}) = \int_0^1 f(\xi)\overline{G}(0,\xi,\bar{t},0)\,d\xi - \int_0^{\bar{t}} u_2(\tau)\frac{\partial\overline{G}}{\partial\xi}(0,1,\bar{t},\tau)\,d\tau$$

$$- \int_0^{\bar{t}} v_1(\tau)\overline{G}(0,0,\bar{t},\tau)\,d\tau + \int_0^{\bar{t}} v_2(\tau)\overline{G}(0,1,\bar{t},\tau)\,d\tau,$$

$$\frac{1}{2}u_2(\bar{t}) = \int_0^1 f(\xi)\overline{G}(1,\xi,\bar{t},0)\,d\xi + \int_0^{\bar{t}} u_1(\tau)\frac{\partial\overline{G}}{\partial\xi}(1,0,\bar{t},\tau)\,d\tau$$

$$- \int_0^{\bar{t}} v_1(\tau)\overline{G}(1,0,\bar{t},\tau)\,d\tau + \int_0^{\bar{t}} v_2(\tau)\overline{G}(1,1,\bar{t},\tau)\,d\tau,$$

$$(6.76)$$

where we have used the fact that

$$\frac{\partial\overline{G}}{\partial\xi}(\bar{x},\xi,\bar{t},\tau) = \frac{(\bar{x}-\xi)}{2(\bar{t}-\tau)}\overline{G}(\bar{x},\xi,\bar{t},\tau), \tag{6.77}$$

so that $\frac{\partial\overline{G}}{\partial\xi}$ vanishes when $\xi = \bar{x}$, thereby eliminating integrals involving $\frac{\partial\overline{G}}{\partial\xi}(0,0,\bar{t},\tau)$ and $\frac{\partial\overline{G}}{\partial\xi}(1,1,\bar{t},\tau)$.

For a well posed heat conduction problem we have two prescribed boundary conditions which leaves only two of the functions $u_1(\bar{t})$, $u_2(\bar{t})$, $v_1(\bar{t})$ and $v_2(\bar{t})$ unknown, and we also have the two equations (6.76) for those unknowns. For prescribed surface temperature or temperature gradient, we are given two of the functions $u_1(\bar{t})$, $u_2(\bar{t})$, $v_1(\bar{t})$ and $v_2(\bar{t})$ and we find the other two by solving (6.76), and having done this we determine $\overline{T}(\bar{x},\bar{t})$ from (6.73). Strictly speaking, Newton cooling conditions do not eliminate unknowns but rather increase the number of equations for determining the unknowns, but for all practical purposes the difference is immaterial. For example, given the boundary conditions

$$\overline{T}(0,\bar{t}) = 0, \quad \frac{\partial\overline{T}}{\partial\bar{x}}(1,\bar{t}) + \overline{T}(1,\bar{t}) = 1,$$

for problem (6.71) we have

$$u_1(\bar{t}) = 0, \quad v_2(\bar{t}) = 1 - u_2(\bar{t}),$$

and on substitution into (6.76) we obtain the two integral equations

$$\int_0^{\bar{t}} v_1(\tau)\overline{G}(0,0,\bar{t},\tau)\,d\tau \;-\; \int_0^{\bar{t}} (1 - u_2(\tau))\overline{G}(0,1,\bar{t},\tau)\,d\tau$$

$$+\; \int_0^{\bar{t}} u_2(\tau)\frac{\partial \overline{G}}{\partial \xi}(0,1,\bar{t},\tau)\,d\tau \;=\; \int_0^1 f(\xi)\overline{G}(0,\xi,\bar{t},0)\,d\xi,$$

$$\int_0^{\bar{t}} v_1(\tau)\overline{G}(1,0,\bar{t},\tau)\,d\tau \;-\; \int_0^{\bar{t}} (1 - u_2(\bar{t}))\overline{G}(1,1,\bar{t},\tau)\,d\tau$$

$$=\; -\frac{1}{2}u_2(\bar{t}) \;+\; \int_0^1 f(\xi)\overline{G}(1,\xi,\bar{t},0)\,d\xi,$$

for the undetermined functions $u_2(\bar{t})$ and $v_1(\bar{t})$. Once we determine the functions $u_2(\bar{t})$ and $v_1(\bar{t})$ (and hence by a trivial calculation, $v_2(\bar{t})$) from these two equations, we can determine $\overline{T}(\bar{x},\bar{t})$ for $0 \le \bar{x} \le 1$ from (6.73). Unfortunately, it is in general impossible to solve the resulting system of equations analytically, and a numerical procedure is necessary. Any method of solving a heat conduction problem based solving (6.76) to determine the boundary behaviour of the temperature and temperature gradient, and (6.73) to find $\overline{T}(\bar{x},\bar{t})$ is referred to as a *boundary integral method*. We present a simple boundary integral method in the following.

To obtain numerical solutions of (6.76) we proceed as follows. We divide the interval $[0,\bar{t}]$ into N equal time steps of length $\Delta\bar{t} = \bar{t}/N$, and we express the time integrals occurring in (6.73) and (6.76) in the form

$$\int_0^{\bar{t}} \phi(\tau)w(\bar{x},\xi,\bar{t},\tau)\,d\tau \;=\; \sum_{i=1}^{N} \int_{(i-1)\Delta\bar{t}}^{i\Delta\bar{t}} \phi(\tau)w(\bar{x},\xi,\bar{t},\tau)\,d\tau,$$

where $\phi(\tau)$ represents any of the four functions $u_1(\bar{t})$, $u_2(\bar{t})$, $v_1(\bar{t})$ or $v_2(\bar{t})$ and w is either $\overline{G}$ or $\frac{\partial \overline{G}}{\partial \xi}$. Since for fixed $\bar{x}$, ξ and $\bar{t}$, neither $\overline{G}$ nor $\frac{\partial \overline{G}}{\partial \xi}$ changes sign as τ varies, we can apply the mean value theorem to obtain

$$\int_{(i-1)\Delta\bar{t}}^{i\Delta\bar{t}} \phi(\tau)w(\bar{x},\xi,\bar{t},\tau)\,d\tau \;=\; \phi(\bar{t}_i)\int_{(i-1)\Delta\bar{t}}^{i\Delta\bar{t}} w(\bar{x},\xi,\bar{t},\tau)\,d\tau, \qquad (6.78)$$

where $\bar{t}_i$ is a number such that $(i-1)\Delta\bar{t} \le \bar{t}_i \le i\Delta\bar{t}$ and which depends on the functions $\phi(\bar{t})$ and w and the values of $\bar{x}$, ξ and $\bar{t}$. We obtain our approximation by assuming that the $\bar{t}_i$ are independent of any of $\phi(\bar{t})$, w, $\bar{x}$, ξ, $\bar{t}$, but that it does satisfy $(i-1)\Delta\bar{t} \le \bar{t}_i \le i\Delta\bar{t}$. In what follows we will assume that

$$\bar{t}_i = i\Delta\bar{t}, \qquad (6.79)$$

which is equivalent to assuming that the boundary temperatures and temperature gradients are piecewise constant. The precise value chosen for $\bar{t}_i$ is somewhat immaterial with regard to setting up the boundary integral approximation, although it does affect some of the details and the accuracy of the solutions obtained (see Problem 12 for the case $\bar{t}_i = (i - \frac{1}{2})\Delta\bar{t}$). This assumption reduces (6.76) to two equations in $4N$ variables, namely $u_1(\bar{t}_i)$, $u_2(\bar{t}_i)$, $v_1(\bar{t}_i)$ and $v_2(\bar{t}_i)$ ($1 \le i \le N$). Having made this approximation, however, we can no longer expect these equations to hold for all times between zero and $\bar{t}$, but rather we require them to hold only at the times $\bar{t}_i$ ($1 \le i \le N$). In this way we reduce (6.76) to a system of $2N$ equations in $4N$ variables, namely

$$\sum_{j=1}^{N} \left\{ -A_{ij}(0,1)u_2(\bar{t}_j) - B_{ij}(0,0)v_1(\bar{t}_j) + B_{ij}(0,1)v_2(\bar{t}_j) \right\}$$

$$= \frac{1}{2}u_1(\bar{t}_i) - \int_0^1 f(\xi)\overline{G}(0,\xi,\bar{t}_i,0)\,\mathrm{d}\xi,$$

$$1 \le i \le N, \quad (6.80)$$

$$\sum_{j=1}^{N} \left\{ A_{ij}(1,0)u_1(\bar{t}_j) - B_{ij}(1,0)v_1(\bar{t}_j) + B_{ij}(1,1)v_2(\bar{t}_j) \right\}$$

$$= \frac{1}{2}u_2(\bar{t}_i) - \int_0^1 f(\xi)\overline{G}(1,\xi,\bar{t}_i,0)\,\mathrm{d}\xi,$$

and where

$$A_{ij}(\bar{x},\xi) = \int_{(j-1)\Delta\bar{t}}^{j\Delta\bar{t}} \frac{\partial\overline{G}}{\partial\xi}(\bar{x},\xi,\bar{t}_i,\tau)\,\mathrm{d}\tau,$$

$$(6.81)$$

$$B_{ij}(\bar{x},\xi) = \int_{(j-1)\Delta\bar{t}}^{j\Delta\bar{t}} \overline{G}(\bar{x},\xi,\bar{t}_i,\tau)\,\mathrm{d}\tau.$$

In a similar manner we obtain the approximation

$$\lambda\tilde{T}(\bar{x},\bar{t}_i) = \int_0^1 f(\xi)\overline{G}(\bar{x},\xi,\bar{t}_i,0)\,\mathrm{d}\xi$$

$$(6.82)$$

$$+ \sum_{j=1}^{N} \left\{ A_{ij}(\bar{x},0)u_1(\bar{t}_j) - A_{ij}(\bar{x},1)u_2(\bar{t}_j) - B_{ij}(\bar{x},0)v_1(\bar{t}_j) + B_{ij}(\bar{x},1)v_2(\bar{t}_j) \right\},$$

for the temperature at $0 \le \bar{x} \le 1$ and time $\bar{t}_i$, and where λ is defined by (6.75). Note that, unlike the finite difference method, we can in theory find the temperature directly at any point $0 \le \bar{x} \le 1$. The spatial integrals in (6.80) and (6.82) can be done either analytically if possible, or numerically using any one of a range of numerical integration methods available, and as such will not concern us greatly.

Before proceeding, we require expressions for $A_{ij}(\overline{x}, \xi)$ and $B_{ij}(\overline{x}, \xi)$, namely we have to evaluate the integrals

$$A_{ij}(\overline{x}, \xi) = \frac{(\overline{x} - \xi)}{4\sqrt{\pi}} \int_{(j-1)\Delta\overline{t}}^{j\Delta\overline{t}} \frac{H(i\Delta\overline{t} - \tau)}{(i\Delta\overline{t} - \tau)^{\frac{3}{2}}} \exp \frac{-(\overline{x} - \xi)^2}{4(i\Delta\overline{t} - \tau)} \, d\tau,$$

$$B_{ij}(\overline{x}, \xi) = \frac{1}{2\sqrt{\pi}} \int_{(j-1)\Delta\overline{t}}^{j\Delta\overline{t}} \frac{H(i\Delta\overline{t} - \tau)}{\sqrt{i\Delta\overline{t} - \tau}} \exp \frac{-(\overline{x} - \xi)^2}{4(i\Delta\overline{t} - \tau)} \, d\tau.$$

It is clear that for $j > i$ both $A_{ij}(\overline{x}, \xi)$ and $B_{ij}(\overline{x}, \xi)$ must vanish because the Heaviside step function is zero over the entire range of integration in this case. For $i \geq j$ it is not difficult to establish by a simple change of variables $\sigma = (\overline{x} - \xi)/2\sqrt{i\Delta\overline{t} - \tau}$ that $A_{ij}(\overline{x}, \xi)$ is given by

$$A_{ij}(\overline{x}, \xi) = \begin{cases} \frac{1}{2}\,\mathrm{erf}\!\left(\frac{\overline{x} - \xi}{2\sqrt{\Delta\overline{t}(i - j)}}\right) - \frac{1}{2}\,\mathrm{erf}\!\left(\frac{\overline{x} - \xi}{2\sqrt{\Delta\overline{t}(i + 1 - j)}}\right), & i > j, \\[2ex] \frac{1}{2}\,\mathrm{sgn}(\overline{x} - \xi) - \frac{1}{2}\,\mathrm{erf}\!\left(\frac{\overline{x} - \xi}{2\sqrt{\Delta\overline{t}}}\right), & i = j, \\[2ex] 0, & i < j, \end{cases} \tag{6.83}$$

where the function $\mathrm{sgn}(\overline{x})$ is defined by

$$\mathrm{sgn}(\overline{x}) = \begin{cases} 1, & \overline{x} > 0, \\ 0, & \overline{x} = 0, \\ -1, & \overline{x} < 0. \end{cases}$$

We note that from the definition of the integral it is clear that $A_{ij}(\overline{x}, \xi)$ is anti-symmetric in $\overline{x}$ and ξ, in the sense that

$$A_{ij}(\overline{x}, \xi) = -A_{ij}(\xi, \overline{x}).$$

It is somewhat more difficult to obtain expressions for $B_{ij}(\overline{x}, \xi)$, but it can be shown (see Problem 11) that

$$
B_{ij}(\overline{x},\xi) = \begin{cases}
\left\{\begin{aligned}
&\sqrt{\tfrac{\Delta\overline{t}}{\pi}(i+1-j)}\,\exp\frac{-(\overline{x}-\xi)^2}{4\Delta\overline{t}(i+1-j)} - \frac{|\overline{x}-\xi|}{2}\,\mathrm{erfc}\!\left(\frac{|\overline{x}-\xi|}{2\sqrt{\Delta\overline{t}(i+1-j)}}\right) \\
&-\sqrt{\tfrac{\Delta\overline{t}}{\pi}(i-j)}\,\exp\frac{-(\overline{x}-\xi)^2}{4\Delta\overline{t}(i-j)} + \frac{|\overline{x}-\xi|}{2}\,\mathrm{erfc}\!\left(\frac{|\overline{x}-\xi|}{2\sqrt{\Delta\overline{t}(i-j)}}\right)
\end{aligned}\right\}, & i > j \\[2ex]
\sqrt{\dfrac{\Delta\overline{t}}{\pi}}\,\exp\dfrac{-(\overline{x}-\xi)^2}{4\Delta\overline{t}} - \dfrac{|\overline{x}-\xi|}{2}\,\mathrm{erfc}\!\left(\dfrac{|\overline{x}-\xi|}{2\sqrt{\Delta\overline{t}}}\right), & i = j, \\[2ex]
0, & i < j.
\end{cases}
\tag{6.84}
$$

We note that the matrix $\mathbf{B}(\overline{x},\xi)$ is symmetric in $\overline{x}$ and ξ in the sense that

$$
\mathbf{B}(\overline{x},\xi) = \mathbf{B}(\xi,\overline{x}),
$$

and that the matrices $\mathbf{A}(\overline{x},\xi) = \big(A_{ij}(\overline{x},\xi)\big)$ and $\mathbf{B}(\overline{x},\xi) = \big(B_{ij}(\overline{x},\xi)\big)$ are both lower triangular matrices since $A_{ij}(\overline{x},\xi) = B_{ij}(\overline{x},\xi) = 0$ for $j > i$. Moreover, for $i \geq j$ we have

$$
A_{i+1\,j+1}(\overline{x},\xi) = A_{ij}(\overline{x},\xi), \quad B_{i+1\,j+1}(\overline{x},\xi) = B_{ij}(\overline{x},\xi),
$$

so that the matrices $\mathbf{A}(\overline{x},\xi)$ and $\mathbf{B}(\overline{x},\xi)$ are fully determined by the values $A_{i1}(\overline{x},\xi)$ and $B_{i1}(\overline{x},\xi)$ $(1 \leq i \leq N)$ respectively. Specifically we can write

$$
\mathbf{A}(\overline{x},\xi) = \begin{pmatrix}
A_{11}(\overline{x},\xi) & 0 & \cdots & 0 \\
A_{21}(\overline{x},\xi) & A_{11}(\overline{x},\xi) & & 0 \\
A_{31}(\overline{x},\xi) & A_{21}(\overline{x},\xi) & & 0 \\
\vdots & \vdots & & \vdots \\
A_{N1}(\overline{x},\xi) & A_{(N-1)1}(\overline{x},\xi) & \cdots & A_{11}(\overline{x},\xi)
\end{pmatrix},
$$

$$
\mathbf{B}(\overline{x},\xi) = \begin{pmatrix}
B_{11}(\overline{x},\xi) & 0 & \cdots & 0 \\
B_{21}(\overline{x},\xi) & B_{11}(\overline{x},\xi) & & 0 \\
B_{31}(\overline{x},\xi) & B_{21}(\overline{x},\xi) & & 0 \\
\vdots & \vdots & & \vdots \\
B_{N1}(\overline{x},\xi) & B_{(N-1)1}(\overline{x},\xi) & \cdots & B_{11}(\overline{x},\xi)
\end{pmatrix}.
\tag{6.85}
$$

By way of illustration, we consider the boundary integral solution to (6.71) subject to the prescribed boundary temperatures

$$
\overline{T}(0,\overline{t}) = g_1(\overline{t}), \quad \overline{T}(1,\overline{t}) = g_2(\overline{t}),
\tag{6.86}
$$

where $g_1(\bar{t})$ and $g_2(\bar{t})$ are known functions of time $\bar{t}$. Using (6.86) we can substitute $u_1(\bar{t}) = g_1(\bar{t})$ and $u_2(\bar{t}) = g_2(\bar{t})$ so that (6.80) becomes

$$\sum_{j=1}^{N} \left\{ -B_{ij}(0,0)v_1(\bar{t}_j) + B_{ij}(0,1)v_2(\bar{t}_j) \right\} = \tfrac{1}{2}g_1(\bar{t}_i) + \sum_{j=1}^{N} A_{ij}(0,1)g_2(\bar{t}_j)$$

$$- \int_0^1 f(\xi)\overline{G}(0,\xi,\bar{t}_i,0)\,\mathrm{d}\xi,$$

$$1 \le i \le N,$$

$$\sum_{j=1}^{N} \left\{ -B_{ij}(1,0)v_1(\bar{t}_j) + B_{ij}(1,1)v_2(\bar{t}_j) \right\} = \tfrac{1}{2}g_2(\bar{t}_i) - \sum_{j=1}^{N} A_{ij}(1,0)g_1(\bar{t}_j)$$

$$- \int_0^1 f(\xi)\overline{G}(1,\xi,\bar{t}_i,0)\,\mathrm{d}\xi,$$

giving us $2N$ equations for the $2N$ unknown boundary temperature gradient values $v_1(\bar{t}_j)$ and $v_2(\bar{t}_j)$ $(1 \le j \le N)$. We can write these as a coupled system of matrix equations

$$\left. \begin{aligned} -\mathbf{C}\underline{v}_1 + \mathbf{D}\underline{v}_2 &= \underline{b}_1, \\ -\mathbf{D}\underline{v}_1 + \mathbf{C}\underline{v}_2 &= \underline{b}_2, \end{aligned} \right\} \tag{6.87}$$

where the matrices $\mathbf{C}$ and $\mathbf{D}$ are given by

$$\mathbf{C} = \mathbf{B}(0,0) = \mathbf{B}(1,1), \quad \mathbf{D} = \mathbf{B}(0,1) = \mathbf{B}(1,0), \tag{6.88}$$

and the vectors $\underline{v}_1$ and $\underline{v}_2$ are given by

$$\underline{v}_1 = \begin{pmatrix} v_1(\bar{t}_1) \\ \vdots \\ v_1(\bar{t}_N) \end{pmatrix}, \quad \underline{v}_2 = \begin{pmatrix} v_2(\bar{t}_1) \\ \vdots \\ v_2(\bar{t}_N) \end{pmatrix},$$

and the ith components of vectors $\underline{b}_1$ and $\underline{b}_2$ are given respectively by

$$\tfrac{1}{2}g_1(\bar{t}_i) + \sum_{j=1}^{N} A_{ij}(0,1)g_2(\bar{t}_j) - \int_0^1 f(\xi)\overline{G}(0,\xi,\bar{t}_i,0)\,\mathrm{d}\xi,$$

$$\tfrac{1}{2}g_2(\bar{t}_i) - \sum_{j=1}^{N} A_{ij}(1,0)g_1(\bar{t}_j) - \int_0^1 f(\xi)\overline{G}(1,\xi,\bar{t}_i,0)\,\mathrm{d}\xi.$$

Since both matrices $\mathbf{C}$ and $\mathbf{D}$ are lower triangular matrices (see (6.85)) we can solve (6.87) quite effectively by forward substitution (see Problem 13). Once the vectors $\underset{\sim}{v}_1$ and $\underset{\sim}{v}_2$ are determined, we find our approximate temperature $\tilde{T}(\bar{x}, \bar{t}_i)$ using (6.82).

<u>Example 6.8</u> Describe a boundary integral method to solve the heat flow problem on [0, 1] with initial temperature $u_0 = 1$ and zero prescribed temperature at the endpoints $\bar{x} = 0$ and $\bar{x} = 1$.

Since in this case we have zero prescribed temperature at $\bar{x} = 0$ and at $\bar{x} = 1$ and initial temperature unity, that is $u_0(\bar{t}) = u_1(\bar{t}) = 0$ and $f(\bar{x}) = 1$, the approximation (6.82) for the temperature becomes

$$\tilde{T}(\bar{x}, \bar{t}_i) = \int_0^1 \overline{G}(\bar{x}, \xi, \bar{t}_i, 0) \, d\xi + \sum_{j=1}^{N} \left\{ -B_{ij}(\bar{x}, 0)v_1(\bar{t}_j) + B_{ij}(\bar{x}, 1)v_2(\bar{t}_j) \right\}, \quad (6.89)$$

where λ is defined by (6.75) and where the vectors of approximate boundary temperature gradients $\underset{\sim}{v}_1$ and $\underset{\sim}{v}_2$ have to be determined from (6.87), and where the ith components of the vectors $\underset{\sim}{b}_1$ and $\underset{\sim}{b}_2$ in (6.87) are given respectively by

$$-\int_0^1 \overline{G}(0, \xi, \bar{t}_i, 0) \, d\xi, \quad -\int_0^1 \overline{G}(1, \xi, \bar{t}_i, 0) \, d\xi.$$

These spatial integrals can be evaluated analytically, and we find that

$$\int_0^1 \overline{G}(\bar{x}, \xi, \bar{t}, 0) \, d\xi = \frac{1}{2}\left[\text{erf}\left(\frac{\bar{x}}{2\sqrt{t}}\right) + \text{erf}\left(\frac{1 - \bar{x}}{2\sqrt{t}}\right)\right], \quad (6.90)$$

which can be implemented more efficiently than a numerical integration of the Green's function. In view of (6.85), it is only necessary to evaluate $B_{i1}(0, 0)$ and $B_{i1}(1, 0)$ $(1 \le i \le N)$, using (6.84), to determine the matrices $\mathbf{C}$ and $\mathbf{D}$ given by (6.88). Equation (6.87) can then be solved by forward substitution to determine $\underset{\sim}{v}_1$ and $\underset{\sim}{v}_2$. Once these vectors are determined, we can evaluate the approximate temperature $\tilde{T}(\bar{x}, \bar{t}_i)$ which can be determined from (6.89) using (6.84) and (6.90). In Figure 6.14 we present a FORTRAN program which carries out this procedure.

In Table 6.8 we compare the exact solution to this problem with two boundary integral solutions (using 10 and 20 time steps) for time $\bar{t} = 0.1$. The solutions generated are more accurate than either finite difference or finite element solutions with comparable time steps, which is due mainly to the large degree to which the structure of the solution is already inherent in the boundary integral formulation. Thus although the initial analysis required to establish the boundary integral approximation is greater than that required to set up either finite difference or finite element methods, the extra work is rewarded by a scheme which is more efficient and accurate. In particular the facility to take relatively large time steps while retaining accuracy is an advantage of the method. Another advantage which is not apparent in one spatial dimension, and which the boundary integral approach has in common with the finite element method is the facility to easily accommodate complicated spatial geometries. The boundary integral approach also has the property that the dimensionality of the resulting integral equations is one less than the dimensionality of the original problem, which is a significant advantage when working in two and three spatial dimensions. Even in one dimension it is apparent that for the boundary integral method it is only necessary to solve one system of equations once to determine all the information necessary to construct the approximate temperature, whereas for the finite difference and finite element methods it is necessary to solve a system of equations for each time step.

$\bar{x}$	$T(\bar{x}, \bar{t})$	$N = 40$	$N = 20$	$N - 10$
0.0	0.0000	0.0000	0.0000	0.0000
0.1	0.1468	0.1478	0.1490	0.1517
0.2	0.2790	0.2802	0.2828	0.2868
0.3	0.3839	0.3864	0.3888	0.3939
0.4	0.4513	0.4541	0.4569	0.4625
0.5	0.4745	0.4774	0.4803	0.4862

Table 6.8. Comparison of the exact solution with boundary integral solutions of Example 6.8 for three values of $N = 10, 20, 40$ at time $\bar{t} = 0.1$.

6.6 Sample programs

In this section we give the FORTRAN code for the main subroutines used to generate the numerical solutions given in this chapter. To avoid unnecessary details we have omitted the code concerned with the input of initial and boundary data and the output of the calculated solutions, except in the boundary integral program given in Figure 6.14. All subroutines have been tested on a SPERRY 1100/72 computer and all are written in standard FORTRAN 77, except the boundary integral algorithm which uses the non-standard functions ERF and ERFC. These functions are not standard FORTRAN 77 functions and many FORTRAN compilers will not recognize them as intrinsic functions so that the reader will have to either write their own functions to calculate erf(x) and erfc(x) or obtain such functions from a library of mathematical routines. The rational approximations given in Abramowitz and Stegun (1972), namely Abramowitz and Stegun 7.1.26 and 7.1.28 are good approximations to use. As well we have used two routines, SPTSL and SGTSL, from the LINPACK linear algebra package to solve the tridiagonal matrix equations which arise in the implicit finite difference and in the finite element solutions. The source code for LINPACK and the package itself are currently widely available.

First we consider a subroutine to solve the heat conduction problem for the finite region $0 \le \bar{x} \le 1$ with specified initial temperature and constant prescribed temperature at $\bar{x} = 0$ and $\bar{x} = 1$, using an explicit finite difference method. Such a subroutine is shown in Figure 6.7, and it is a generalization of the algorithm described in Example 6.1. The numerical temperature $\tilde{T}_{ij}$ $0 \le i \le N$ is stored as an $N + 1$ dimensional array. The initial temperature is passed to the subroutine in the array T, the number N of space intervals in the integer argument N, the constant prescribed temperatures at $\bar{x} = 0$ and $\bar{x} = 1$ in the real arguments TX0 and TX1 respectively, the ratio δ in the real argument DELTA and the number of time steps necessary in the integer argument LOOPS. The subroutine uses two internal arrays TOLD and TNEW which hold the current and updated temperatures respectively. The first task is to copy the initial temperature as passed in T into the array TOLD, and this is done in lines 15–16. The subroutine enters its main time stepping loop at line 18. At each execution this loop performs one time step to find TNEW from TOLD, and then copies TNEW into TOLD, ready for the next time step. The inner loop in lines 21–22 performs the explicit finite difference calculations (6.7) to determine TNEW(I) for $1 \le I \le$ N-1. Then the boundary temperatures TNEW(0) and TNEW(N) are set to their prescribed values, TX0 and TX1 respectively. This completes one single time step, and the loop in lines 28–29 copies the updated temperature TNEW into the array TOLD, ready for the next time step. The entire outer loop is executed LOOPS times. On leaving the main loop the updated temperature is copied back into the array T in lines 33–34 and the subroutine returns with the updated temperature stored in the array T.

```
 1:          SUBROUTINE XPLICT( T,N,TX0,TX1,DELTA,LOOPS  )
 2:          DIMENSION T(0:N),TNEW(0:200),TOLD(0:200)
 3:* T(0:N)  is the array containing the initial temperature
 4:* N+1     is the number of nodes in the finite difference
 5:*         grid. N must be less than 201
 6:* TX0     is the constant fixed temperature at the end X=0
 7:* TX1     is the constant fixed temperature at the end X=1
 8:* DELTA   is the ratio of time step to square of space step
 9:* LOOPS  is the number of time steps to be taken
10:* TOLD     is an array to hold the OLD temperature
11:* TNEW     is an array to hold the NEW temperature as it is
12:*          calculated at each time step.
13:*
14:* First initialize TOLD to the initial temperature
15:        DO 100 I=0,N
16:100        TOLD(I)=T(I)
17:* Loop for LOOPS time steps
18:        DO 400 I=1,LOOPS
19* Use the explicit finite difference method to calculate
20:* the NEW temperature from the OLD temperature
21:          DO 200 J=1,N-1
22:200         TNEW(J)=TOLD(J)+DELTA*(TOLD(J+1)+TOLD(J-1)-2*TOLD(J))
23:* Set the boundary temperatures to their prescribed values
24:          TNEW(0)-TX0
25:          TNEW(N)=TX1
26:* Once the NEW temperature has been completely determined,it
27:* becomes the OLD temperature for the next time step
28:          DO 300 J=0,N
29:300         TOLD(J)=TNEW(J)
30:400     CONTINUE
31:* Copy the calculated temperature back into T and return after
32:* completing LOOPS time steps
33:        DO 500 I=0,N
34:500        T(I)=TOLD(I)
35:        RETURN
36:        END
```

Figure 6.7. FORTRAN subroutine illustrating the explicit finite difference method for prescribed temperature boundary conditions.

This subroutine is easily adapted to a wide variety of constant boundary conditions, such as constant prescribed temperature gradient or Newton cooling into media at constant temperatures, by modifying the code in lines 24–25 which determines the boundary temperature at $\bar{x} = 0$ and $\bar{x} = 1$ respectively. Thus, for example, the boundary conditions

$$-\frac{\partial \overline{T}}{\partial \bar{x}}(0,\bar{t}) + \overline{T}(0,\bar{t}) = A, \quad \frac{\partial \overline{T}}{\partial \bar{x}}(1,\bar{t}) + \overline{T}(1,\bar{t})^4 = B,$$

which lead to the finite difference equations

$$\tilde{T}_{0j+1} = \tilde{T}_{0j} + 2\delta\left(\tilde{T}_{1j} - (1 + \Delta\bar{x})\tilde{T}_{0j} + A\,\Delta\bar{x}\right),$$

$$\tilde{T}_{Nj+1} = \tilde{T}_{Nj} + 2\delta\left(\tilde{T}_{N-1j} - \tilde{T}_{Nj} + \Delta\bar{x}(B - \tilde{T}_{Nj}^4)\right),$$

where A and B are constants are easily accommodated by replacing lines 24–25 by the following two lines

```
24:TNEW(0)=TOLD(0)+2*DELTA*(TOLD(1)-(1+1.0/N)*TOLD(0)+TX0/N)
25:TNEW(N)=TOLD(0)+2*DELTA*(TOLD(N-1)-TOLD(N)+(TX1-TOLD(N)**4)/N)    .
```

Here the variables `TX0` and `TX1` are now used to carry the values of A and B respectively, and we have used the fact that the step length $\Delta\bar{x}$ is given by $1/N$.

In Figure 6.8 we present a subroutine which solves the heat equation posed on the finite interval $0 \le \bar{x} \le 1$ with specified initial temperature and constant prescribed boundary temperatures at $\bar{x} = 0$ and $\bar{x} = 1$, by using a fully implicit finite difference algorithm. This subroutine performs each time step by solving the matrix equation (6.25) using the symmetric tridiagonal solver routine `SPTSL` from the LINPACK library, which is more efficient than inverting the matrix (6.26) and using (6.27) for each time step. This is a generalization of the algorithm described in Example 6.3. As for the explicit finite difference method, we store the temperature as an array of $N + 1$ nodal values, and the arguments passed to the subroutine are the same as those in the previous figure. The symmetric tridiagonal matrix given by (6.26) is stored as two arrays, `DIAG` which holds the $N - 1$ diagonal elements (each given by $1 + 2\delta$) and `OFFDIA` which holds the $N - 2$ non-zero off diagonal elements (each given by $-\delta$). Because of the behaviour of the LINPACK routine `SPTSL`, only one internal array, `TNEW`, is needed to hold the temperature.

The subroutine works as follows. In lines 11–12 the array `TNEW` is set to the initial temperature passed in the array `T` and the time stepping loop begins at line 14.

Inside this loop the arrays DIAG and OFFDIA are initialized (lines 16–19), which is necessary at each time step because SPTSL scrambles them on invocation. The array TNEW is set equal to $\underset{\sim}{T}{}^{j-1} + \delta \underset{\sim}{U}$ in lines 21–22 by adding $\mathrm{DELTA*TX0}$ and

```
 1:          SUBROUTINE  IMPLIC(  T,N,TX0,TX1,DELTA,LOOPS   )
 2:          DIMENSION  T(0:N),TNEW(0:200)
 3:          DIMENSION  DIAG(200),OFFDIA(200)
 4:*  TNEW       is  the  array  to  hold  the  temperature
 5:*  DIAG       is  the  array  to  hold  the  diagonal  elements  of  the
 6:*             symmetric  tridiagonal  matrix
 7:*  OFFDIA     is  the  array  to  hold  the  non-zero  off  diagonal  elements
 8:*             of  the  symmetric  tridiagonal  matrix
 9:*
10:*  First  initialize  TNEW  to  the  initial  temperature
11:          DO  100  I=0,N
12:100       TNEW(I)=T(I)
13:*  Loop  for  LOOPS  time  steps
14:          DO  400  I=1,LOOPS
15:*  Set  up  the  diagonal  and  off  diagonal  elements  of  the  matrix
16:          DO  200  J=1,N-1
17:             DIAG(J)=1+2*DELTA
18:             OFFDIA(J)=-DELTA
19:200       CONTINUE
20:*  Add  the  extra  components  to  TNEW(1)  and  TNEW(N-1)
21:          TNEW(1)=TNEW(1)+DELTA*TX0
22:          TNEW(N-1)=TNEW(N-1)+DELTA*TX1
23:*  Use  the  LINPACK  routine  SPTSL  to  solve  the  system  of  equations
24:          CALL  SPTSL(  N-1,DIAG,OFFDIA,TNEW(1)   )
25:*  Set  the  end  temperatures  to  their  prescribed  values
26:          TNEW(0)=TX0
27:          TNEW(N)=TX1
28:400    CONTINUE
29:*  Copy  the  NEW  temperature  back  into  T  and  return  after  completing
30:*  LOOPS  time  steps
31:          DO  500  I=0,N
32:500       T(I)=TNEW(I)
33:       RETURN
34:       END
```

Figure 6.8. FORTRAN subroutine illustrating the fully implicit finite difference method for prescribed temperature boundary conditions.

DELTA*TX1 respectively to TNEW(1) and TNEW(N-1), and then the LINPACK tridiagonal solver SPTSL is invoked at line 24 to solve the system of linear equations. The subroutine passes the dimension of the system, N-1, the symmetric tridiagonal matrix in DIAG and OFFDIA, and the first element of the known array TNEW, TNEW(1), to SPTSL which returns the solution in TNEW(1) ... TNEW(N-1). Finally in the loop TNEW(0) and TNEW(N) are set to their prescribed constant values TX0 and TX1 respectively in lines 26–27. This time stepping loop is executed LOOPS times, and then the updated temperature TNEW is copied back into the array T in lines 31–32. The subroutine ends and the updated temperature is passed back to the calling program in T.

The Crank-Nicolson solver shown in Figure 6.9 is a generalization of the algorithm described in Example 6.4 and is intended to solve the same heat flow problem as the previous two subroutines. It works in a similar manner to the implicit solver in Figure 6.8, except that two arrays TOLD and TNEW are needed to contain the current and updated temperature respectively. The array TOLD is set to the initial temperature passed in the array T in lines 5–6 and the main time stepping loop starts at line 8. In lines 11–15 the arrays DIAG and OFFDIA are initialized to the diagonal and off diagonal of the matrix **C** given in (6.32) and TNEW is set equal to the right hand side of (6.31) at line 14 and in lines 17–18. The subroutine SPTSL is called to solve the system of equations, and returns the result in TNEW(1) ... TNEW(N-1). The boundary temperatures TNEW(0) and TNEW(N) are set to their prescribed values in lines 22–23 and then TNEW is copied into TOLD ready for the next time step. After LOOPS time steps, the updated temperature is copied back into the array T in lines 25–26 and the subroutine returns to the calling program.

In Figures 6.10 and 6.11 we give fully implicit and Crank-Nicolson solvers for the heat equation subject to prescribed initial temperature and the boundary conditions

$$-\frac{\partial \overline{T}}{\partial \overline{x}}(0, \overline{t}) = A, \quad \frac{\partial \overline{T}}{\partial \overline{x}}(1, \overline{t}) + \overline{T}(1, \overline{t}) = B,$$

where A and B are fixed constants represented by A and B respectively in the given listings. The fully implicit finite difference approximations for the temperatures $\tilde{T}_{0j}$

and $\tilde{T}_{Nj}$ are in this case

$$\left.\begin{array}{l}(1 + 2\delta)\tilde{T}_{0j} - 2\delta\tilde{T}_{1j} = \tilde{T}_{0j-1} + 2\delta\,\Delta\overline{x}A\,, \\[2ex] (1 + 2\delta(1 + \Delta\overline{x}))\tilde{T}_{Nj} - 2\delta\tilde{T}_{1j} = \tilde{T}_{Nj-1} + 2\delta\,\Delta\overline{x}B\,,\end{array}\right\} \quad j \geq 1,$$

while for the Crank-Nicolson procedure we have

```
 1:            SUBROUTINE CRANKN( T,N,TX0,TX1,DELTA,LOOPS  )
 2:            DIMENSION T(0:N),TOLD(0:200),TNEW(0:200)
 3:            DIMENSION DIAG(200),OFFDIA(200)
 4:* Set TOLD to the initial temperature
 5:            DO 100 I=0,N
 6:100            TOLD(I)=T(I)
 7:* Loop for LOOPS time steps
 8:            DO 400 I=1,LOOPS
 9:* Set up the diagonal and off diagonal matrix elements and
10:* set up the known temperature vector function
11:            DO 200 J=1,N-1
12:              DIAG(J)=1+DELTA
13:              OFFDIA(J)=-DELTA/2
14:              TNEW(J)=(1-DELTA)*TOLD(J)+DELTA*(TOLD(J-1)+TOLD(J+1))/2
15:200         CONTINUE
16:* Add the extra components to TNEW(1) and TNEW(N-1)
17:            TNEW(1)=TNEW(1)+DELTA*TX0
18:            TNEW(N-1)=TNEW(N-1)+DELTA*TX1
19:* Call LINPACK routine SPTSL to solve the system of equations
20:            CALL SPTSL( N-1,DIAG,OFFDIA,TNEW(1)  )
21:* Set the end temperatures to their prescribed values
22:            TNEW(0)=TX0
23:            TNEW(N)=TX1
24:* Copy TNEW into TOLD
25:            DO 300 J=0,N
26:300            TOLD(J)=TNEW(J)
27:400         CONTINUE
28:*
29:* Copy the updated temperature into T and return
30:         DO 500 I=0,N
31:500            T(I)=TNEW(I)
32:         RETURN
33:         END
```

Figure 6.9. FORTRAN subroutine illustrating the Crank-Nicolson finite difference method for prescribed temperature boundary conditions.

$$(1 + \delta)\tilde{T}_{0j+1} - \delta\tilde{T}_{1j+1} = (1 - \delta)\tilde{T}_{0j} + \delta\tilde{T}_{1j} + 2\delta\,\Delta\overline{x}A ,$$

$$(1 + \delta(1 + \Delta\overline{x}))\tilde{T}_{Nj+1} - \delta\tilde{T}_{N-1j+1}$$
$$= (1 - \delta(1 + \Delta\overline{x}))\tilde{T}_{Nj} + \delta\tilde{T}_{N-1j} + 2\delta\,\Delta\overline{x}B ,$$

$$\left. \right\} \quad j \geq 0.$$

Since the tridiagonal matrices associated with the fully implicit and Crank-Nicolson finite difference schemes for this problem are no longer symmetric, it is necessary to store them as three arrays; the N non-zero subdiagonal elements in RLOWER, the

```
 1:         SUBROUTINE IMPLC2( T,N,A,B,DELTA,LOOPS )
 2:         DIMENSION T(0:N),TNEW(0:200)
 3:         DIMENSION DIAG(0:200),UPPER(0:200),RLOWER(0:200)
 4:* Set TNEW equal to the initial temperature
 5:         DO 100 I=0,N
 6:100         TNEW(I)=T(I)
 7:* Enter time stepping loop
 8:         DO 400 I=1,LOOPS
 9:* Set up diagonal, sub and super diagonals of the matrix
10:         DO 200 J=0,N
11:            DIAG(J)=1+2*DELTA
12:            RLOWER(J)=-DELTA
13:            UPPER(J)=-DELTA
14:200         CONTINUE
15:* Adjust ends of the diagonal, sub and super diagonal arrays
16:         UPPER(0)=-2*DELTA
17:         DIAG(N)=1+2*DELTA*(1.0+1.0/N)
18:         RLOWER(N)=-2*DELTA
19:* Adjust end temperature values to accommodate boundary
20:* conditions.
21:         TNEW(0)=TNEW(0)+2*DELTA*A/N
22:         TNEW(N)=TNEW(N)+2*DELTA*B/N
23:* Call the general tridiagonal solver SGTSL
24:         CALL SGTSL( N+1,RLOWER,DIAG,UPPER,TNEW,INFO)
25:400     CONTINUE
26:* Copy TNEW back into T and leave the subroutine
27:         DO 500 I=0,N
28:500         T(I)=TNEW(I)
29:         RETURN
30:         END
```

Figure 6.10.   FORTRAN subroutine illustrating the fully implicit finite difference method for problems involving temperature gradient boundary conditions.

$N + 1$ diagonal elements in DIAG and the N non-zero superdiagonal elements in UPPER. The LINPACK general purpose tridiagonal solver SGTSL is used to solve the linear system at each time step.

```
 1:          SUBROUTINE CRANK2( T,N,A,B,DELTA,LOOPS )
 2:          DIMENSION T(0:N),TNEW(0:200),TOLD(0:200)
 3:          DIMENSION DIAG(0:200),UPPER(0:200),RLOWER(0:200)
 4:* Set TOLD equal to the initial temperature
 5:          DO 100 I=0,N
 6:100          TOLD(I)=T(I)
 7:* Enter the time stepping loop
 8:          DO 400 I=1,LOOPS
 9:* Set up the diagonal, sub and super diagonals of the
10:* matrix and set up TNEW
11:          DO 200 J=1,N-1
12:             DIAG(J)=1+DELTA
13:             RLOWER(J)=-DELTA/2
14:             UPPER(J)=-DELTA/2
15:             TNEW(J)=(1-DELTA)*TOLD(J)+DELTA*(TOLD(J-1)+TOLD(J+1))/2
16:200          CONTINUE
17:* Adjust ends of the diagonal, sub and super diagonal arrays
18:          UPPER(0)=-DELTA
19:          DIAG(0)=1+DELTA
20:          RLOWER(N)=-DELTA
21:          DIAG(N)=1+DELTA*(1.0+1.0/N)
22:* Adjust end values of TNEW to accommodate boundary conditions
23:          TNEW(0)=(1-DELTA)*TOLD(0)+DELTA*TOLD(1)
24:          TNEW(0)=TNEW(0)+2*DELTA*A/N
25:          TNEW(N)=(1-DELTA*(1.0+1.0/N))*TOLD(N)+DELTA*TOLD(N-1)
26:          TNEW(N)=TNEW(N)+2*DELTA*B/N
27:* Call the general tridiagonal solver SGTSL
28:          CALL SGTSL( N+1,RLOWER,DIAG,UPPER,TNEW,INFO)
29:* Copy the updated temperature in TNEW back into TOLD
30:          DO 300 J=0,N
31:300          TOLD(J)=TNEW(J)
32:400       CONTINUE
33:* Copy updated temperature into T and exit the subroutine
34:          DO 500 I=0,N
35:500          T(I)=TNEW(I)
36:          RETURN
37:          END
```

Figure 6.11. FORTRAN subroutine illustrating the Crank-Nicolson finite difference method for problems involving temperature gradient boundary conditions.

These two subroutines work in the same manner as the previous two subroutines. Both copy the initial temperature from the array T into an internal array and enter the time stepping loop. Here the arrays representing the tridiagonal matrices are set up (SGTSL scrambles them at each invocation) and TNEW is set equal the known vector in the linear system. The linear system is solved by SGTSL which is given the dimension of the linear system, N+1, the subdiagonal RLOWER, diagonal DIAG and superdiagonal UPPER of the matrix and the known vector in TNEW. It returns the solution in TNEW and an integer diagnostic in INFO. In the case of the Crank-Nicolson routine, TNEW is then copied into TOLD ready for the next time step. After LOOPS time steps the updated temperature is copied back into T and the subroutine finishes.

In Figure 6.12 we give a finite element program which solves the heat equation on [0, 1] subject to prescribed initial temperature and fixed boundary temperatures at $\bar{x} = 0$ and $\bar{x} = 1$. The initial temperature is passed in T, the finite element mesh in XNODE and the number of nodes in NODES. Thus T(I) represents the initial temperature at nodal position XNODE(I). The fixed boundary temperatures at $\bar{x} = 0$ and $\bar{x} = 1$ are passed in TX0 and TX1 respectively, the length of the time step in TSTEP, the number of time steps required in LOOPS and θ in THETA. In lines 6–7 the routine calculates the length of the elements, then in lines 9–18 it sets up the *full* $N \times N$ capacitance and stiffness matrices using (6.50) and stores them as arrays of diagonal and off diagonal elements. The arrays CD and CO store the diagonal and off diagonal elements, respectively, of the capacitance matrix while the real arrays KD and KO contain the diagonal and off diagonal elements of the stiffness matrix. Once these matrices have been generated the program calculates the matrices **D** and **E**, defined by (6.57), and stores the respective diagonal elements in DD and ED and the off diagonal elements in DOFF and EOFF, in lines 20–25. Once **D** and **E** are determined the capacitance matrix is unnecessary and the arrays holding it can be used for temporary storage.

```
 1:           SUBROUTINE FINIT1(T,XNODE,NODES,TX0,TX1,TSTEP,LOOPS,THETA)
 2:           REAL T(NODES),XNODE(NODES),TOLD(200),TNEW(200)
 3:           REAL HSIZE(200),CD(200),CO(200),KD(200),KO(200)
 4:           REAL DD(200),DOFF(200),ED(200),EOFF(200)
 5:* Calculate size of elements
 6:           DO 100 I=1,NODES-1
 7:100            HSIZE(I)=XNODE(I+1)-XNODE(I)
 8:* Use these to generate the FULL N*N matrices
 9:           CD(1)=HSIZE(1)/3.0
10:           KD(1)=1.0/HSIZE(1)
11:           DO 110 I=2,NODES-1
12:               CD(I)=( HSIZE(I-1)+HSIZE(I) )/3.0
```

```
13:110        KD(I)=1.0/HSIZE(I-1)+1.0/HSIZE(I)
14:      CD(NODES)=HSIZE(NODES-1)/3.0
15:      KD(NODES)=1.0/HSIZE(NODES-1)
16:      DO 120 I=1,NODES-1
17:         CO(I)=HSIZE(I)/6.0
18:120      KO(I)=-1.0/HSIZE(I)
19:* Now generate the time step / finite element matrices
20:      DO 130 I=2,NODES-1
21:         DD(I)=CD(I)+TSTEP*THETA*KD(I)
22:130      ED(I)=CD(I)-TSTEP*(1-THETA)*KD(I)
23:      DO 140 I=2,NODES-2
24:         DOFF(I)=CO(I)+TSTEP*THETA*KO(I)
25:140      EOFF(I)=CO(I)-TSTEP*(1-THETA)*KO(I)
26:* Set up the initial temperature
27:      N=NODES-1
28:      DO 150 I=1,NODES
29:150      TOLD(I)=T(I)
30:* Now enter the solver loop
31:      DO 500 LOOP = 1,LOOPS
32:* Set up load vector
33:         TNEW(1)=TX0
34:         TNEW(2)=ED(2)*TOLD(2)+EOFF(2)*TOLD(3)
35:         TNEW(N)=ED(N)*TOLD(N)+EOFF(N-1)*TOLD(N-1)
36:         TNEW(2)=TNEW(2)-TSTEP*KO(1)*TX0
37:         TNEW(N)=TNEW(N)-TSTEP*KO(N)*TX1
38:         TNEW(NODES)=TX1
39:         DO 190 I=3,NODES-2
40:190      TNEW(I)=ED(I)*TOLD(I)+EOFF(I)*TOLD(I+1)+EOFF(I-1)*TOLD(I-1)
41:* Prepare to solve
42:         DO 200 I-1,NODES
43:            CD(I)=DD(I)
44:200         CO(I)=DOFF(I)
45:         CALL SPTSL( NODES-2,CD(2),CO(2),TNEW(2)   )
46:         DO 210 I=1,NODES
47:210         TOLD(I)=TNEW(I)
48:         TOLD(1)=TX0
49:         TOLD(NODES)=TX1
50:500   CONTINUE
51:* Copy back updated temperature into T
52:      DO 600 I=1,NODES
53:600      T(I)=TOLD(I)
54:      RETURN
55:      END
```

Figure 6.12. FORTRAN subroutine illustrating the finite element method for prescribed temperature boundary conditions.

After setting up the matrices, the initial temperature in T is copied into the array TOLD in lines 28–29 and the time stepping procedure begins at line 31. At each time step TNEW is set equal to the load vector (the right hand side of (6.56)) in lines 33–40, and the arrays DD and DOFF are copied into CD and CO which are used to pass the matrix **D** to the routine SPTSL rather than passing and recalculating DD and DOFF at each time step. The linear system is solved by calling SPTSL which returns the updated temperature in TNEW. This is then copied back into TOLD ready for the next time step and after LOOPS time steps the temperature is copied into the array T and returned to the calling program.

In Figure 6.13 we present a finite element program which solves the heat equation subject to prescribed initial temperature and the natural boundary conditions

$$-\frac{\partial \overline{T}}{\partial \overline{x}}(0, \overline{t}) + \mu_0 \overline{T}(0, \overline{t}) = A, \quad \frac{\partial \overline{T}}{\partial \overline{x}}(1, \overline{t}) + \mu_1 \overline{T}(1, \overline{t}) = B.$$

This subroutine functions in much the same way as the previous subroutine. The quantities A, B, μ_0 and μ_1 are passed in the arguments A, B, R0 and R1 respectively, and otherwise the arguments are the same as in the previous figure. The element lengths are calculated and these are used to find the stiffness and capacitance matrices which are stored in the same way as in the previous subroutine. The matrices **D** and **E** are calculated and the initial temperature copied into TOLD. In the main time stepping loop, TNEW is set equal to the load vector (that is, the right hand side of (6.67)) and the arrays holding **D** are copied into the arrays originally used for the capacitance matrix and the linear system is solved by SPTSL which returns the solution in TNEW. The updated temperature TNEW is copied back into TOLD ready for the next time step. After LOOPS time steps the updated temperature is copied back into T and returned to the calling routine.

```
 1:           SUBROUTINE  FINIT2(T,XNODE,NODES,A,R0,B,R1,TSTEP,LOOPS,THETA)
 2:           REAL  T(NODES),XNODE(NODES),TOLD(200),TNEW(200)
 3:           REAL  HSIZE(200),CD(200),CO(200),KD(200),KO(200)
 4:           REAL  DD(200),DOFF(200),ED(200),EOFF(200)
 5:* Calculate  size  of  elements
 6:           DO 100  I=1,NODES-1
 7:100            HSIZE(I)=XNODE(I+1)-XNODE(I)
 8:* Use  these  to  generate  the  FULL N*N  matrices
 9:           CD(1)=HSIZE(1)/3.0
10:           KD(1)=1.0/HSIZE(1)+R0
11:           DO 110  I=2,NODES-1
12:              CD(I)=(  HSIZE(I-1)+HSIZE(I)  )/3.0
13:110            KD(I)=1.0/HSIZE(I-1)+1.0/HSIZE(I)
```

```
14:          CD(NODES)=HSIZE(NODES-1)/3.0
15:          KD(NODES)=1.0/HSIZE(NODES-1)+R1
16:          DO 120 I=1,NODES-1
17:              CO(I)=HSIZE(I)/6.0
18:120           KO(I)=-1.0/HSIZE(I)
19:* Now generate the time step / finite element matrices
20:          DO 130 I=1,NODES
21:              DD(I)=CD(I)+TSTEP*THETA*KD(I)
22:130           ED(I)=CD(I)-TSTEP*(1-THETA)*KD(I)
23:          DO 140 I=1,NODES-1
24:              DOFF(I)=CO(I)+TSTEP*THETA*KO(I)
25:140           EOFF(I)=CO(I)-TSTEP*(1-THETA)*KO(I)
26:* Set up the initial temperature
27:          DO 150 I=1,NODES
28:150           TOLD(I)=T(I)
29:* Now enter the solver loop
30:          DO 500 LOOP = 1,LOOPS
31:* Set up load vector
32:          DO 190 I=2,NODES-1
33:190         TNEW(I)=EOFF(I)*TOLD(I+1)+EOFF(I-1)*TOLD(I-1)+ED(I)*TOLD(I)
34:          TNEW(1)=ED(1)*TOLD(1)+EOFF(1)*TOLD(2)
35:          TNEW(1)=TNEW(1)+TSTEP*A
36:          TNEW(NODES)=ED(NODES)*TOLD(NODES)+EOFF(NODES-1)*TOLD(NODES-1)
37:          TNEW(NODES)=TNEW(NODES)+TSTEP*B
38:* Prepare to solve
39:          DO 200 I=1,NODES
40:              CD(I)=DD(I)
41:200           CO(I)=DOFF(I)
42:          CALL SPTSL( NODES,CD,CO,TNEW )
43:          DO 210 I=1,NODES
44:210           TOLD(I)=TNEW(I)
45:500      CONTINUE
46:* Copy updated temperature back into T
47:          DO 600 I=1,NODES
48:600           T(I)=TOLD(I)
49:          RETURN
50:          END
```

Figure 6.13. FORTRAN subroutine illustrating the finite element method for problems involving temperature gradient boundary conditions.

Finally in Figure 6.14 we present a complete boundary integral program for the heat equation subject to constant initial temperature and constant prescribed boundary temperatures at $\bar{x} = 0$ and $\bar{x} = 1$. The code given is in fact more sophisticated than is required for such a simple problem, and the reader should have little difficulty in modifying it for more general boundary and initial conditions. The arrays C, D, E and F are used to hold the first column of matrices of the form $\mathbf{A}(\bar{x}, \xi)$ and $\mathbf{B}(\bar{x}, \xi)$ for various combinations of $\bar{x} = 0$, $\bar{x} = 1$ and $\xi = 0$, $\xi = 1$. In view of the structure of these matrices (see (6.85)), it is only necessary to know the first column to fully determine the matrix, and we urge the reader to bear this in mind when examining some of the matrix manipulations in the program. The arrays U1 and U2 are used to store the nodal boundary temperature values at $\bar{x} = 0$ and $\bar{x} = 1$ respectively while the arrays V1 and V2 are used for the nodal boundary temperature gradients at $\bar{x} = 0$ and $\bar{x} = 1$. The arrays B1 and B2 are used to store the vectors $\underline{b}_1$ and $\underline{b}_2$ which occur in (6.87). Initially the program prompts the user for the initial temperature value (T0), the boundary temperatures at $\bar{x} = 0$ and $\bar{x} = 1$ (TX0 and TX1 respectively), the time $\bar{t}$ (TIME) and the number of time steps to use (NSTEP).

The program then determines the matrices $\mathbf{B}(0, 0)$ and $\mathbf{B}(0, 1)$ which are saved in C and D, corresponding to the matrices $\mathbf{C}$ and $\mathbf{D}$ in (6.87). Similarly E and F are used to determine the matrices $\mathbf{A}(0, 1)$ and $\mathbf{A}(1, 0)$. The functions AMAT and BMAT are used to calculate the matrix elements $A_{ij}(\bar{x}, \xi)$ and $B_{ij}(\bar{x}, \xi)$ using (6.83) and (6.84) respectively. Notice that both routines adopt the strategy of testing the size of the arguments to the error and complementary error functions before evaluating the difference of two such functions. This is because, for large x and y it is better to evaluate

$$\mathrm{erf}(x) - \mathrm{erf}(y), \tag{6.91}$$

as

$$\mathrm{erfc}(y) - \mathrm{erfc}(x), \tag{6.92}$$

since for such values, $\mathrm{erf}(x)$ and $\mathrm{erf}(y)$ are of very nearly equal value and many (if not all) significant figures can be lost in the difference (6.91), whereas the difference (6.92) does not suffer from this problem to the same extent. For similar reasons it is best to evaluate (6.92) as (6.91) if x and y are small. This strategy has been adopted throughout this program.

The arrays U1 and U2 are set equal to the prescribed boundary temperatures TX0 at $\bar{x} = 0$ and TX1 at $\bar{x} = 1$ respectively, and then the arrays B1 and B2 corresponding to the vectors $\underline{b}_1$ and $\underline{b}_2$ in (6.87) are initialized. To do this we use the function SPACEI to evaluate the spatial integrals. The function SPACEI uses the exact analytical representation (6.90) to evaluate the spatial integral, although for more general initial conditions a numerical integration would be necessary. Once all the known arrays are set up the subroutine $SOLVE$ is invoked to solve the system (6.87) by using the forward substitution method outlined in Problem 13. Subroutine $SOLVE$ returns the solution for $\underset{\sim}{v}_1$ and $\underset{\sim}{v}_2$, the boundary temperature gradient values, in arrays $V1$ and $V2$. With $V1$ and $V2$ determined, the approximate temperature can be calculated for any $0 \le x \le 1$ using function $TEMP$, which is a simple implementation of equation (6.82). The program uses $TEMP$ to find and write out the temperature at the points $\bar{x} = 0.0, 0.1, \ldots, 0.9, 1.0$.

```
 1:            PROGRAM BI
 2:            REAL C(100),D(100),E(100),F(100),B1(100)
 3:            REAL U1(100),U2(100),V1(100),V2(100),B2(100)
 4:* Prompt user for initial and boundary data
 5:100       WRITE(6,*) 'Initial temp,temp at x=0,temp at x=1 '
 6:            READ(5,*,END=999,ERR=100) T0,TX0,TX1
 7:101       WRITE(6,*) 'Time,number of time steps '
 8:            READ(5,*,END=999,ERR=101) TIME,NSTEP
 9:* Set up matrix and known vector arrays
10:            TSTEP=TIME/NSTEP
11:            DO 130 I=1,NSTEP
12:               C(I)=BMAT(I,1,0.0,0.0,TSTEP)
13:               D(I)=BMAT(I,1,0.0,1.0,TSTEP)
14:               E(I)=AMAT(I,1,0.0,1.0,TSTEP)
15:               F(I)=AMAT(I,1,1.0,0.0,TSTEP)
16:               U1(I)=TX0
17:130            U2(I)=TX1
18:* Form B1 and B2 vectors
19:            DO 150 I=1,NSTEP
20:               B1(I)=0.5*U1(I)-SPACEI(T0,0.0,I*TSTEP)
21:               B2(I)=0.5*U2(I)-SPACEI(T0,1.0,I*TSTEP)
22:               DO 140 J=1,I
23:                  B1(I)=B1(I)+U2(I+1-J)*E(J)
24:140               B2(I)=B2(I)-U1(I+1-J)*F(J)
25:150       CONTINUE
26:* Solve system
27:            CALL SOLVE(V1,V2,C,D,B1,B2,NSTEP)
28:* And write out values at x=0,0.1,...,0.9,1.0
29:            DO 300 I=0,10
30:               XX=FLOAT(I)/10.0
31:300            WRITE(6,*) XX,TEMP(XX,TSTEP,U1,U2,V1,V2,NSTEP,T0)
32:            GOTO 100
33:999       STOP
34:            END
```

```
35:        FUNCTION AMAT(I,J,X,XI,TSTEP)
36:        IF (X.NE.XI) THEN
37:           SIGN=ABS(X-XI)/(X-XI)
38:        ELSE
39:           SIGN=0
40:        ENDIF
41:        IF (J.GT.I) THEN
42:           AMAT=0
43:        ELSEIF (I.EQ.J) THEN
44:           AMAT=SIGN*ERFC(ABS(X-XI)/SQRT(4*TSTEP))/2
45:        ELSE
46:           THING1=ABS((X-XI)/SQRT(4*TSTEP*(I-J)))
47:           THING2=ABS((X-XI)/SQRT(4*TSTEP*(I+1-J)))
48:           IF ((THING1.GT.1).OR.(THING2.GT.1))  THEN
49:              ANS=ERFC(THING2)-ERFC(THING1)
50:           ELSE
51:              ANS=ERF(THING1)-ERF(THING2)
52:           ENDIF
53:           AMAT=SIGN*ANS/2
54:        ENDIF
55:        RETURN
56:        END


57:        FUNCTION BMAT(I,J,X,XI,TS)
58:        PARAMETER (PI=4.0*DATAN(1.0))
59:        IF (I.LT.J) THEN
60:           BMAT=0
61:        ELSEIF (I.EQ.J) THEN
62:           BIT1=2*SQRT(TS/PI)*EXP(-(X-XI)**2/(4*TS))
63:           BIT2=ABS(X-XI)*ERFC(ABS(X-XI)/(2*SQRT(TS)))
64:           BMAT=(BIT1-BIT2)/2
65:        ELSE
66:           BIT1=2*SQRT(TS/PI*(I+1-J))
67:           BIT1=BIT1*EXP(-(X-XI)**2/(4*TS*(I+1-J)))
68:           BIT1A=2*SQRT(TS/PI*(I-J))
69:           BIT1A=BIT1A*EXP(-(X-XI)**2/(4*TS*(I-J)))
70:           BIT1=BIT1-BIT1A
71:           THING1=ABS(X-XI)/SQRT(4*TS*(I+1-J))
72:           THING2=ABS(X-XI)/SQRT(4*TS*(I-J))
73:           IF ((THING1.GT.1).OR.(THING2.GT.1))  THEN
74:              BIT2=ABS(X-XI)*(ERFC(THING1)-ERFC(THING2))
75:           ELSE
76:              BIT2=ABS(X-XI)*(ERF(THING2)-ERF(THING1))
77:           ENDIF
78:           BMAT=(BIT1-BIT2)/2
79:        ENDIF
80:        RETURN
81:        END
```

```
 82:         FUNCTION SPACEI(T0,X,TIME)
 83:         IF (TIME.LE.0) THEN
 84:           SPACEI=T0
 85:         ELSE
 86:           THING1=X/SQRT(4*TIME)
 87:           THING2=(X-1)/SQRT(4*TIME)
 88:           IF ((ABS(THING1).GT.1).OR.(ABS(THING2).GT.1))THEN
 89:             BIT=ERFC(THING2)-ERFC(THING1)
 90:           ELSE
 91:             BIT=ERF(THING1)-ERF(THING2)
 92:           ENDIF
 93:           SPACEI=T0*BIT/2
 94:         ENDIF
 95:         RETURN
 96:         END


 97:         FUNCTION TEMP(X,T,U1,U2,V1,V2,N,T0)
 98:         DIMENSION U1(N),U2(N),V1(N),V2(N)
 99:         SUM=SPACEI(T0,X,N*T)
100:         DO 100 J=1,N
101:           SUM=SUM+AMAT(N,J,X,0.0,T)*U1(J)
102:           SUM=SUM-AMAT(N,J,X,1.0,T)  *U2(J)
103:           SUM=SUM-BMAT(N,J,X,0.0,T)*V1(J)
104:           SUM=SUM+BMAT(N,J,X,1.0,T)  *V2(J)
105:100     CONTINUE
106:         IF ((X.EQ.0).OR.(X.EQ.1))  THEN
107:           TEMP=2*SUM
108:         ELSE
109:           TEMP=SUM
110:         ENDIF
111:         RETURN
112:         END
```

Figure 6.14a. FORTRAN program, subroutines and functions for the boundary integral method.

```
113:          SUBROUTINE SOLVE(X,Y,A,B,C,D,N)
114:* Solve the coupled lower triangular matrix equations
115:          DIMENSION X(N),Y(N),A(N),B(N),C(N),D(N)
116:          DIMENSION C1(100),D1(100)
117:          DO 100 I=1,N
118:             C1(I)=C(I)
119:             D1(I)=D(I)
120:100       CONTINUE
121:          DET=B(1)**2-A(1)**2
122:          DO 120 I=1,N
123:            DO 110 J=1,I-1
124:              C1(I)=C1(I)+A(I+1-J)*X(J)-B(I+1-J)*Y(J)
125:110           D1(I)=D1(I)+B(I+1-J)*X(J)-A(I+1-J)*Y(J)
126:            X(I)=(A(1)*C1(I)-B(1)*D1(I))/DET
127:            Y(I)=(B(1)*C1(I)-A(1)*D1(I))/DET
128:120       CONTINUE
129:          RETURN
130:          END
```

Figure 6.14b. FORTRAN program, subroutines and functions for the boundary integral method.

PROBLEMS

1. Show that in terms of the non-dimensional variables defined by (6.1), the dimensional boundary conditions

$$\textbf{(i)} \quad T(0, t) = g(t),$$

$$\textbf{(ii)} \quad -k \frac{\partial T}{\partial x}(0, t) = h(t),$$

$$\textbf{(iii)} \quad -k \frac{\partial T}{\partial x}(0, t) + h T(0, t) = h T_0(t),$$

become respectively

$$\textbf{(i)} \quad \overline{T}(0, \bar{t}) = \overline{g}(\bar{t}),$$

$$\textbf{(ii)} \quad -\frac{\partial \overline{T}}{\partial \bar{x}}(0, \bar{t}) = \overline{h}(\bar{t}),$$

$$\textbf{(iii)} \quad -\frac{\partial \overline{T}}{\partial \bar{x}}(0, \bar{t}) + \overline{\mu} \overline{T}(0, \bar{t}) = \overline{T}_0(\bar{t}),$$

where the constant $\overline{\mu}$ and functions $\overline{g}(\bar{t})$, $\overline{h}(\bar{t})$ and $\overline{T}_0(\bar{t})$ are given by

$$\overline{\mu} = \frac{\ell h}{k}, \quad \overline{g}(\bar{t}) = \frac{g(t)}{T_0}, \quad \overline{h}(\bar{t}) = \frac{\ell}{k T_0} h(t), \quad \overline{T}_0(\bar{t}) = \frac{h \ell}{k T_0} T_0(t).$$

2. In this problem a method for obtaining the eigenvalues of symmetric tridiagonal matrices is outlined, which enables the determination of stability restrictions which apply to finite difference schemes.

(i) Let $\mathbf{A}$ denote the $N \times N$ symmetric tridiagonal matrix

$$\mathbf{A} = \begin{pmatrix} a_1 & b_1 & 0 & 0 & \cdots & & 0 \\ b_1 & a_2 & b_2 & 0 & \cdots & & 0 \\ & & & & & & \vdots \\ 0 & b_2 & a_3 & b_3 & & & \vdots \\ \vdots & & & & & & \vdots \\ \vdots & & & & a_{N-2} & b_{N-2} & 0 \\ 0 & & & & b_{N-2} & a_{N-1} & b_{N-1} \\ 0 & \cdots & 0 & 0 & & b_{N-1} & a_N \end{pmatrix}.$$

Associated with $\mathbf{A}$ is the *Sturm sequence* of polynomials $(p_i(x))_{i=1}^{N}$ defined recursively by

$$p_0(x) = 1, \quad p_1(x) = a_1 - x,$$

$$p_i(x) = (a_i - x)p_{i-1}(x) - b_{i-1}^2 p_{i-2}(x), \quad 2 \leq i \leq N.$$

Show by induction that $p_N(x)$ is the characteristic polynomial of $\mathbf{A}$, that is

$$\det|\mathbf{A} - x\mathbf{I}| = p_N(x).$$

(ii)   Suppose that $a_i = 0 \ (1 \le i \le N)$ and $b_i = \frac{1}{2} \ (1 \le i \le N - 1)$ in the matrix $\mathbf{A}$ and show by induction that the Sturm sequence for $\mathbf{A}$ then has the property that

$$p_i(\cos\theta) = \frac{(-1)^i}{2^i} \frac{\sin(i+1)\theta}{\sin\theta}, \quad 1 \le i \le N,$$

and hence that the eigenvalues of $\mathbf{A}$ are

$$\lambda_i = \cos\left(\frac{i\pi}{N+1}\right), \quad 1 \le i \le N.$$

(iii)   Show that if $\mathbf{B}$ is an $N \times N$ matrix with eigenvalues $\lambda_i \ (1 \le i \le N)$ and $\mathbf{I}$ the $N \times N$ identity matrix, then the eigenvalues of the matrix $(\alpha\mathbf{B} + \beta\mathbf{I})$ are $(\alpha\lambda_i + \beta) \ (1 \le i \le N)$ for any real numbers α and β.

(iv)   Using the results of parts **(ii)** and **(iii)**, show that the eigenvalues of the $(N-1) \times (N-1)$ symmetric tridiagonal matrix $\mathbf{A}$ given by (6.16) are given by

$$\lambda_i = 1 - 2\delta\left(1 - \cos\left(\frac{i\pi}{N}\right)\right) = 1 - 4\delta\sin^2\left(\frac{i\pi}{2N}\right), \quad 1 \le i \le N - 1.$$

3.   Show that the explicit finite difference scheme for the heat flow problem

$$
\left.
\begin{aligned}
\frac{\partial \overline{T}}{\partial \overline{t}} &= \frac{\partial^2 \overline{T}}{\partial \overline{x}^2}, \quad 0 < \overline{x} < 1, \\[2mm]
-\frac{\partial \overline{T}}{\partial \overline{x}}(0,\overline{t}) &= 1, \quad \frac{\partial \overline{T}}{\partial \overline{x}}(1,\overline{t}) + \overline{T}(1,\overline{t}) = 0, \\[2mm]
\overline{T}(\overline{x},0) &= 0,
\end{aligned}
\right\}
$$

can be written in the form

$$\underline{T}^{j+1} = \mathbf{A}\underline{T}^j + \delta\underline{U}, \quad \underline{T}^0 = \underline{0},$$

where the $(N+1)$ dimensional vectors $\underline{T}^j$ and $\underline{U}$ are given by

$$
\underline{T}^j = \begin{pmatrix} T_{0j} \\ T_{1j} \\ \vdots \\ \vdots \\ T_{N-1\,j} \\ T_{Nj} \end{pmatrix}, \quad
\underline{U} = \begin{pmatrix} 2\Delta\overline{x} \\ 0 \\ \vdots \\ 0 \\ 0 \end{pmatrix},
$$

and the $(N + 1) \times (N + 1)$ tridiagonal asymmetric matrix $\mathbf{A}$ is given by

$$\mathbf{A} = \begin{pmatrix} 1-2\delta & 2\delta & 0 & 0 & \cdots & & 0 \\ \delta & 1-2\delta & \delta & 0 & \cdots & & 0 \\ & & & & & & \vdots \\ 0 & \delta & 1-2\delta & \delta & & & \\ \vdots & & & & & & \vdots \\ & & & 1-2\delta & \delta & & 0 \\ 0 & \cdots & 0 & \delta & 1-2\delta & & \delta \\ 0 & \cdots & 0 & 0 & 2\delta & & 1-2(\delta+\Delta\overline{x}) \end{pmatrix}.$$

4. Using the results of parts **(ii)** and **(iii)** of Problem 2, show that the eigenvalues of the $(N - 1) \times (N - 1)$ symmetric tridiagonal matrix $\mathbf{B}$ given by (6.26) are

$$\lambda_i = 1 + 2\delta\left(1 - \cos\left(\frac{i\pi}{N}\right)\right) = 1 + 4\delta\sin^2\left(\frac{i\pi}{2N}\right), \quad 1 \le i \le N - 1.$$

5. Prove the convergence and stability of the Crank-Nicolson scheme (6.31) for prescribed boundary temperatures. To prove the stability of (6.31) determine the eigenvalues of $\mathbf{C}^{-1}\mathbf{D}$, where $\mathbf{C}$ and $\mathbf{D}$ are given by (6.32), as follows. Write $\mathbf{C}$ and $\mathbf{D}$ in the form

$$\mathbf{C} = \mathbf{I} + \delta\mathbf{E}, \quad \mathbf{D} = \mathbf{I} - \delta\mathbf{E},$$

where $\mathbf{I}$ is the $(N - 1) \times (N - 1)$ identity matrix and $\mathbf{E}$ is the $(N - 1) \times (N - 1)$ tridiagonal symmetric matrix

$$\mathbf{E} = \begin{pmatrix} 1 & -\frac{1}{2} & 0 & 0 & \cdots & & 0 \\ -\frac{1}{2} & 1 & -\frac{1}{2} & 0 & \cdots & & 0 \\ & & & & & & \vdots \\ 0 & -\frac{1}{2} & 1 & -\frac{1}{2} & & & \\ \vdots & & & & & & \vdots \\ & & & -\frac{1}{2} & 1 & -\frac{1}{2} & 0 \\ 0 & \cdots & 0 & \frac{1}{2} & 1 & & -\frac{1}{2} \\ 0 & \cdots & 0 & 0 & -\frac{1}{2} & & 1 \end{pmatrix}.$$

Then, if λ_i is an eigenvalue of $\mathbf{C}^{-1}\mathbf{D}$ with eigenvector $\underset{\sim}{x}_i$ show that

$$(\mathbf{D} - \lambda_i\mathbf{C})\underset{\sim}{x}_i = 0,$$

from which it follows that

$$\mathbf{E}\underset{\sim}{x}_i = \frac{1}{\delta}\left(\frac{1 - \lambda_i}{1 + \lambda_i}\right)\underset{\sim}{x}_i,$$

so that $\frac{1}{\delta}\left(\frac{1-\lambda_i}{1+\lambda_i}\right)$ is an eigenvalue of $\mathbf{E}$. Determine all the eigenvalues of $\mathbf{E}$ using parts **(ii)** and **(iii)** of Problem 2 and hence show that the eigenvalues of $\mathbf{C}^{-1}\mathbf{D}$ are

$$\lambda_i = \frac{1 - 2\delta \sin^2\left(\frac{i\pi}{2N}\right)}{1 + 2\delta \sin^2\left(\frac{i\pi}{2N}\right)}, \quad 1 \le i \le N - 1.$$

6. Assume that there is some constant θ such that $0 \le \theta \le 1$ and

$$\frac{\partial \overline{T}}{\partial \overline{t}}(i\Delta\overline{x}, (j + \theta)\Delta\overline{t}) = \frac{1}{\Delta\overline{t}}\left(\overline{T}_{ij+1} - \overline{T}_{ij}\right), \quad 0 \le i \le N, j \ge 0,$$

and approximate $\overline{T}(i\Delta\overline{x}, (j + \theta)\Delta\overline{t})$ by the weighted average

$$\overline{T}(i\Delta\overline{x}, (j + \theta)\Delta\overline{t}) \approx \theta\overline{T}_{ij+1} + (1 - \theta)\overline{T}_{ij},$$

so that

$$\frac{\partial \overline{T}}{\partial \overline{x}}(i\Delta\overline{x}, (j + \theta)\Delta\overline{t}) \approx \frac{\theta}{2\Delta\overline{x}}\left(\overline{T}_{i+1j+1} - \overline{T}_{i-1j+1}\right) + \frac{1 - \theta}{2\Delta\overline{x}}\left(\overline{T}_{i+1j} - \overline{T}_{i-1j}\right),$$

$$\frac{\partial^2 \overline{T}}{\partial \overline{x}^2}(i\Delta\overline{x}, (j + \theta)\Delta\overline{t}) \approx \frac{\theta}{(\Delta\overline{x})^2}\left(\overline{T}_{i+1j+1} + \overline{T}_{i-1j+1} - 2\overline{T}_{ij+1}\right)$$

$$+ \frac{(1 - \theta)}{(\Delta\overline{x})^2}\left(\overline{T}_{i+1j} + \overline{T}_{i-1j} - 2\overline{T}_{ij}\right).$$

Deduce that the heat equation is approximated by the difference equations

$$\tilde{T}_{ij+1} - \theta\delta\left(\tilde{T}_{i+1j+1} + \tilde{T}_{i-1j+1} - 2\tilde{T}_{ij+1}\right)$$
$$= \tilde{T}_{ij} + (1 - \theta)\delta\left(\tilde{T}_{i+1j} + \tilde{T}_{i-1j} - 2\tilde{T}_{ij}\right), \quad j \ge 0,$$

where $\delta = \Delta\overline{t}/(\Delta\overline{x})^2$. Show that the explicit, Crank-Nicolson and fully implicit finite difference schemes are special cases of this finite difference scheme, corresponding to the values $\theta = 0$, $\frac{1}{2}$ and 1 respectively. Assuming prescribed temperature boundary conditions, write this finite difference scheme in matrix form and using the method outlined in the previous question, show that the scheme is stable for $0 < \delta \le \frac{1}{2(1 - 2\theta)}$ if $0 \le \theta < \frac{1}{2}$ and $0 < \delta$ if $\frac{1}{2} \le \theta \le 1$.

$\left[\text{Hint, to determine the eigenvalues of a matrix of the form } \mathbf{A}^{-1}\mathbf{B} \text{ use the method}\right.$
outlined in the previous question and show that these eigenvalues are given by

$$\lambda_i = \frac{1 - 4(1 - \theta)\delta \sin^2\frac{i\pi}{2N}}{1 + 4\theta\delta \sin^2\frac{i\pi}{2N}}, \quad 1 \le i \le N - 1. \Big]$$

7. Continuation. The non-dimensional cylindrical or spherical radially symmetric heat equation can be written in the form

$$\frac{\partial \overline{T}}{\partial \overline{t}} = \frac{\partial^2 \overline{T}}{\partial \overline{r}^2} + \frac{m}{\overline{r}} \frac{\partial \overline{T}}{\partial \overline{r}}, \quad 0 < \overline{r} < 1,$$

$$\frac{\partial \overline{T}}{\partial \overline{r}}(0, \overline{t}) = 0,$$

where $m = 1$ corresponds to the cylindrical case and $m = 2$ to the spherical case and the zero temperature gradient condition at the origin is necessary to obtain finite solutions. Divide the interval $[0, 1]$ into N equal steps each of length $\Delta \overline{r} = 1/N$ and let

$$\overline{T}_{ij} = \overline{T}(i\Delta \overline{r}, j\Delta \overline{t}), \quad 0 \le i \le N, j \ge 0.$$

Derive the finite difference approximations

$$\tilde{T}_{ij+1} - \theta \delta \left[\left(1 + \frac{m}{2i}\right) \tilde{T}_{i+1j+1} + \left(1 - \frac{m}{2i}\right) \tilde{T}_{i-1j+1} - 2\tilde{T}_{ij+1} \right]$$

$$1 \le i \le N,$$

$$= \tilde{T}_{ij} + (1 - \theta)\delta \left[\left(1 + \frac{m}{2i}\right) \tilde{T}_{i+1j} + \left(1 - \frac{m}{2i}\right) \tilde{T}_{i-1j} - 2\tilde{T}_{ij} \right],$$

for this problem for $1 \le i \le N$ by assuming that

$$\frac{\partial \overline{T}}{\partial \overline{t}}(i\Delta \overline{r}, (j + \theta)\Delta \overline{t}) = \frac{1}{\Delta \overline{t}} \left(\overline{T}_{ij+1} - \overline{T}_{ij} \right),$$

$$\overline{T}(i\Delta \overline{r}, (j + \theta)\Delta \overline{t}) \approx \theta \overline{T}_{ij+1} + (1 - \theta)\overline{T}_{ij},$$

for $0 \le \theta \le 1$. Using l'Hopital's rule to treat the term $\frac{m}{\overline{r}} \frac{\partial \overline{T}}{\partial \overline{r}}$ as $\overline{r}$ tends to zero, and by approximating the zero temperature gradient condition at the origin by

$$\frac{1}{2\Delta \overline{r}} \left(\overline{T}_{1j} - \overline{T}_{-1j} \right) = 0, \quad j \ge 0,$$

where $\overline{T}_{-1j}$ is a fictitious temperature, deduce the finite difference expression

$$\tilde{T}_{0j+1} - 2\theta \delta (1 + m)\left(\tilde{T}_{1j+1} - \tilde{T}_{0j+1} \right)$$

$$j \ge 0,$$

$$= \tilde{T}_{0j} + 2(1 - \theta)\delta(1 + m)\left(\tilde{T}_{1j} - \tilde{T}_{0j} \right),$$

for the temperature at the origin. $\left[\text{Notice that the special values } \theta = 0, \frac{1}{2} \text{ and } 1 \right.$ correspond to the explicit, Crank-Nicolson and fully implicit finite difference methods for this problem respectively. $\Big]$

8.   Consider the finite element solution of the prescribed boundary temperature problem (6.6). Assuming a constant element length h, show that the capacitance and stiffness matrices $\mathbf{C}$ and $\mathbf{K}$, defined by (6.53) become the $(N-2) \times (N-2)$ symmetric tridiagonal matrices

$$
\mathbf{C} = \frac{h}{6}
\begin{bmatrix}
4 & 1 & 0 & 0 & \cdots & 0 \\
1 & 4 & 1 & 0 & \cdots & 0 \\
 & & & & \vdots & \\
0 & 1 & 4 & 1 & & \vdots \\
\vdots & & & & & \\
\vdots & & 0 & 1 & 4 & 1 \\
0 & \cdots & 0 & 0 & 1 & 4
\end{bmatrix}, \quad
\mathbf{K} = \frac{1}{h}
\begin{bmatrix}
2 & -1 & 0 & 0 & \cdots & 0 \\
-1 & 2 & 1 & 0 & \cdots & 0 \\
 & & & & \vdots & \\
0 & -1 & 2 & 1 & & \vdots \\
\vdots & & & & & \\
\vdots & & 0 & -1 & 2 & 1 \\
0 & \cdots & 0 & 0 & -1 & 2
\end{bmatrix}.
$$

Hence show that the matrices $\mathbf{D}$ and $\mathbf{E}$ defined by (6.57) are given by the $(N-2) \times (N-2)$ symmetric tridiagonal matrices

$$
\mathbf{D} = \frac{h}{6}
\begin{bmatrix}
4+12\theta\delta & 1-6\theta\delta & 0 & 0 & \cdots & 0 \\
1-6\theta\delta & 4+12\theta\delta & 1-6\theta\delta & 0 & \cdots & 0 \\
 & & & & & \vdots \\
0 & 1-6\theta\delta & 4+12\theta\delta & 1-6\theta\delta & & \vdots \\
0 & 0 & 1-6\theta\delta & 4+12\theta\delta & & 0 \\
\vdots & & & & & \\
\vdots & & 0 & 1-6\theta\delta & 4+12\theta\delta & 1-6\theta\delta \\
0 & \cdots & 0 & 0 & 1-6\theta\delta & 4+12\theta\delta
\end{bmatrix},
$$

$$
\mathbf{E} = \frac{h}{6}
\begin{bmatrix}
4-12(1-\theta)\delta & 1+6(1-\theta)\delta & 0 & 0 & \cdots & 0 \\
1+6(1-\theta)\delta & 4-12(1-\theta)\delta & 1-6\theta\delta & 0 & \cdots & 0 \\
 & & & & & \vdots \\
0 & 1+6(1-\theta)\delta & 4-12(1-\theta)\delta & 1-6\theta\delta & & \vdots \\
0 & 0 & 1+6(1-\theta)\delta & 4-12(1-\theta)\delta & & 0 \\
\vdots & & & & & \\
\vdots & & 0 & 1+6(1-\theta)\delta & 4-12(1-\theta)\delta & 1-6\theta\delta \\
0 & \cdots & 0 & 0 & 1+6(1-\theta)\delta & 4-12(1-\theta)\delta
\end{bmatrix},
$$

where $\delta = \Delta\bar{t}/h^2$ is as usual the ratio of the time step to the square of the space step. In order that the finite element equation (6.56) be stable, we require the spectral radius of $\mathbf{D}^{-1}\mathbf{E}$ be less than or equal to unity. Using the method outlined in Problem 5, show that the eigenvalues of $\mathbf{D}^{-1}\mathbf{E}$ are given by

$$\lambda_i = \frac{1 - 6(1 - \theta)\delta \sin^2 \frac{i\pi}{2(N-1)}}{1 + 6\theta\delta \sin^2 \frac{i\pi}{2(N-1)}}, \quad 1 \leq i \leq N - 2,$$

so that (6.56) is stable if $0 < \delta \leq \frac{1}{3(1-2\theta)}$ if $0 \leq \theta < \frac{1}{2}$ and for all δ if $\frac{1}{2} \leq \theta \leq 1$.

9. Consider the cylindrical or spherical radially symmetric heat equation

$$\bar{r}^m \frac{\partial \bar{T}}{\partial \bar{t}} = \frac{\partial}{\partial \bar{r}}\left(\bar{r}^m \frac{\partial \bar{T}}{\partial \bar{r}}\right), \quad 0 < \bar{r} < 1,$$

$$\frac{\partial \bar{T}}{\partial \bar{r}}(0, \bar{t}) = 0,$$

where $m = 1$ for the cylindrical geometry and $m = 2$ for the spherical geometry. Show that if $\phi(\bar{r})$ in any test function in $H[0, 1]$ then $\bar{T}(\bar{r}, \bar{t})$ satisfies

$$\int_0^1 \bar{r}^m \left[\phi(\bar{r})\frac{\partial \bar{T}}{\partial \bar{t}} + \frac{\partial \phi}{\partial \bar{r}} \frac{\partial \bar{T}}{\partial \bar{r}}\right] d\bar{r} = \phi(1)\frac{\partial \bar{T}}{\partial \bar{r}}(1, \bar{t}), \quad \bar{t} > 0.$$

If the initial temperature $\bar{T}(\bar{r}, 0)$ is prescribed by

$$\bar{T}(\bar{r}, 0) = f(\bar{r}), \quad 0 \leq \bar{r} \leq 1,$$

describe the weak formulation of the cylindrical or spherical heat equation subject to:

(i) the essential boundary condition $\bar{T}(1, \bar{t}) = g(\bar{t})$,

(ii) the natural boundary condition $\dfrac{\partial \bar{T}}{\partial \bar{r}}(1, \bar{t}) + \mu \bar{T}(1, \bar{t}) = \bar{T}_0(\bar{t})$, where $\mu \geq 0$ is constant.

10. Let $\phi(\bar{x}, \bar{t})$ be continuous for $0 \leq \bar{x} \leq 1$ and $\bar{t} \geq 0$. With $\bar{G}(\bar{x}, \xi, \bar{t}, \tau)$ defined by (6.74) and λ by (6.75) we can show that

$$-\lambda\phi(\bar{x}, \bar{t}) = \int_0^{\bar{t}} \int_0^1 \phi(\xi, \tau)\left(\frac{\partial \bar{G}}{\partial \tau} + \frac{\partial^2 \bar{G}}{\partial \xi^2}\right) d\xi \, d\tau,$$

as follows.

(i) Show that if $(\bar{x}, \bar{t}) \neq (\xi, \tau)$ then

$$\frac{\partial \bar{G}}{\partial \tau} + \frac{\partial^2 \bar{G}}{\partial \xi^2} = 0.$$

268   **Numerical solutions**

(ii)   Using **(i)** show that for $0 < \bar{x} < 1$

$$\int_0^{\bar{t}} \int_0^1 \phi\left(\frac{\partial \overline{G}}{\partial \tau} + \frac{\partial^2 \overline{G}}{\partial \xi^2}\right) d\xi \, d\tau = \int_{\bar{t}-\epsilon}^{\bar{t}} \int_{\bar{x}-\delta}^{\bar{x}+\delta} \phi\left(\frac{\partial \overline{G}}{\partial \tau} + \frac{\partial^2 \overline{G}}{\partial \xi^2}\right) d\xi \, d\tau,$$

for all ϵ and δ such that $0 < \epsilon \le \bar{t}$ and $0 \le \bar{x} - \delta < \bar{x} + \delta \le 1$. Hence from the mean value theorem, there are constants $\bar{x}'$ and $\bar{t}'$ such that $\bar{t} - \epsilon \le \bar{t}' \le \bar{t}$ and $\bar{x} - \delta \le \bar{x}' \le \bar{x} + \delta$ and

$$\int_{\bar{t}-\epsilon}^{\bar{t}} \int_{\bar{x}-\delta}^{\bar{x}+\delta} \phi\left(\frac{\partial \overline{G}}{\partial \tau} + \frac{\partial^2 \overline{G}}{\partial \xi^2}\right) d\xi \, d\tau = \phi(\bar{x}',\bar{t}') \int_{\bar{t}-\epsilon}^{\bar{t}} \int_{\bar{x}-\delta}^{\bar{x}+\delta} \left(\frac{\partial \overline{G}}{\partial \tau} + \frac{\partial^2 \overline{G}}{\partial \xi^2}\right) d\xi \, d\tau.$$

(iii)   Show that for any $\epsilon > 0$ and $\delta > 0$

$$\int_{\bar{t}-\epsilon}^{\bar{t}} \int_{\bar{x}-\delta}^{\bar{x}+\delta} \left(\frac{\partial \overline{G}}{\partial \tau} + \frac{\partial^2 \overline{G}}{\partial \xi^2}\right) d\xi \, d\tau$$

$$= -\int_{\bar{x}-\delta}^{\bar{x}+\delta} \overline{G}(\bar{x},\xi,\bar{t},\bar{t}-\epsilon) \, d\xi + \int_{\bar{t}-\epsilon}^{\bar{t}} \left[\frac{\partial \overline{G}}{\partial \xi}\right]_{\xi=\bar{x}-\delta}^{\xi=\bar{x}+\delta} d\tau$$

$$= -1,$$

by evaluating the single integrals on the right hand side of this expression.

(iv)   Hence using parts **(ii)** and **(iii)** and allowing ϵ and δ to tend to zero show that for $0 < \bar{x} < 1$

$$-\phi(\bar{x},\bar{t}) = \int_0^{\bar{t}} \int_0^1 \phi(\xi,\tau)\left(\frac{\partial \overline{G}}{\partial \tau} + \frac{\partial^2 \overline{G}}{\partial \xi^2}\right) d\xi \, d\tau.$$

(v)   For the case $\bar{x} = 0$ show that

$$\int_0^{\bar{t}} \int_0^1 \phi\left(\frac{\partial \overline{G}}{\partial \tau} + \frac{\partial^2 \overline{G}}{\partial \xi^2}\right) d\xi \, d\tau = \int_{\bar{t}-\epsilon}^{\bar{t}} \int_0^{\delta} \phi\left(\frac{\partial \overline{G}}{\partial \tau} + \frac{\partial^2 \overline{G}}{\partial \xi^2}\right) d\xi \, d\tau$$

$$= \phi(\bar{x}',\bar{t}') \int_{\bar{t}-\epsilon}^{\bar{t}} \int_0^{\delta} \left(\frac{\partial \overline{G}}{\partial \tau} + \frac{\partial^2 \overline{G}}{\partial \xi^2}\right) d\xi \, d\tau$$

$$= -\frac{1}{2}\phi(\bar{x}',\bar{t}'),$$

for any $0 < \epsilon \le \bar{t}$ and $0 < \delta \le 1$ and constants $\bar{x}'$ and $\bar{t}'$ such that $\bar{t} - \epsilon \le \bar{t}' \le \bar{t}$ and $0 \le \bar{x}' \le \delta$. Letting ϵ and δ tend to zero show that for $\bar{x} = 0$

$$\int_0^{\bar{t}} \int_0^1 \phi \left(\frac{\partial \overline{G}}{\partial \tau} + \frac{\partial^2 \overline{G}}{\partial \xi^2} \right) d\xi \, d\tau = -\frac{1}{2} \phi(0, \bar{t}).$$

(vi) Repeat **(v)** for $\bar{x} = 1$ and hence deduce the original assertion.

(vii) Derive (6.72) by considering the integral

$$\int_0^{\bar{t}} \int_0^1 \left\{ \overline{T}(\xi, \tau) \left(\frac{\partial \overline{G}}{\partial \tau} + \frac{\partial^2 \overline{G}}{\partial \xi^2} \right) + \overline{G}(\bar{x}, \xi, \bar{t}, \tau) \left(\frac{\partial \overline{T}}{\partial \tau} - \frac{\partial^2 \overline{T}}{\partial \xi^2} \right) \right\} d\xi \, d\tau,$$

where $\overline{T}(\bar{x}, \bar{t})$ is a solution of the non-dimensional heat equation.

11. Let the function $f(\bar{t})$ be defined by

$$f(\bar{t}) = \frac{1}{2\sqrt{\pi}} \int_0^{\bar{t}} \frac{1}{\sqrt{\bar{t} - \tau}} \exp \frac{-(\bar{x} - \xi)^2}{4(\bar{t} - \tau)} \, d\tau.$$

Show by making the change of variable $\sigma = |\bar{x} - \xi|/2\sqrt{\bar{t} - \tau}$ that if $\bar{x} \ne \xi$ then

$$f(\bar{t}) = \frac{|\bar{x} - \xi|}{2\sqrt{\pi}} \int_{|\bar{x} - \xi|/2\sqrt{\bar{t}}}^{\infty} \frac{1}{\sigma^2} e^{-\sigma^2} \, d\sigma,$$

and using (2.41) show that the Laplace transform of $f(t)$ is

$$\hat{f}(p) = \frac{1}{2} \frac{e^{-|\bar{x} - \xi|q}}{pq},$$

where here $q = \sqrt{p}$. Hence deduce that

$$f(\bar{t}) = \sqrt{\frac{\bar{t}}{\pi}} \exp \frac{-(\bar{x} - \xi)^2}{4\bar{t}} - \frac{|\bar{x} - \xi|}{2} \operatorname{erfc} \left(\frac{|\bar{x} - \xi|}{2\sqrt{\bar{t}}} \right).$$

Show that for $\bar{t} > \bar{t}_0$ we have

$$f(\bar{t} - \bar{t}_0) = \frac{1}{2\sqrt{\pi}} \int_{\bar{t}_0}^{\bar{t}} \frac{1}{\sqrt{\bar{t} - \tau}} \exp \frac{-(\bar{x} - \xi)^2}{4(\bar{t} - \tau)} \, d\tau$$

$$= \sqrt{\frac{\bar{t} - \bar{t}_0}{\pi}} \exp \frac{-(\bar{x} - \xi)^2}{4(\bar{t} - \bar{t}_0)} - \frac{|\bar{x} - \xi|}{2} \operatorname{erfc} \left(\frac{|\bar{x} - \xi|}{2\sqrt{\bar{t} - \bar{t}_0}} \right),$$

and hence derive (6.84).

12. In formulating the boundary integral numerical approximation suppose $\bar{t}_i$ to be the midpoint of the ith time interval, that is

$$\bar{t}_i = (i - \tfrac{1}{2})\Delta\bar{t}, \quad 1 \le i \le N,$$

and show that the matrix elements $A_{ij}(\bar{x}, \xi)$ and $B_{ij}(\bar{x}, \xi)$ defined by (6.81) are given by

$$A_{ij}(\bar{x}, \xi) = \begin{cases} \dfrac{1}{2}\,\text{erf}\!\left(\dfrac{\bar{x} - \xi}{2\sqrt{\Delta\bar{t}(i - \frac{1}{2} - j)}}\right) - \dfrac{1}{2}\,\text{erf}\!\left(\dfrac{\bar{x} - \xi}{2\sqrt{\Delta\bar{t}(i + \frac{1}{2} - j)}}\right), & i > j, \\[20pt] \dfrac{1}{2}\,\text{sgn}(\bar{x} - \xi) - \dfrac{1}{2}\,\text{erf}\!\left(\dfrac{\bar{x} - \xi}{\sqrt{2\Delta\bar{t}}}\right), & i = j, \\[16pt] 0, & i < j, \end{cases}$$

$$B_{ij}(\bar{x}, \xi) = \begin{cases} \left\{ \sqrt{\dfrac{\Delta\bar{t}}{\pi}(i + \tfrac{1}{2} - j)}\,\exp\dfrac{-(\bar{x} - \xi)^2}{4\Delta\bar{t}(i + \frac{1}{2} - j)} - \dfrac{|\bar{x} - \xi|}{2}\,\text{erfc}\!\left(\dfrac{|\bar{x} - \xi|}{2\sqrt{\Delta\bar{t}(i + \frac{1}{2} - j)}}\right) \right. \\[18pt] \left. \quad - \sqrt{\dfrac{\Delta\bar{t}}{\pi}(i - \tfrac{1}{2} - j)}\,\exp\dfrac{-(\bar{x} - \xi)^2}{4\Delta\bar{t}(i - \frac{1}{2} - j)} + \dfrac{|\bar{x} - \xi|}{2}\,\text{erfc}\!\left(\dfrac{|\bar{x} - \xi|}{2\sqrt{\Delta\bar{t}(i - \frac{1}{2} - j)}}\right) \right\}, & i > j \\[20pt] \sqrt{\dfrac{\Delta\bar{t}}{2\pi}}\,\exp\dfrac{-(\bar{x} - \xi)^2}{2\Delta\bar{t}} - \dfrac{|\bar{x} - \xi|}{2}\,\text{erfc}\!\left(\dfrac{|\bar{x} - \xi|}{\sqrt{2\Delta\bar{t}}}\right), & i = j, \\[16pt] 0, & i < j, \end{cases}$$

rather than by (6.83) and (6.84) respectively. Apart from these changes the boundary integral procedure is essentially unchanged by this choice of $\bar{t}_i$. In general, how would you expect this change to affect the accuracy of the method? $\Big[$Hint, assuming that $w(\tau)$ does not change sign over the range of integration, decide which of the following two approximations is likely to be more accurate

$$\int_{(i-1)\Delta\bar{t}}^{i\Delta\bar{t}} \phi(\tau)w(\tau)\,\mathrm{d}\tau \approx \phi(i\Delta\bar{t}) \int_{(i-1)\Delta\bar{t}}^{i\Delta\bar{t}} w(\tau)\,\mathrm{d}\tau,$$

$$\text{or} \quad \int_{(i-1)\Delta\bar{t}}^{i\Delta\bar{t}} \phi(\tau)w(\tau)\,\mathrm{d}\tau \approx \phi\big((i - \tfrac{1}{2})\Delta\bar{t}\big) \int_{(i-1)\Delta\bar{t}}^{i\Delta\bar{t}} w(\tau)\,\mathrm{d}\tau.\Big]$$

13. In general the boundary integral method leads to a system of $2N$ equations of the form

$$\mathbf{A}\underset{\sim}{x} + \mathbf{B}\underset{\sim}{y} = \underset{\sim}{a},$$

$$\mathbf{C}\underset{\sim}{x} + \mathbf{D}\underset{\sim}{y} = \underset{\sim}{b},$$

where $\underset{\sim}{x}$, y, $\underset{\sim}{a}$ and $\underset{\sim}{b}$ are N dimensional vectors and $\mathbf{A}$, $\mathbf{B}$, $\mathbf{C}$ and $\mathbf{D}$ are $N \times N$ lower triangular matrices with non-zero diagonal elements. Specifically, the components A_{ij}, B_{ij}, C_{ij} and D_{ij} ($1 \le i,j \le N$) of the matrices $\mathbf{A}$, $\mathbf{B}$, $\mathbf{C}$ and $\mathbf{D}$ satisfy the relations

$$A_{ij} = 0, \quad B_{ij} = 0, \quad C_{ij} = 0, \quad D_{ij} = 0, \quad \text{if } i < j,$$

$$A_{ii} \neq 0, \quad B_{ii} \neq 0, \quad C_{ii} \neq 0, \quad D_{ii} \neq 0, \quad 1 \le i \le N.$$

Show that such a system can be solved by forward substitution, that is by determining the first components of $\underset{\sim}{x}$ and y, x_1 and y_1 and using these to determine the second components x_2 and y_2 and so forth.

$\Bigg[$ Hint, show that x_1 and y_1 satisfy the equations

$$A_{11}x_1 + B_{11}y_1 = a_1,$$
$$C_{11}x_1 + C_{11}y_1 = b_1,$$

and that

$$A_{22}x_2 + B_{22}y_2 = a_2 - A_{21}x_1 - B_{21}y_1,$$
$$C_{22}x_2 + D_{22}y_2 = b_2 - C_{21}x_1 - D_{21}y_1,$$

and so forth. $\Bigg]$

Chapter Seven

Melting or freezing moving boundary problems

7.1 Introduction

Many important heat conduction problems in science, engineering and industry involve a phase change. Thus, for example, the processes of casting metals or plastics, freezing or thawing of foods and ice production all involve a material either melting or freezing (solidifying). The surface separating frozen and unfrozen regions of the material changes with time and the essential aspect of such heat conduction problems is the mathematical determination of the position of this surface as a function of time. Mathematically similar processes also arise in chemical engineering where a chemical reaction is taking place and a moving reaction surface separates reacted and unreacted regions. Phase change problems are examples of what are loosely referred to as *moving boundary problems* and their study presents one of the more exciting and challenging areas of current applied mathematical research. The existence of a moving boundary generally means that the problem does not admit a simple closed form analytical solution and accordingly much research has focussed on approximate solution techniques, both analytical and numerical. In this chapter we mainly consider only the simplest moving boundary problems which involve the melting or freezing of a semi-infinite half-space and we present some of the simpler approximate solution techniques.

Heat conduction moving boundary problems are associated with Josef Stefan (1835-1893) who, in 1889, appears to have been the first to make an extensive study of them, although such a problem was certainly considered by Gabriel Lame (1795-1870) and Benoit Clapeyron (1799-1864) as early as 1831. However, heat conduction moving boundary problems are now widely referred to as *Stefan problems*. In 1889 Stefan considered the problem of a semi-infinite half-space $0 \leq x < \infty$ consisting of a pure material which can exist in either a liquid or solid phase. Initially, the material is assumed in the liquid phase at a known uniform temperature T_ℓ. At time $t = 0$ the surface $x = 0$ is cooled and subsequently maintained at a given constant temperature T_0 which is below the constant freezing temperature T_f of the liquid. Immediately the liquid along $x = 0$ freezes and subsequently a freezing front moves progressively through the liquid such that behind the front the material is in the solid phase while ahead of the front the material is in the liquid phase. After time t the problem is to determine the thickness $X(t)$ of the solid layer and the temperatures $T_1(x,t)$ and $T_2(x,t)$ in the solid and liquid phases respectively (see Figure 7.1). Clearly the temperatures satisfy the inequalities

$$T_0 \leq T_1(x,t) \leq T_f \leq T_2(x,t) \leq T_\ell. \tag{7.1}$$

Assuming constant thermal properties in both phases and no density change on solidification these temperatures are determined by solving heat conduction problems in domains which vary with time, thus

$$\left. \begin{array}{l} \dfrac{\partial T_1}{\partial t} = \kappa_1 \dfrac{\partial^2 T_1}{\partial x^2}, \quad 0 < x < X(t), \\[2ex] T_1(0,t) = T_0, \quad T_1(X(t),t) = T_f, \quad t > 0, \end{array} \right\} \text{(solid)} \tag{7.2}$$

and

$$\left. \begin{array}{l} \dfrac{\partial T_2}{\partial t} = \kappa_2 \dfrac{\partial^2 T_2}{\partial x^2}, \quad X(t) < x < \infty, \\[2ex] T_2(X(t),t) = T_f, \quad T_2(\infty,t) = T_\ell, \quad t > 0, \\[2ex] T_2(x,0) = T_\ell, \quad 0 \leq x \leq \infty, \end{array} \right\} \text{(liquid)} \tag{7.3}$$

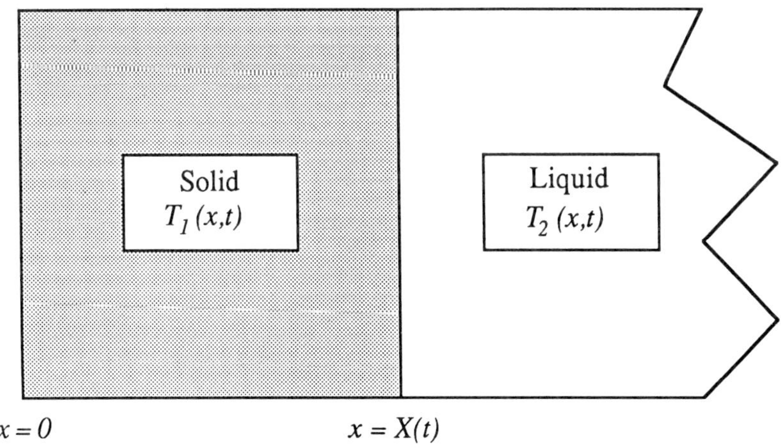

Figure 7.1. Freezing liquid with a moving interface separating solid and liquid regions.

where κ_1 and κ_2 are the constant thermal diffusivities of the solid and liquid phases respectively, and we note that there is no initial condition imposed on $T_1(x, t)$, since initially there is no solid (that is $X(0) = 0$).

Since $X(t)$ is unknown and variable, an additional equation is required for its determination. This equation is known as the *Stefan condition* which states that the freezing front moves in such a way that its velocity is proportional to the jump in heat flux across the front. In general, pure substances change phase isothermally and there is a *latent heat* associated with the phase change. The latent heat of a given phase change is the quantity of heat liberated when a unit mass undergoes that phase change completely and isothermally. This isothermal release of energy is due to the large and discontinuous rearrangement of the molecular structure which accompanies a phase change. In particular, the latent heat of fusion is the quantity of heat liberated when a unit mass of pure liquid completely freezes at its fusion temperature. Conversely, the latent heat of fusion is also the quantity of heat which must be added to melt a unit mass of pure solid at its fusion temperature. The latent heat has dimensions of [energy]/[mass], and is usually given in units of joule/gram. For physically stable situations the latent heat of fusion is always a positive quantity. Hence if L denotes the latent heat of fusion for the material under consideration and ρ the constant density of its solid and liquid phases, then as the freezing front moves a distance dX during time dt, a quantity of heat $\rho L\, dX$ is released from the freezing liquid and must be removed by conduction. During the time interval dt we have

$$k_2 \frac{\partial T_2}{\partial x}(X(t), t)\, dt,$$

units of heat flowing *into* the phase change region from the liquid and

$$k_1 \frac{\partial T_1}{\partial x}(X(t), t)\, dt,$$

units of heat flowing *out of* the phase change zone into the solid (see Figure 7.2), where k_1 and k_2 are the thermal conductivities of the solid and liquid respectively. The difference must equal the liberated latent heat so we have

$$k_1 \frac{\partial T_1}{\partial x}(X(t), t)\, \mathrm{d}t \;-\; k_2 \frac{\partial T_2}{\partial x}(X(t), t)\, \mathrm{d}t \;=\; \rho L\, \mathrm{d}X,$$

and thus across the phase change front there is a discontinuity in heat flux due to the liberation of this latent heat and we have the Stefan condition

$$k_1 \frac{\partial T_1}{\partial x}(X(t), t) \;-\; k_2 \frac{\partial T_2}{\partial x}(X(t), t) \;=\; \rho L \frac{\mathrm{d}X}{\mathrm{d}t}. \tag{7.4}$$

Altogether equations (7.2)–(7.4) together with the initial condition

$$X(0) = 0, \tag{7.5}$$

constitute the mathematical prescription for the determination of the boundary position $x = X(t)$ and the temperatures $T_1(x, t)$ and $T_2(x, t)$ in the solid and liquid regions.

In the case of Newton cooling on the surface $x = 0$ the condition $(7.2)_2$ is replaced by

$$k_1 \frac{\partial T_1}{\partial x}(0, t) = h\,[T_1(0, t) - T_0], \tag{7.6}$$

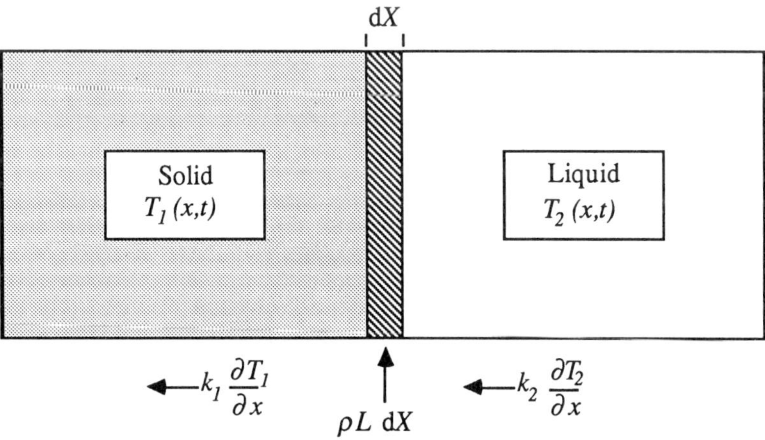

Figure 7.2. Energy balance or Stefan condition (7.4) at the moving interface between solid and liquid regions in a freezing liquid.

while for the case of genuine radiation on the surface $(7.2)_2$ is replaced by

$$k_1 \frac{\partial T_1}{\partial x}(0, t) = \sigma E \left[T_1^4(0, t) - T_0^4 \right],$$
(7.7)

where h, σ, E and T_0 have their usual meanings (see Section 1.5). We remark that in either of these latter two cases solidification along $x = 0$ occurs only after a finite time, since the surface temperature does not immediately reach the melting temperature T_f in these cases, and accordingly a standard heat conduction problem must first be solved to determine this finite time.

In most of this chapter we completely avoid this complexity and indeed the necessity of solving two heat conduction problems by making the assumption that the liquid is initially everywhere at the constant fusion temperature T_f, that is $T_\ell = T_f$. For this situation $T_2(x, t) = T_f$ is the appropriate solution of problem (7.3) (as there is always zero temperature gradient in the liquid) and the heat conduction problem in the solid phase with Newton cooling on the surface $x = 0$, becomes

$$\begin{aligned} \frac{\partial T_1}{\partial t} &= \kappa_1 \frac{\partial^2 T_1}{\partial x^2}, \quad 0 < x < X(t), \\[2mm] k_1 \frac{\partial T_1}{\partial x}(0, t) &= h[T_1(0, t) - T_0], \quad T_1(X(t), t) = T_f, \quad t > 0, \\[2mm] k_1 \frac{\partial T_1}{\partial x}(X(t), t) &= \rho L \frac{dX}{dt}, \quad X(0) = 0, \end{aligned}$$
(7.8)

which is the main problem we consider throughout this chapter. As there is only non-trivial heat conduction in the one (solid) phase, we refer to problem (7.8) as a *single phase problem*, whereas the original problem with genuine heat conduction in both phases is known as a *two phase problem*. For this problem it is convenient to introduce non-dimensional variables so that the essential parameters of the problem emerge. We therefore introduce non-dimensional variables and constants α and β defined by

$$\begin{aligned} \bar{x} &= \frac{x}{a}, \quad \overline{X}(\bar{t}) = \frac{X(t)}{a}, \quad \bar{t} = \frac{\kappa_1 t}{a^2}, \\[2mm] \overline{T}(\bar{x}, \bar{t}) &= \frac{T_f - T_1(x, t)}{T_f - T_0}, \\[2mm] \alpha &= \frac{L}{c_1(T_f - T_0)}, \quad \beta = \frac{k_1}{ha}, \end{aligned}$$
(7.9)

and problem (7.8) becomes

$$
\begin{array}{l}
\dfrac{\partial \overline{T}}{\partial \overline{t}} = \dfrac{\partial^2 \overline{T}}{\partial \overline{x}^2}, \quad 0 < \overline{x} < \overline{X}(\overline{t}), \\[2mm]
\overline{T}(0,\overline{t}) - \beta \dfrac{\partial \overline{T}}{\partial \overline{x}}(0,\overline{t}) = 1, \quad \overline{T}(\overline{X}(\overline{t}),\overline{t}) = 0, \\[2mm]
\dfrac{\partial \overline{T}}{\partial \overline{x}}(\overline{X}(\overline{t}),\overline{t}) = -\alpha \dfrac{\mathrm{d}\overline{X}}{\mathrm{d}\overline{t}}, \quad \overline{X}(0) = 0.
\end{array}
\right\} \qquad (7.10)
$$

In the above a denotes any convenient length scale, such as the boundary position at some non-zero time, c_1 is the specific heat of the solid and we have used the relation $\kappa_1 = k_1/\rho c_1$. The constant α, for which specific numerical values are given in Table 7.1, is known as the *Stefan number* or the phase change parameter and is the ratio of the latent heat to the sensible heat of the solidified phase while the constant β^{-1} is sometimes known as the *Biot modulus* and denoted by Bi. For small values of α (for example, for most metals freezing $\alpha < 1$) the solidification process is very rapid while for large values of α (for example, under laboratory conditions for water $\alpha \approx 8.0$) the solidification process is much slower. We observe that the case β zero corresponds to no Newton cooling at the surface $\overline{x} = 0$, that is the surface temperature is prescribed. Further we note that the parameter β, if non-zero, can be completely removed from

Substance	L	T_f	c_1	α
	J/g	K	J/g-K	
Aluminium	370	933	0.90	0.65
Copper	204	1356	0.39	0.49
Gold	65	1336	0.13	0.48
Iron	251	1813	0.45	0.37
Lead	23	600	0.13	0.59
Nickel	307	1723	0.44	0.49
Silver	118	1336	0.23	0.49
Tin	60	505	0.23	1.26
Water	333	273	4.20	7.93

Table 7.1. Latent heat L (joule/gram), fusion temperature T_f (kelvin), specific heat c_1 (joule/gram-kelvin) and Stefan number α for a variety of materials at constant surface temperature $T_0 = 297$K, except for Water where $T_0 = 263$K.

problem (7.10) by using the variables $\bar{x}/\beta$, $\overline{X}/\beta$ and $\bar{t}/\beta^2$. However, the above formulation is preferable so that the case β zero can be recovered since this case admits an exact similarity solution which is described in the following section. The case β non-zero is not solvable exactly and subsequent sections of this chapter develop various standard analytical approximations and the enthalpy formulation which is particularly suited to the numerical solution of heat conduction moving boundary problems.

7.2 Exact solution for prescribed surface temperature

For prescribed surface temperature (β zero), problem (7.10) becomes

$$
\left.
\begin{aligned}
&\frac{\partial \overline{T}}{\partial \bar{t}} = \frac{\partial^2 \overline{T}}{\partial \bar{x}^2}, \quad 0 < \bar{x} < \overline{X}(\bar{t}), \\[2mm]
&\overline{T}(0,\bar{t}) = 1, \quad \overline{T}(\overline{X}(\bar{t}),\bar{t}) = 0, \\[2mm]
&\frac{\partial \overline{T}}{\partial \bar{x}}(\overline{X}(\bar{t}),\bar{t}) = -\alpha \frac{\mathrm{d}\overline{X}}{\mathrm{d}\bar{t}}, \quad \overline{X}(0) = 0.
\end{aligned}
\right\}
\tag{7.11}
$$

It is a simple matter to verify that this problem remains unchanged under the stretching transformation of variables

$$
\bar{x}^* = \lambda\bar{x}, \quad \bar{t}^* = \lambda^2\bar{t}, \quad \overline{X}^* = \lambda\overline{X}, \quad \overline{T}^* = \overline{T},
\tag{7.12}
$$

which has invariants

$$
\frac{\bar{x}^*}{\sqrt{\bar{t}^*}} = \frac{\bar{x}}{\sqrt{\bar{t}}}, \quad \frac{\overline{X}^*}{\sqrt{\bar{t}^*}} = \frac{\overline{X}}{\sqrt{\bar{t}}}, \quad \overline{T}^* = \overline{T}.
$$

Therefore we look for a solution of (7.11) of the form

$$
\overline{T}(\bar{x},\bar{t}) = \psi\!\left(\frac{\bar{x}}{\sqrt{\bar{t}}}\right), \quad \overline{X}(\bar{t}) = \sqrt{2\gamma\bar{t}},
\tag{7.13}
$$

which is automatically invariant under (7.12), where ψ denotes an arbitrary twice differentiable function and γ is a positive constant.

With $\zeta = \bar{x}/\sqrt{\bar{t}}$ and using the partial derivatives

$$
\frac{\partial \overline{T}}{\partial \bar{t}} = \frac{-\zeta\psi'(\zeta)}{2\bar{t}}, \quad \frac{\partial \overline{T}}{\partial \bar{x}} = \frac{\psi'(\zeta)}{\sqrt{\bar{t}}}, \quad \frac{\partial^2 \overline{T}}{\partial \bar{x}^2} = \frac{\psi''(\zeta)}{\bar{t}},
$$

we find that (7.11) becomes

$$\psi''(\zeta) + \frac{\zeta}{2}\psi'(\zeta) = 0,$$
$$\psi(0) = 1, \quad \psi\left(\sqrt{2\gamma}\right) = 0, \quad \psi'\left(\sqrt{2\gamma}\right) = -\alpha\sqrt{\frac{\gamma}{2}}, \tag{7.14}$$

where primes denote differentiation with respect to ζ. A first integration of $(7.14)_1$ yields

$$\psi'(\zeta) = C_1 e^{-\zeta^2/4},$$

so that on a further integration we have

$$\psi(\zeta) = C_1 \int_0^\zeta e^{-\xi^2/4}\, d\xi + C_2,$$

where C_1 and C_2 denote integration constants. From $\psi(0) = 1$ and $\psi(\sqrt{2\gamma}) = 0$ we may deduce

$$\psi(\zeta) = 1 - \frac{\int_0^\zeta e^{-\xi^2/4}\, d\xi}{\int_0^{\sqrt{2\gamma}} e^{-\xi^2/4}\, d\xi},$$

which, on making the change of variable $\eta = \xi/2$ gives

$$\psi(\zeta) = 1 - \mathrm{erf}\left(\frac{\zeta}{2}\right)\Big/ \mathrm{erf}\left(\sqrt{\frac{\gamma}{2}}\right), \tag{7.15}$$

where the error function $\mathrm{erf}(x)$ is defined by (2.11). Thus, altogether the exact solution of the problem (7.11) becomes

$$\overline{T}(\overline{x},\overline{t}) = 1 - \mathrm{erf}\left(\frac{\overline{x}}{2\sqrt{\overline{t}}}\right)\Big/ \mathrm{erf}\left(\sqrt{\frac{\gamma}{2}}\right), \quad \overline{X}(\overline{t}) = \sqrt{2\gamma\overline{t}}, \tag{7.16}$$

where the positive constant γ is determined as a root of the transcendental equation

$$\alpha e^{\frac{\gamma}{2}}\sqrt{\frac{\pi\gamma}{2}}\,\mathrm{erf}\left(\sqrt{\frac{\gamma}{2}}\right) = 1, \tag{7.17}$$

which follows immediately from $(7.14)_4$ and (7.15). The transcendental equation (7.17) must be solved numerically for the root γ (see Table 7.2). For small α (rapid solidification process), γ is large while for large α (slow solidification process), γ is small and the reader will note that these statements are consistent with the solution $\overline{X(t)} = \sqrt{2\gamma t}$ for the boundary motion.

<u>Example 7.1</u> Show that for non-zero α the transcendental equation (7.17) becomes

$$\frac{1}{\alpha} = \gamma \sum_{n=0}^{\infty} \frac{n!(2\gamma)^n}{(2n + 1)!}, \tag{7.18}$$

that is

$$\frac{1}{\alpha} = \gamma\left(1 + \frac{\gamma}{3} + \frac{\gamma^2}{15} + \frac{\gamma^3}{105} + \cdots\right). \tag{7.19}$$

From this equation deduce that

$$\frac{1}{\gamma} = \alpha + \frac{1}{3} - \frac{2}{45\alpha} + \frac{16}{945\alpha^2} + \cdots,$$

which is a useful approximate formula for large α.

From the definition (2.11) of $\mathrm{erf}(x)$ we can rearrange (7.17) to read

$$\frac{1}{\alpha} = \gamma \int_0^1 e^{\gamma(1 - \xi^2)/2}\, d\xi, \tag{7.20}$$

α	γ	$\frac{1}{\gamma}$	$\alpha + \frac{1}{3}$
0.001	10.9706	0.0912	0.3343
0.005	8.0664	0.1240	0.3383
0.010	6.8520	0.1459	0.3433
0.050	4.1902	0.2387	0.3833
0.100	3.1600	0.3165	0.4333
0.500	1.2819	0.7801	0.8333
1.000	0.7690	1.3005	1.3333
5.000	0.1878	5.3251	5.3333
10.000	0.0968	10.3291	10.3333
50.000	0.0199	50.3325	50.3333
100.000	0.0100	100.3332	100.3333
500.000	0.0020	500.3333	500.3333
1000.000	0.0010	1000.3333	1000.3333

Table 7.2. Values of the solution γ of (7.17) and comparison of the upper and lower bounds (7.23) for $\frac{1}{\gamma}$, for various values of the Stefan number α.

so that we may expand the exponential in a Taylor series, and since the Taylor series converges absolutely, we can then interchange the orders of summation and integration thus

$$\frac{1}{\alpha} = \gamma \int_0^1 \sum_{n=0}^{\infty} \frac{\gamma^n}{n!} \frac{(1 - \xi^2)^n}{2^n} \, d\xi = \gamma \sum_{n=0}^{\infty} \frac{\gamma^n}{n! 2^n} \int_0^1 \left(1 - \xi^2\right)^n d\xi. \tag{7.21}$$

If we set

$$I_n = \int_0^1 \left(1 - \xi^2\right)^n d\xi,$$

then we have by integration by parts

$$I_n = 2n \int_0^1 \xi^2 \left(1 - \xi^2\right)^{n-1} d\xi$$

$$= 2n \int_0^1 \left(1 - \xi^2\right)^{n-1} d\xi - 2n \int_0^1 \left(1 - \xi^2\right)^n d\xi$$

$$= 2n I_{n-1} - 2n I_n,$$

so that

$$I_n = \frac{(2n)}{(2n + 1)} I_{n-1},$$

and therefore

$$I_n = \frac{(2n).(2n - 2) \cdots .4.2.}{(2n + 1).(2n - 1) \cdots .5.3.1} I_0.$$

But clearly $I_0 = 1$ and thus

$$I_n = \frac{\left(2^n n!\right)^2}{(2n + 1)!},$$

and the desired first result (7.18) follows immediately from (7.2).

For the second result, note that the left hand side of (7.18) is a monotonically increasing function of γ, so that for any given α, there is at most one γ which can satisfy (7.18). Hence in principle it is possible to uniquely invert (7.18). Suppose that, for large α at least, we can write

$$\frac{1}{\gamma} = \alpha + A + \frac{B}{\alpha} + \frac{C}{\alpha^2} + O\left(\frac{1}{\alpha^3}\right), \tag{7.22}$$

where A, B and C denote constants to be determined by substitution into (7.19). Now

$$\gamma = \frac{1}{\alpha}\left[1 + \frac{A}{\alpha} + \frac{B}{\alpha^2} + \frac{C}{\alpha^3} + O\left(\frac{1}{\alpha^4}\right)\right]^{-1},$$

and expanding binomially we may deduce

$$\gamma = \frac{1}{\alpha}\left[1 - \frac{A}{\alpha} + \frac{(A^2 - B)}{\alpha^2} + O\left(\frac{1}{\alpha^3}\right)\right].$$

Now, from this equation, (7.19) and (7.22) we have

$$\frac{1}{\alpha\gamma} = 1 + \frac{A}{\alpha} + \frac{B}{\alpha^2} + \frac{C}{\alpha^3} + O\left(\frac{1}{\alpha^4}\right)$$

$$= 1 + \frac{1}{3\alpha}\left[1 - \frac{A}{\alpha} + \frac{(A^2 - B)}{\alpha^2}\right] + \frac{1}{15\alpha^2}\left[1 - \frac{2A}{\alpha}\right] + \frac{1}{105\alpha^3} + O\left(\frac{1}{\alpha^4}\right),$$

and on equating coefficients of α^{-1}, α^{-2} and α^{-3} we obtain

$$A = \frac{1}{3}, \quad B = \frac{1}{15} - \frac{A}{3}, \quad C = \frac{1}{105} - \frac{2A}{15} + \frac{(A^2 - B)}{3},$$

from which the given result can be deduced.

<u>Example 7.2</u> Show that the positive roots γ of (7.17) satisfy the inequalities

$$\alpha \le \frac{1}{\gamma} \le \alpha + \frac{1}{3}. \tag{7.23}$$

In order to verify these inequalities we first observe that γ is positive for α positive. This follows since the integral in (7.20) is positive for any γ. Next, to show $\alpha\gamma \le 1$ we suppose the contrary, namely $\alpha\gamma > 1$. Then from (7.19) we have

$$\frac{1}{\alpha\gamma} = \left(1 + \frac{\gamma}{3} + \frac{\gamma^2}{15} + \frac{\gamma^3}{105} + \cdots\right) < 1,$$

which is clearly absurd for $\gamma > 0$, therefore we have established a contradiction and the first inequality of (7.23) follows. For the second inequality of (7.23) we again suppose the converse holds, that is that $\frac{1}{\gamma} > \alpha + \frac{1}{3}$ and since α is positive it follows that we must have $\gamma < 3$. Now, from the assumption $\frac{1}{\gamma} > \alpha + \frac{1}{3}$ we also have

$$\frac{\gamma}{\left(1 - \frac{\gamma}{3}\right)} < \frac{1}{\alpha},$$

which on expanding the left hand side as a geometric series (valid since $0 < \frac{\gamma}{3} < 1$) and using (7.18) for the right hand side gives

$$\sum_{n=0}^{\infty} \left(\frac{\gamma}{3}\right)^n < \sum_{n=0}^{\infty} \frac{n!(2\gamma)^n}{(2n + 1)!},$$

which is again clearly absurd for $\gamma > 0$ since evidently

$$\left(\tfrac{1}{3}\right)^n \geq \frac{n!\,2^n}{(2n+1)!} > 0,$$

with equality for n zero and one and strict inequality for $n \geq 2$. Thus a contradiction is again established and the second inequality of (7.23) follows. We note that the $\tfrac{1}{3}$ in (7.23) is the best constant that can be achieved since as is apparent from the large α formula for $\tfrac{1}{\gamma}$ given in the previous example, $\tfrac{1}{\gamma}$ tends to $\alpha + \tfrac{1}{3}$ as α tends to infinity. This and the validity of the inequalities (7.23) are confirmed numerically in Table 7.2.

This exact solution and its generalization to the case when the liquid is initially not at its freezing point (see Problems 11 and 12) are usually credited to Franz Neumann (1798-1895), as given in unpublished lectures delivered in Konigsberg in the early 1860's. As mentioned previously, this is despite the fact that as early as 1831 Lame and Clapeyron studied the problem of determining the thickness of the solid crust generated by the cooling of a liquid under a constant surface temperature and gave the exact solution (7.16) with the square root time thickness.

7.3 Pseudo steady state and large α approximations

In this section we introduce three important issues, the pseudo steady state approximation, the expansion of the solution in powers of α^{-1} and the idea of boundary fixing transformations which are frequently used for both analytical and numerical solutions of moving boundary problems. These issues can be seen to arise in connection with the exact solution given in the previous section as follows. If we introduce the *boundary fixing transformation* ω defined by

$$\omega = \frac{\bar{x}}{\bar{X}(\bar{t})}, \tag{7.24}$$

then the interval of interest $0 \leq \bar{x} \leq \bar{X}(\bar{t})$ transforms to the fixed unit interval $0 \leq \omega \leq 1$ and, moreover, the exact solution (7.16) becomes

$$\bar{T}(\bar{x},\bar{t}) = 1 - \text{erf}\left(\sqrt{\tfrac{\gamma}{2}}\,\omega\right)\Big/\text{erf}\left(\sqrt{\tfrac{\gamma}{2}}\right).$$

Now for small γ we may approximate this temperature as follows

$$\overline{T}(\overline{x},\overline{t}) = 1 - \left(\int_0^{\sqrt{\frac{\gamma}{2}}\omega} e^{-\xi^2}\, d\xi \right)\Big/\left(\int_0^{\sqrt{\frac{\gamma}{2}}} e^{-\xi^2}\, d\xi \right)$$

$$\approx 1 - \left(\int_0^{\sqrt{\frac{\gamma}{2}}\omega} \left(1 - \xi^2\right) d\xi \right)\Big/\left(\int_0^{\sqrt{\frac{\gamma}{2}}} \left(1 - \xi^2\right) d\xi \right)$$

$$= 1 - \omega\left(1 - \frac{\gamma\omega^2}{6}\right)\Big/\left(1 - \frac{\gamma}{6}\right)$$

$$\approx 1 - \omega\left(1 - \frac{\gamma\omega^2}{6}\right)\left(1 + \frac{\gamma}{6}\right)$$

$$\approx 1 - \omega - \frac{\gamma\omega}{6}\left(1 - \omega^2\right).$$

Thus, on using the result of Example 7.1, namely $\gamma \approx \frac{1}{\alpha}$, we may deduce

$$\overline{T}(\overline{x},\overline{t}) = 1 - \frac{\overline{x}}{\overline{X}(\overline{t})} - \frac{\overline{x}}{6\alpha\,\overline{X}(\overline{t})}\left[1 - \frac{\overline{x}^2}{\overline{X}(\overline{t})^2}\right] + O\left(\frac{1}{\alpha^2}\right). \qquad (7.25)$$

The first term of this equation is referred to as the pseudo steady state temperature estimate of the temperature profile and is denoted by $\overline{T}_{pss}(\overline{x},\overline{t})$. It is in fact the solution to the problem obtained by neglecting $\frac{\partial \overline{T}}{\partial \overline{t}}$ in $(7.11)_1$, that is the solution of

$$\left.\begin{array}{l} \dfrac{\partial^2 \overline{T}_{pss}}{\partial \overline{x}^2} = 0, \quad 0 < \overline{x} < \overline{X}(\overline{t}), \\[2ex] \overline{T}_{pss}(0,\overline{t}) = 1, \quad \overline{T}_{pss}(\overline{X}(\overline{t}),\overline{t}) = 0. \end{array}\right\}$$

Thus the pseudo steady state approximation is the limiting large α solution and represents a good estimate of the actual temperature profile for slow solidification processes (when heat conduction proceeds very much more rapidly than phase change, that is when the latent heat L is very much greater than the sensible heat $c_1(T_f - T_0)$). Similarly using $\frac{1}{\gamma} \approx \alpha + \frac{1}{3}$ we may deduce for the exact boundary motion $\overline{X}(\overline{t}) = \sqrt{2\gamma\overline{t}}$ the following approximation

$$\overline{X}(\overline{t}) = \sqrt{\frac{2\overline{t}}{\alpha}}\left[1 - \frac{1}{6\alpha} + O\left(\frac{1}{\alpha^2}\right)\right]. \qquad (7.26)$$

The first term of this equation is the pseudo steady state boundary motion which alternatively arises by integrating the Stefan condition

$$\frac{\partial \overline{T}_{\text{pss}}}{\partial \overline{x}}\left(\overline{X}(\overline{t}), \overline{t}\right) = -\alpha \frac{\mathrm{d}\overline{X}}{\mathrm{d}\overline{t}}, \quad \overline{X}(0) = 0.$$

<u>Example 7.3</u> For β non-zero show that the pseudo steady state estimates for the temperature and boundary motion for problem (7.10) are given respectively by

$$\overline{T}_{\text{pss}}(\overline{x}, \overline{t}) = \left(\frac{\overline{X}(\overline{t}) - \overline{x}}{\overline{X}(\overline{t}) + \beta}\right), \quad \overline{t}_{\text{pss}}(\overline{X}) = \frac{\alpha}{2}\overline{X}\left(\overline{X} + 2\beta\right). \tag{7.27}$$

The pseudo steady state estimates are obtained by solving

$$\left.\begin{array}{c}
\dfrac{\partial^2 \overline{T}_{\text{pss}}}{\partial \overline{x}^2} = 0, \quad 0 < \overline{x} < \overline{X}(\overline{t}), \\[4mm]
\overline{T}_{\text{pss}}(0, \overline{t}) - \beta \dfrac{\partial \overline{T}_{\text{pss}}}{\partial \overline{x}}(0, \overline{t}) = 1, \quad \overline{T}_{\text{pss}}(\overline{X}(\overline{t}), \overline{t}) = 0, \\[4mm]
\dfrac{\partial \overline{T}_{\text{pss}}}{\partial \overline{x}}(\overline{X}(\overline{t}), \overline{t}) = -\alpha \dfrac{\mathrm{d}\overline{X}}{\mathrm{d}\overline{t}}, \quad \overline{X}(0) = 0.
\end{array}\right\} \tag{7.28}$$

On integrating $(7.28)_1$ we readily obtain

$$\overline{T}_{\text{pss}}(\overline{x}, \overline{t}) = A(\overline{t}) + B(\overline{t})\overline{x},$$

where $A(\overline{t})$ and $B(\overline{t})$ denote arbitrary functions of non-dimensional time $\overline{t}$. From $(7.28)_2$ and $(7.28)_3$ we have

$$A(\overline{t}) - \beta B(\overline{t}) = 1, \quad A(\overline{t}) + B(\overline{t})\overline{X}(\overline{t}) = 0,$$

and therefore

$$A(\overline{t}) = \frac{\overline{X}(\overline{t})}{\overline{X}(\overline{t}) + \beta}, \quad B(\overline{t}) = \frac{-1}{\overline{X}(\overline{t}) + \beta},$$

and the given expression is now immediately apparent. From $(7.27)_1$ and $(7.28)_4$ we have

$$\frac{1}{\overline{X} + \beta} = \alpha \frac{\mathrm{d}\overline{X}}{\mathrm{d}\overline{t}},$$

which on integration and using $\overline{X}(0) = 0$ gives $(7.27)_2$.

In order to derive the α^{-1} correction for β non-zero either of two methods may employed. The first method involves introducing a new time scale $\tau = \bar{t}/\alpha$ while the second utilizes the boundary fixing transformation (7.24) and employs the boundary position $\overline{X}(\bar{t})$ as the time variable.

<u>Method 1</u> Make the following transformations

$$\tau = \frac{\bar{t}}{\alpha}, \quad Y(\tau) = \overline{X}(\bar{t}), \quad \Phi(\bar{x},\tau) = \overline{T}(\bar{x},\bar{t}),$$

so that problem (7.10) becomes

$$\left.\begin{aligned}
\frac{1}{\alpha}\frac{\partial \Phi}{\partial \tau} &= \frac{\partial^2 \Phi}{\partial \bar{x}^2}, \quad 0 < \bar{x} < Y(\tau), \\[2mm]
\Phi(0,\tau) - \beta\frac{\partial \Phi}{\partial \bar{x}}(0,\tau) &= 1, \quad \Phi(Y(\tau),\tau) = 0, \\[2mm]
\frac{\partial \Phi}{\partial \bar{x}}(Y(\tau),\tau) &= -\frac{dY}{d\tau}, \quad Y(0) = 0.
\end{aligned}\right\}
\tag{7.29}$$

We now assume that $\Phi(\bar{x},\tau)$ may be expanded in the regular perturbation series

$$\Phi(\bar{x},\tau) = \Phi_0(\bar{x},\tau) + \frac{1}{\alpha}\Phi_1(\bar{x},\tau) + O\left(\frac{1}{\alpha^2}\right),
\tag{7.30}$$

so that on substituting (7.30) into (7.29) the order one and order $1/\alpha$ equations for $\Phi_0(\bar{x},\tau)$ and $\Phi_1(\bar{x},\tau)$ are as follows

$$\left.\begin{aligned}
\frac{\partial^2 \Phi_0}{\partial \bar{x}^2} &= 0, \quad 0 < \bar{x} < Y(\tau), \\[2mm]
\Phi_0(0,\tau) - \beta\frac{\partial \Phi_0}{\partial \bar{x}}(0,\tau) &= 1, \quad \Phi_0(Y(\tau),\tau) = 0,
\end{aligned}\right\}
\tag{7.31}$$

and

$$\left.\begin{aligned}
\frac{\partial^2 \Phi_1}{\partial \bar{x}^2} &= \frac{\partial \Phi_0}{\partial \tau}, \quad 0 < \bar{x} < Y(\tau), \\[2mm]
\Phi_1(0,\tau) - \beta\frac{\partial \Phi_1}{\partial \bar{x}}(0,\tau) &= 0, \quad \Phi_1(Y(\tau),\tau) = 0.
\end{aligned}\right\}
\tag{7.32}$$

Evidently $\Phi_0(\bar{x},\tau)$ is the pseudo steady state estimate and is given by

$$\Phi_0(\bar{x},\tau) = \frac{Y(\tau) - \bar{x}}{Y(\tau) + \beta}.$$

From this equation we may deduce

$$\frac{\partial \Phi_0}{\partial \tau} = \frac{(\overline{x} + \beta)}{(Y + \beta)^2} \frac{dY}{d\tau},$$

and therefore $\Phi_1(\overline{x}, \tau)$ is obtained by integrating the equation

$$\frac{\partial^2 \Phi_1}{\partial \overline{x}^2} = \frac{(\overline{x} + \beta)}{(Y + \beta)^2} \frac{dY}{d\tau},$$

thus

$$\frac{\partial \Phi_1}{\partial \overline{x}} = \frac{(\overline{x}^2 + 2\beta\overline{x})}{2(Y + \beta)^2} \frac{dY}{d\tau} + C(\tau),$$

and

$$\Phi_1(\overline{x}, \tau) = \frac{(\overline{x}^3 + 3\beta\overline{x}^2)}{6(Y + \beta)^2} \frac{dY}{d\tau} + C(\tau)\overline{x} + D(\tau),$$

where $C(\tau)$ and $D(\tau)$ denote arbitrary functions of τ. From the boundary conditions $(7.32)_2$ and $(7.32)_3$ we can deduce

$$C(\tau) = -\frac{(Y^3 + 3\beta Y^2)}{6(Y + \beta)^3} \frac{dY}{d\tau}, \quad D(\tau) = \frac{-\beta(Y^3 + 3\beta Y^2)}{6(Y + \beta)^3} \frac{dY}{d\tau},$$

and therefore $\Phi_1(\overline{x}, \tau)$ becomes

$$\Phi_1(\overline{x}, \tau) = \frac{(\overline{x} - Y)}{6(Y + \beta)^3} \frac{dY}{d\tau}\left[(Y + \beta)\overline{x}^2 + (Y + \beta)(Y + 3\beta)\overline{x} + (Y + 3\beta)\beta Y\right].$$

From the above expressions for Φ_0 and Φ_1 we find from (7.30) that the Stefan condition $(7.29)_4$ becomes

$$\frac{-1}{(Y + \beta)} + \frac{\left[Y^3 + 3\beta Y^2 + 3\beta^2 Y\right]}{3\alpha(Y + \beta)^3} \frac{dY}{d\tau} + O\left(\frac{1}{\alpha^2}\right) = -\frac{dY}{d\tau},$$

which may be rearranged to give

$$d\tau = (Y + \beta)dY + \frac{1}{3\alpha}\left[(Y + \beta) - \frac{\beta^3}{(Y + \beta)^2}\right]dY + O\left(\frac{1}{\alpha^2}\right).$$

On integration and using $Y(0) = 0$ we may deduce, after some simplification

$$\tau = \frac{Y}{2}(Y + 2\beta) + \frac{Y^2}{6\alpha}\left(\frac{Y + 3\beta}{Y + \beta}\right) + O\left(\frac{1}{\alpha^2}\right).$$

In terms of $\bar{t}$ and $\overline{X}(\bar{t})$ we may summarize the above as follows

$$\overline{T}(\bar{x},\bar{t}) = \left(\frac{\overline{X} - \bar{x}}{\overline{X} + \beta}\right) \times$$

$$\left\{1 - \frac{[(\overline{X} + \beta)\bar{x}^2 + (\overline{X} + \beta)(\overline{X} + 3\beta)\bar{x} + (\overline{X} + 3\beta)\beta\overline{X}]}{6\alpha(\overline{X} + \beta)^3} + O\left(\frac{1}{\alpha^2}\right)\right\}, \quad (7.33)$$

$$\bar{t} = \frac{\alpha}{2}\overline{X}(\overline{X} + 2\beta) + \frac{\overline{X}^2(\overline{X} + 3\beta)}{6(\overline{X} + \beta)} + O\left(\frac{1}{\alpha}\right), \quad (7.34)$$

and we may confirm that in the limit as β tends to zero these expressions are consistent with (7.25) and (7.26) respectively.

<u>Method 2</u> Make the following boundary fixing transformation, with the boundary position $\overline{X}(\bar{t})$ replacing the time variable, thus

$$\omega = \frac{\bar{x}}{\overline{X}(\bar{t})}, \quad \chi = \overline{X}(\bar{t}), \quad \Psi(\omega,\chi) = \overline{T}(\bar{x},\bar{t}),$$

then on using the partial derivatives

$$\frac{\partial\overline{T}}{\partial\bar{x}} = \frac{1}{\chi}\frac{\partial\Psi}{\partial\omega}, \quad \frac{\partial^2\overline{T}}{\partial\bar{x}^2} = \frac{1}{\chi^2}\frac{\partial^2\Psi}{\partial\omega^2},$$

$$\frac{\partial\overline{T}}{\partial\bar{t}} = \frac{-1}{\chi}\left(\omega\frac{\partial\Psi}{\partial\omega} - \chi\frac{\partial\Psi}{\partial\chi}\right)\frac{\mathrm{d}\chi}{\mathrm{d}\bar{t}},$$

we can verify that problem (7.10) becomes

$$\left.\begin{array}{c}\dfrac{\partial^2\Psi}{\partial\omega^2} = \dfrac{1}{\alpha}\dfrac{\partial\Psi}{\partial\omega}(1,\chi)\left(\omega\dfrac{\partial\Psi}{\partial\omega} - \chi\dfrac{\partial\Psi}{\partial\chi}\right), \quad 0 < \omega < 1, \\[3mm] \beta\dfrac{\partial\Psi}{\partial\omega}(0,\chi) = \chi[\Psi(0,\chi) - 1], \quad \Psi(1,\chi) = 0, \\[3mm] \dfrac{\partial\Psi}{\partial\omega}(1,\chi) = -\alpha\chi\dfrac{\mathrm{d}\chi}{\mathrm{d}\bar{t}}, \quad \chi(0) = 0.\end{array}\right\}$$

Equations (7.33) and (7.34) may now be derived assuming that $\Psi(\omega,\chi)$ can be expanded in the regular perturbation series

$$\Psi(\omega,\chi) = \Psi_0(\omega,\chi) + \frac{1}{\alpha}\Psi_1(\omega,\chi) + O\left(\frac{1}{\alpha^2}\right),$$

so that $\Psi_0(\omega,\chi)$ and $\Psi_1(\omega,\chi)$ are obtained respectively as solutions of

$$\frac{\partial^2 \Psi_0}{\partial \omega^2} = 0, \quad 0 < \omega < 1,$$

$$\beta \frac{\partial \Psi_0}{\partial \omega}(0,\chi) = \chi[\Psi_0(0,\chi) - 1], \quad \Psi_0(1,\chi) = 0,$$

and

$$\frac{\partial^2 \Psi_1}{\partial \omega^2} = \frac{\partial \Psi_0}{\partial \omega}(1,\chi)\left(\omega \frac{\partial \Psi_0}{\partial \omega} - \chi \frac{\partial \Psi_0}{\partial \chi}\right), \quad 0 < \omega < 1,$$

$$\beta \frac{\partial \Psi_1}{\partial \omega}(0,\chi) = \chi \Psi_1(0,\chi), \quad \Psi_1(1,\chi) = 0.$$

The analysis of these problems is similar to that of Method 1 and accordingly only the final results are noted, namely

$$\Psi_0(\omega,\chi) = \frac{\chi(1 - \omega)}{(\chi + \beta)},$$

$$\Psi_1(\omega,\chi) = \frac{-\chi^2(1 - \omega)}{6(\chi + \beta)^4}\left[\chi(\chi + \beta)\omega^2 + (\chi + \beta)(\chi + 3\beta)\omega + (\chi + 3\beta)\beta\right].$$

From these expressions, the Stefan condition

$$d\bar{t} = -\alpha\chi\left\{\frac{\partial \Psi_0}{\partial \omega}(1,\chi) + \frac{1}{\alpha}\frac{\partial \Psi_1}{\partial \omega}(1,\chi) + O\left(\frac{1}{\alpha^2}\right)\right\}^{-1} d\chi,$$

and expanding in powers of α^{-1} we may deduce the given equations (7.33) and (7.34) for the temperature and boundary motion respectively.

7.4 Integral formulation, bounds and integral iteration

In this section we present an integral formulation of (7.10) from which an integral expression may be deduced for the boundary motion. From this latter equation and simple bounds for the temperature we may deduce upper and lower bounds for the boundary motion. In addition, using the integral formulation we may construct an integral iteration scheme which generates further approximations to the temperature profile from any given approximation. For the problem of planar freezing defined by (7.10) we have on integrating (7.10)$_1$ from an arbitrary position $\bar{x}$ to the moving boundary $\overline{X}(\overline{t})$

$$\int_{\overline{x}}^{\overline{X}(\overline{t})} \frac{\partial \overline{T}}{\partial \overline{t}}(\xi,\overline{t})\,d\xi = -\alpha\frac{d\overline{X}(\overline{t})}{d\overline{t}} - \frac{\partial \overline{T}}{\partial \overline{x}}(\overline{x},\overline{t}), \tag{7.35}$$

where we have used the Stefan condition $(7.10)_4$ for the temperature gradient at the moving boundary. On repeating this integration for (7.35) and changing the order of integration we obtain

$$\int_{\overline{x}}^{\overline{X}(\overline{t})}\int_{\eta}^{\overline{X}(\overline{t})} \frac{\partial \overline{T}}{\partial \overline{t}}(\xi,\overline{t})\,d\xi\,d\eta = \int_{\overline{x}}^{\overline{X}(\overline{t})}\int_{\overline{x}}^{\xi} \frac{\partial \overline{T}}{\partial \overline{t}}(\xi,\overline{t})\,d\eta\,d\xi$$

$$= \int_{\overline{x}}^{\overline{X}(\overline{t})}(\xi - \overline{x})\frac{\partial \overline{T}}{\partial \overline{t}}(\xi,\overline{t})\,d\xi,$$

and on noting $(7.10)_3$ we find

$$\int_{\overline{x}}^{\overline{X}(\overline{t})}(\xi - \overline{x})\frac{\partial \overline{T}}{\partial \overline{t}}(\xi,\overline{t})\,d\xi = -\alpha\frac{d\overline{X}}{d\overline{t}}(\overline{X} - \overline{x}) + \overline{T}(\overline{x},\overline{t}),$$

which can be rearranged to give the integro-partial differential equation

$$\overline{T}(\overline{x},\overline{t}) = \frac{\partial}{\partial \overline{t}}\int_{\overline{x}}^{\overline{X}(\overline{t})}(\xi - \overline{x})\left[\alpha + \overline{T}(\xi,\overline{t})\right]d\xi. \tag{7.36}$$

We observe that in deriving (7.36) we have utilized all the equations of (7.10) except the surface condition $(7.10)_2$ and the initial condition $(7.10)_5$. From $(7.10)_2$ and (7.36) we can deduce

$$\frac{d}{d\overline{t}}\int_{0}^{\overline{X}(\overline{t})}(\xi + \beta)\left[\alpha + \overline{T}(\xi,\overline{t})\right]d\xi = 1, \tag{7.37}$$

which evidently integrates immediately so that together with $(7.10)_5$ we conclude

$$\overline{t} = \int_{0}^{\overline{X}(\overline{t})}(\xi + \beta)\left[\alpha + \overline{T}(\xi,\overline{t})\right]d\xi. \tag{7.38}$$

Equations (7.36) and (7.38) constitute the basic integral formulation for planar freezing.

Although (7.38) is not an explicit equation for the motion of the boundary, it nevertheless is an important result. We observe that (7.38) is equivalent to the Stefan condition $(7.10)_4$. This equality is by no means obvious. Equation (7.38) is an integral involving temperature values while $(7.10)_4$ involves heat flux values at the moving boundary. In order to see that these equations are indeed equivalent we differentiate (7.38) with respect to time so that we have

$$1 = \alpha(\overline{X} + \beta)\frac{\mathrm{d}\overline{X}}{\mathrm{d}\bar{t}} + \int_0^{\overline{X}} (\xi + \beta)\frac{\partial^2\overline{T}}{\partial\xi^2}(\xi,\bar{t})\,\mathrm{d}\xi,$$

where we have used the heat equation $(7.10)_1$ to eliminate the time partial derivative of $\overline{T}(\bar{x},\bar{t})$. On integrating by parts this equation becomes

$$1 = (\overline{X} + \beta)\left\{\alpha\frac{\mathrm{d}\overline{X}}{\mathrm{d}\bar{t}} + \frac{\partial\overline{T}}{\partial\bar{x}}(\overline{X}(\bar{t}),\bar{t})\right\} + \overline{T}(0,\bar{t}) - \beta\frac{\partial\overline{T}}{\partial\bar{x}}(0,\bar{t}),$$

which on using the surface condition yields precisely the Stefan condition. Clearly (7.38) embodies more information than does $(7.10)_4$ and for this reason may be used to advantage in conjunction with other approximate procedures.

Example 7.4 Verify, for β zero, that the exact Neumann solution (7.16) satisfies equations (7.36) and (7.38).

From $(7.16)_1$ the integral in (7.36) becomes

$$\alpha\int_{\overline{X}}^{\overline{X}(\bar{t})} (\xi - \bar{x})\left\{1 + \gamma\int_{\xi/\overline{X}(\bar{t})}^1 e^{\gamma(1-\eta^2)/2}\,\mathrm{d}\eta\right\}\mathrm{d}\xi$$

$$= \alpha\overline{X}^2\int_\omega^1 (\Omega - \omega)\left\{1 + \gamma\int_\omega^1 e^{\gamma(1-\eta^2)/2}\,\mathrm{d}\eta\right\}\mathrm{d}\Omega$$

$$= \frac{\alpha\overline{X}^2}{2}\left\{(1 - \omega)^2 + \gamma\int_\omega^1 (\eta - \omega)^2 e^{\gamma(1-\eta)^2/2}\,\mathrm{d}\eta\right\}$$

$$= \frac{\alpha\overline{X}^2}{2}\left\{\omega^2 - \omega e^{\gamma(1-\omega^2)/2} + (1 + \gamma\omega^2)\int_\omega^1 e^{\gamma(1-\eta^2)/2}\,\mathrm{d}\eta\right\},$$

where $\omega = \bar{x}/\overline{X}(\bar{t})$ and $\Omega = \xi/\overline{X}(\bar{t})$ and we have changed orders of integration and used integration by parts in a routine manner. On differentiating this expression partially with respect to time equation (7.36) becomes

$$\overline{T}(\bar{x},\bar{t}) = \alpha\overline{X}\frac{\mathrm{d}\overline{X}}{\mathrm{d}\bar{t}}\int_\omega^1 e^{\gamma(1-\eta^2)/2}\,\mathrm{d}\eta.$$

In a similar manner (7.38) with β zero becomes

$$\bar{t} = \frac{\alpha \bar{X}^2}{2} \int_0^1 e^{\gamma(1-\eta^2)/2} \, d\eta,$$

and these equations are clearly consistent with (7.16) and (7.17).

For β zero or non-zero we may use (7.38) to obtain upper and lower bounds on the boundary motion. From equation (7.9) it is apparent that the non-dimensional temperature $\bar{T}(\bar{x}, \bar{t})$ satisfies the inequalities

$$0 \le \bar{T}(\bar{x}, \bar{t}) \le 1, \quad \frac{\partial \bar{T}}{\partial \bar{t}}(\bar{x}, \bar{t}) \ge 0, \tag{7.39}$$

and the first of these together with (7.38) yields

$$\alpha \int_0^{\bar{X}} (\xi + \beta) \, d\xi \le \bar{t}(\bar{X}) \le (\alpha + 1) \int_0^{\bar{X}} (\xi + \beta) \, d\xi,$$

or

$$\frac{\alpha}{2} \bar{X}(\bar{X} + 2\beta) \le \bar{t}(\bar{X}) \le \frac{(\alpha + 1)}{2} \bar{X}(\bar{X} + 2\beta). \tag{7.40}$$

Thus the pseudo steady state boundary motion $(7.27)_2$ is a lower bound to the actual motion (see Figures 7.3 and 7.4). The upper bound in (7.40) may be substantially improved as described in the following example.

<u>Example 7.5</u> Verify the inequality

$$\bar{T}(\bar{x}, \bar{t}) \le \bar{T}_{\text{pss}}(\bar{x}, \bar{t}), \tag{7.41}$$

where $\bar{T}_{\text{pss}}(\bar{x}, \bar{t})$ is given by $(7.27)_1$ and hence, or otherwise, deduce that

$$\bar{t}(\bar{X}) \le \frac{\alpha \bar{X}}{2} (\bar{X} + 2\beta) + \frac{\bar{X}^2(\bar{X} + 3\beta)}{6(\bar{X} + \beta)}, \tag{7.42}$$

and from which observe that the order one corrected boundary motion given by (7.34) is an upper bound to the actual motion.

On differentiating out (7.36) and (7.37) respectively we may deduce

$$\overline{T}(\overline{x},\overline{t}) = \alpha(\overline{X} - \overline{x})\frac{\mathrm{d}\overline{X}}{\mathrm{d}\overline{t}} + \int_{\overline{x}}^{\overline{X}}(\xi - \overline{x})\frac{\partial\overline{T}}{\partial\overline{t}}(\xi,\overline{t})\,\mathrm{d}\xi,$$

$$1 = \alpha(\overline{X} + \beta)\frac{\mathrm{d}\overline{X}}{\mathrm{d}\overline{t}} + \int_{0}^{\overline{X}}(\xi + \beta)\frac{\partial\overline{T}}{\partial\overline{t}}(\xi,\overline{t})\,\mathrm{d}\xi.$$

Now, eliminating $\alpha\dfrac{\mathrm{d}\overline{X}}{\mathrm{d}\overline{t}}$ from these equations gives

$$\overline{T}(\overline{x},\overline{t}) = \left(\frac{\overline{X} - \overline{x}}{\overline{X} + \beta}\right)\left\{1 - \int_{0}^{\overline{X}}(\xi + \beta)\frac{\partial\overline{T}}{\partial\overline{t}}(\xi,\overline{t})\,\mathrm{d}\xi\right\} + \int_{\overline{x}}^{\overline{X}}(\xi - \overline{x})\frac{\partial\overline{T}}{\partial\overline{t}}(\xi,\overline{t})\,\mathrm{d}\xi,$$

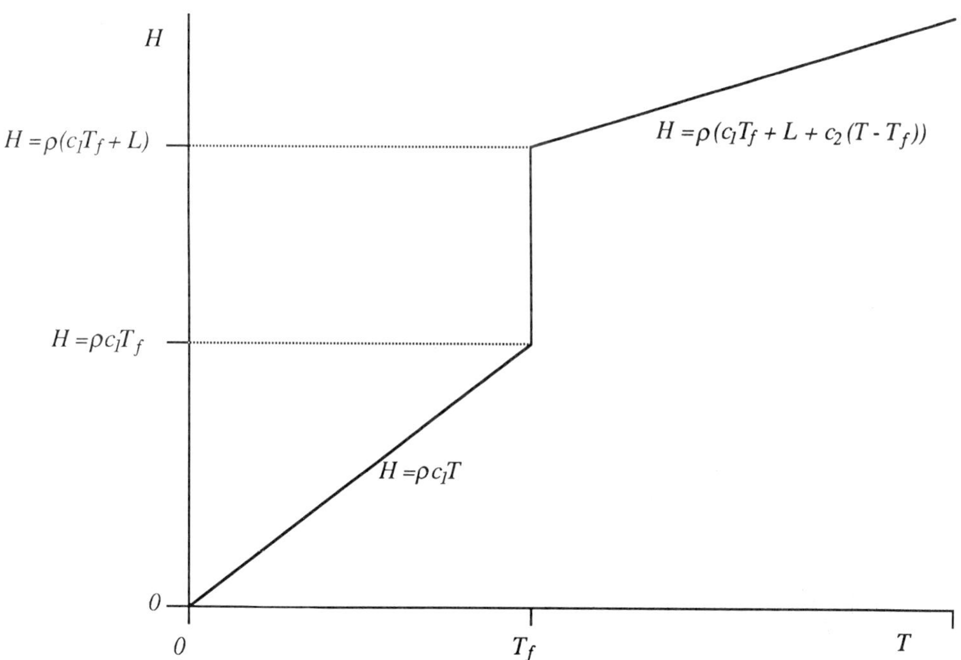

Figure 7.3. Upper and lower bounds for the boundary motion (7.40) ($\cdots$), improved upper bound (7.43) (- - -) and numerical boundary motion (—) for $\alpha = 1.0$ and $\beta = 1.0$.

so that we have

$$\overline{T}(\overline{x},\overline{t}) - \overline{T}_{\mathrm{pss}}(\overline{x},\overline{t})$$

$$= - \int_0^{\overline{X}} \frac{(\overline{X} - \overline{x})(\xi + \beta)}{(\overline{X} + \beta)} \frac{\partial \overline{T}}{\partial \overline{t}}(\xi,\overline{t})\,\mathrm{d}\xi + \int_{\overline{x}}^{\overline{X}} \frac{(\overline{X} + \beta)(\xi - \overline{x})}{(\overline{X} + \beta)} \frac{\partial \overline{T}}{\partial \overline{t}}(\xi,\overline{t})\,\mathrm{d}\xi$$

$$= - \int_0^{\overline{x}} \frac{(\overline{X} - \overline{x})(\xi + \beta)}{(\overline{X} + \beta)} \frac{\partial \overline{T}}{\partial \overline{t}}(\xi,\overline{t})\,\mathrm{d}\xi - \int_{\overline{x}}^{\overline{X}} \frac{(\overline{X} - \xi)(\overline{x} + \beta)}{(\overline{X} + \beta)} \frac{\partial \overline{T}}{\partial \overline{t}}(\xi,\overline{t})\,\mathrm{d}\xi$$

$$\leq 0,$$

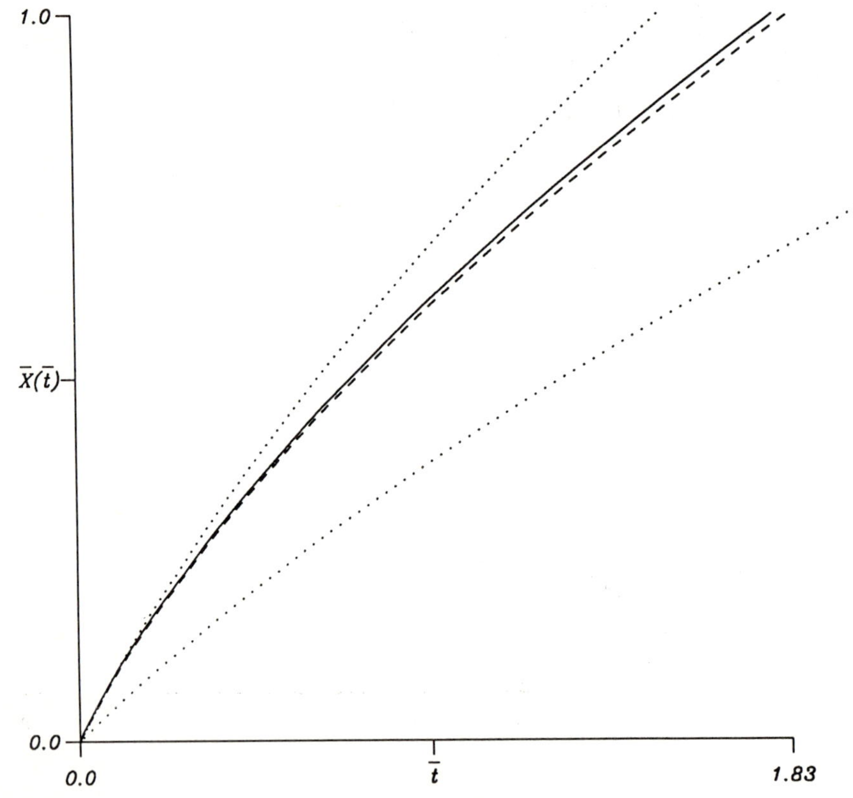

Figure 7.4. Upper and lower bounds for the boundary motion (7.40) ($\cdots$), improved upper bound (7.43) (- - -) and numerical boundary motion (—) for $\alpha = 2.0$ and $\beta = 1.0$.

on noting the equality $(7.39)_2$. Thus we have $\overline{T} \leq \overline{T}_{\mathrm{pss}}$ and therefore from (7.38) we obtain

$$\overline{t}(\overline{X}) \leq \int_0^{\overline{X}} (\xi + \beta)\left[\alpha + \left(\frac{\overline{X} - \xi}{\overline{X} + \beta}\right)\right]\mathrm{d}\xi,$$

and the desired inequality follows by integration.

Thus, altogether from (7.40) and (7.42) we have

$$\frac{\alpha}{2}\overline{X}(\overline{X} + 2\beta) \leq \overline{t}(\overline{X}) \leq \frac{\alpha}{2}\overline{X}(\overline{X} + 2\beta) + \frac{\overline{X}^2(\overline{X} + 3\beta)}{6(\overline{X} + \beta)}, \tag{7.43}$$

which for β zero and the exact Neumann solution $\overline{X}(\overline{t}) = \sqrt{2\gamma\overline{t}}$ yields simply the inequalities (7.23) of Example 7.2. The bounds (7.40) and (7.43) on the moving boundary are shown graphically in Figure 7.3 for $\alpha = 1.0$ and $\beta = 1.0$, and in Figure 7.4 for $\alpha = 2.0$ and $\beta = 1.0$.

Finally in this section we exploit the integral formulation to establish an iterative procedure for improving any given approximation for the temperature. If we use the convention that $\overline{T}^\dagger(\overline{x}, \overline{X})$ denotes the non-dimensional temperature $\overline{T}(\overline{x}, \overline{t})$ but with the boundary position $\overline{X}(\overline{t})$ as a time variable, so that

$$\overline{T}^\dagger(\overline{x}, \overline{X}(\overline{t})) = \overline{T}(\overline{x}, \overline{t}),$$

then on dividing (7.36) by (7.37) we can deduce

$$\overline{T}^\dagger(\overline{x}, \overline{X}) = \frac{\dfrac{\partial}{\partial \overline{X}}\displaystyle\int_{\overline{x}}^{\overline{X}}(\xi - \overline{x})\left[\alpha + \overline{T}^\dagger(\xi, \overline{X})\right]\mathrm{d}\xi}{\dfrac{\partial}{\partial \overline{X}}\displaystyle\int_0^{\overline{X}}(\xi + \beta)\left[\alpha + \overline{T}^\dagger(\xi, \overline{X})\right]\mathrm{d}\xi},$$

from which it is a natural step to formulate the following iterative procedure

$$\overline{T}^\dagger_{n+1}(\overline{x}, \overline{X}) = \frac{\dfrac{\partial}{\partial \overline{X}}\displaystyle\int_{\overline{x}}^{\overline{X}}(\xi - \overline{x})\left[\alpha + \overline{T}^\dagger_n(\xi, \overline{X})\right]\mathrm{d}\xi}{\dfrac{\partial}{\partial \overline{X}}\displaystyle\int_0^{\overline{X}}(\xi + \beta)\left[\alpha + \overline{T}^\dagger_n(\xi, \overline{X})\right]\mathrm{d}\xi}, \tag{7.44}$$

for successive estimates $\overline{T}^\dagger_n(\overline{x}, \overline{X})$ of the temperature $\overline{T}^\dagger(\overline{x}, \overline{X})$. Moreover, for each estimate of the temperature, $\overline{T}^\dagger_n(\overline{x}, \overline{X})$, we may deduce from (7.38) an estimate of the motion of the boundary from the equation

$$\bar{t}_{n+1}(\overline{X}) = \int_0^{\overline{X}} (\xi + \beta)\left[\alpha + \overline{T}_n^\dagger(\xi, \overline{X})\right] d\xi. \tag{7.45}$$

In principle we can continue the above process indefinitely. However, although the actual calculations are straightforward, the expressions obtained rapidly become unmanagable. The following two integrals, obtained by integration by parts, are particularly useful

$$\int_{\bar{x}}^{\overline{X}} (\xi - \bar{x})(\overline{X} - \xi)^n \, d\xi = \frac{(\overline{X} - \bar{x})^{n+2}}{(n+1)(n+2)},$$

$$\int_0^{\overline{X}} (\xi + \beta)(\overline{X} - \xi)^n \, d\xi = \frac{\overline{X}^{n+1}[\overline{X} + (n+2)\beta]}{(n+1)(n+2)},$$

for $n \geq 0$.

Note that we may start the above process with the initial estimate

$$\overline{T}_{-1}^\dagger(\bar{x}, \overline{X}) = 0, \tag{7.46}$$

which is in fact the initial temperature, so that from (7.44) and (7.45) we obtain

$$\overline{T}_0^\dagger(\bar{x}, \overline{X}) = \left(\frac{\overline{X} - \bar{x}}{\overline{X} + \beta}\right), \quad \bar{t}_0(\overline{X}) = \frac{\alpha}{2}\overline{X}(\overline{X} + 2\beta), \tag{7.47}$$

which is simply the pseudo steady state approximation. Using $\overline{T}_0^\dagger$ and the above integrals, we find that (7.44) and (7.45) yield

$$\overline{T}_1^\dagger(\bar{x}, \overline{X}) = \frac{\left\{\alpha\left(\dfrac{\overline{X} - \bar{x}}{\overline{X} + \beta}\right) + \dfrac{1}{2}\left(\dfrac{\overline{X} - \bar{x}}{\overline{X} + \beta}\right)^2 - \dfrac{1}{6}\left(\dfrac{\overline{X} - \bar{x}}{\overline{X} + \beta}\right)^3\right\}}{\left\{\alpha + \dfrac{\overline{X}(\overline{X}^2 + 3\beta\overline{X} + 3\beta^2)}{3(\overline{X} + \beta)^3}\right\}}, \tag{7.48}$$

$$\bar{t}_1(\overline{X}) = \frac{\alpha}{2}\overline{X}(\overline{X} + 2\beta) + \frac{\overline{X}^2(\overline{X} + 3\beta)}{6(\overline{X} + \beta)}, \tag{7.49}$$

and we observe $\bar{t}_1(\overline{X})$ is simply the order one corrected motion (see equation (7.34)). Thus two iterations of (7.46) produce the known lower and upper bounds $\bar{t}_0(\overline{X})$ and

$\bar{t}_1(\overline{X})$, respectively, for the actual motion $\bar{t}(\overline{X})$. A further iteration using $\overline{T}_1^\dagger(\bar{x},\overline{X})$ yields a lengthy expression for $\overline{T}_2^\dagger(\bar{x},\overline{X})$ which we do not give here. However, the corresponding expression for $\bar{t}_2(\overline{X})$ is much simpler and is given by

$$\bar{t}_2(\overline{X}) = \frac{\alpha}{2}\overline{X}(\overline{X}+2\beta) + \frac{\overline{X}^2\left\{\alpha\left(\frac{\overline{X}+3\beta}{\overline{X}+\beta}\right) + \overline{X}\frac{(\overline{X}^2+5\beta\overline{X}+5\beta^2)}{5(\overline{X}+\beta)^3}\right\}}{6\left\{\alpha + \frac{\overline{X}(\overline{X}^2+3\beta\overline{X}+3\beta^2)}{3(\overline{X}+\beta)^3}\right\}}. \qquad (7.50)$$

We observe that $\overline{T}_1^\dagger(\bar{x},\overline{X})$ as given by (7.48) is essentially a cubic in $(\overline{X}-\bar{x})$. We can show that $\overline{T}_1^\dagger$ satisfies the inequalities

$$0 \le \overline{T}_1^\dagger(\bar{x},\overline{X}) \le \overline{T}_0^\dagger(\bar{x},\overline{X}). \qquad (7.51)$$

The left hand inequality follows on inspection, noting $\overline{T}_0^\dagger \le 1$. The right hand inequality follows since

$$\overline{T}_1^\dagger(\bar{x},\overline{X}) - \overline{T}_0^\dagger(\bar{x},\overline{X}) = -\frac{\overline{T}_0^\dagger\left\{(\overline{T}_0^\dagger-1)(\overline{T}_0^\dagger-2) + \frac{2\overline{X}(\overline{X}^2+3\beta\overline{X}+3\beta^2)}{(\overline{X}+\beta)^3}\right\}}{6\left\{\alpha + \frac{\overline{X}(\overline{X}^2+3\beta\overline{X}+3\beta^2)}{3(\overline{X}+\beta)^3}\right\}} \le 0.$$

Having established (7.51), it follows from (7.45) that at least $\bar{t}_2(\overline{X})$ satisfies the known bounds, namely

$$\bar{t}_0(\overline{X}) \le \bar{t}_2(\overline{X}) \le \bar{t}_1(\overline{X}), \qquad (7.52)$$

although, of course the lower bound is immediately apparent from inspection of (7.50). The approximate boundary motion $\bar{t}_2(\overline{X})$ is shown graphically in Figure 7.5 along with the bounding boundary motions $\bar{t}_0(\overline{X})$ and $\bar{t}_1(\overline{X})$ and a numerical estimate obtained using the enthalpy method, as outlined in the final section of this chapter.

We remark that the integral iteration procedure outlined above is particularly suited to computer based algebraic manipulation packages which are now widely available. For the purposes of illustration we give below the expression for $\bar{t}_3(\overline{X})$ obtained using the algebraic manipulation language REDUCE, namely

$$\bar{t}_3(\overline{X}) = \frac{\alpha}{2}\overline{X}(\overline{X} + 2\beta) + \frac{\overline{X}^2}{6}\left\{\frac{\text{Numerator}}{\text{Denominator}}\right\},$$

where

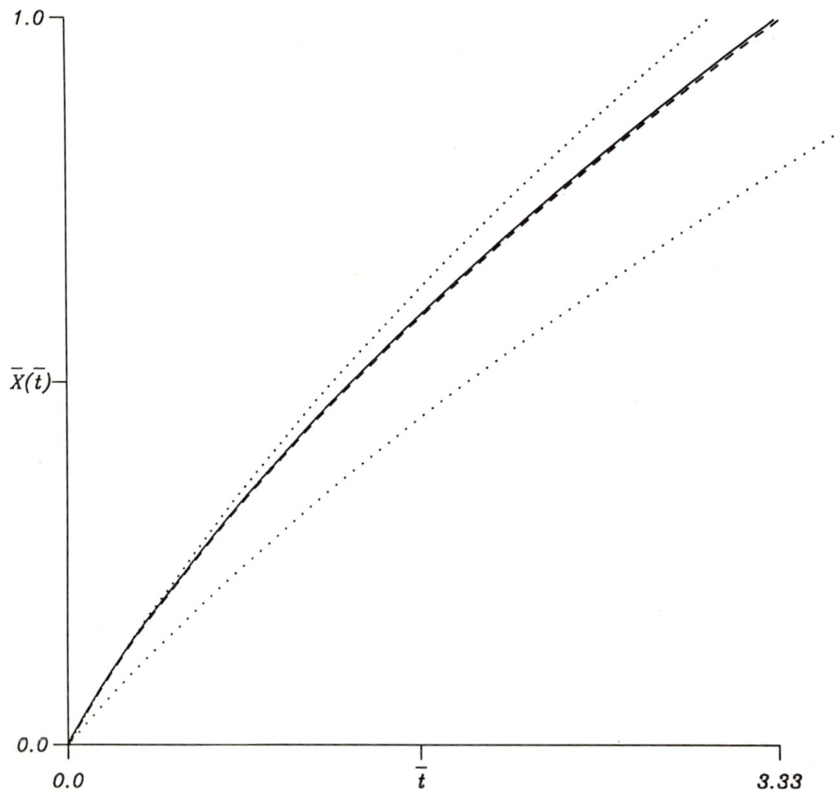

Figure 7.5. Approximate boundary motions $\bar{t}_0(\overline{X})$ $(\cdot\cdot\cdot)$, $\bar{t}_1(\overline{X})$ $(---)$ and $\bar{t}_2(\overline{X})$ $(-\cdot-)$ compared with the numerical boundary motion $(—)$ for $\alpha = 0.1$ and $\beta = 0.5$.

Numerator

$$= \alpha^3\left(\frac{\overline{X} + 3\beta}{\overline{X} + \beta}\right) + \frac{\alpha^2\overline{X}(13\overline{X}^3 + 78\beta\overline{X}^2 + 150\beta^2\overline{X} + 105\beta^3)}{15(\overline{X} + \beta)^4}$$

$$+ \frac{\alpha\overline{X}^2(65\overline{X}^5 + 585\beta\overline{X}^4 + 2067\beta^2\overline{X}^3 + 3696\beta^3\overline{X}^2 + 3339\beta^4\overline{X} + 1260\beta^5)}{315(\overline{X} + \beta)^7}$$

$$+ \frac{\overline{X}^3(\overline{X}^3 + 8\beta\overline{X}^2 + 21\beta^2\overline{X} + 21\beta^3)}{105(\overline{X} + \beta)^6},$$

Denominator

$$= \alpha^3 + \frac{\alpha^2\overline{X}(\overline{X}^2 + 3\beta\overline{X} + 3\beta^2)}{(\overline{X} + \beta)^3}$$

$$+ \frac{\alpha\overline{X}^2(13\overline{X}^4 + 78\beta\overline{X}^3 + 195\beta^2\overline{X}^2 + 225\beta^3\overline{X} + 90\beta^4)}{45(\overline{X} + \beta)^6}$$

$$+ \frac{\overline{X}^3(\overline{X}^3 + 6\beta\overline{X}^2 + 15\beta^2\overline{X} + 15\beta^3)}{45(\overline{X} + \beta)^6}.$$

We point out that the convergence of such iterative schemes are not well understood and such approximations are not necessarily more accurate than preceeding estimates but are certainly more computationally efficient to evaluate than a full numerical solution of the problem.

7.5 Approximate analytical solutions by heat-balance

In this section we obtain approximate analytical solutions of the moving boundary problem (7.10) by the heat-balance method described in Chapter 5. For the purposes of illustration we first consider the case β zero. From equation (7.35) with $x = 0$ we have the heat balance condition

$$\frac{\mathrm{d}}{\mathrm{d}\bar{t}}\left[\int_0^{\overline{X}(\bar{t})} \overline{T}(\xi, \bar{t})\,\mathrm{d}\xi + \alpha\overline{X}(\bar{t})\right] + \frac{\partial\overline{T}}{\partial\bar{x}}(0, \bar{t}) = 0. \tag{7.53}$$

Quadratic temperature profile

We now seek to approximate the solution of (7.10) by an expression of the form

$$\overline{T}(\overline{x},\overline{t}) = A_1(\overline{t})\big(\overline{x} - \overline{X}(\overline{t})\big) + A_2(\overline{t})\big(\overline{x} - \overline{X}(\overline{t})\big)^2,\qquad (7.54)$$

where $A_1(\overline{t})$ and $A_2(\overline{t})$ are determined from known information at the endpoints $\overline{x} = 0$ and $\overline{x} = \overline{X}(\overline{t})$ in the following manner. On differentiating $\overline{T}\big(\overline{X}(\overline{t}),\overline{t}\big) = 0$ we have

$$\frac{\partial \overline{T}}{\partial \overline{t}}\big(\overline{X}(\overline{t}),\overline{t}\big) + \frac{\partial \overline{T}}{\partial \overline{x}}\big(\overline{X}(\overline{t}),\overline{t}\big)\frac{d\overline{X}}{d\overline{t}} = 0,$$

and from this equation and (7.10) we obtain

$$\left(\frac{\partial \overline{T}}{\partial \overline{x}}\big(\overline{X}(\overline{t}),\overline{t}\big)\right)^2 = \alpha\frac{\partial^2 \overline{T}}{\partial \overline{x}^2}\big(\overline{X}(\overline{t}),\overline{t}\big),\qquad (7.55)$$

which is the essential non-linearity in the problem. From (7.54) and (7.55) we obtain the relation

$$A_2(\overline{t}) = \frac{A_1(\overline{t})^2}{2\alpha},$$

which, together with the surface condition $\overline{T}(0,\overline{t}) = 1$ gives the following quadratic equation for the determination of $A_1(\overline{t})$, thus

$$A_1(\overline{t})^2\overline{X}(\overline{t})^2 - 2\alpha A_1(\overline{t})\overline{X}(\overline{t}) - 2\alpha = 0,$$

where the two roots correspond to the situations of freezing in the negative and positive half-planes. This can be seen from the Stefan condition $(7.10)_4$ from which it is apparent that $A_1(\overline{t})$ must be negative. Alternatively since $\overline{x} < \overline{X}(\overline{t})$ it is clear from (7.54) that taking the negative root ensures $\overline{T} \geq 0$. Thus we have

$$A_1(\overline{t}) = \frac{\big[\alpha - \sqrt{\alpha^2 + 2\alpha}\,\big]}{\overline{X}(\overline{t})},\quad A_2(\overline{t}) = \frac{\big[1 + \alpha - \sqrt{\alpha^2 + 2\alpha}\,\big]}{\overline{X}(\overline{t})^2}.\qquad (7.56)$$

Now in order to determine $\overline{X}(\overline{t})$ by the heat-balance method, we simply substitute (7.54) into (7.53), thus

$$\frac{d}{d\overline{t}}\Big[\alpha\overline{X}(\overline{t}) - \tfrac{1}{2}A_1(\overline{t})\overline{X}(\overline{t})^2 + \tfrac{1}{3}A_2(\overline{t})\overline{X}(\overline{t})^3\Big] + A_1(\overline{t}) - 2A_2(\overline{t})\overline{X}(\overline{t}) = 0,$$

and on using (7.56) this equation simplifies to give the correct form of the boundary motion, namely $\overline{X}(\bar{t}) = \sqrt{2\gamma\bar{t}}$ with the constant $\frac{1}{\gamma}$ approximated as follows

$$\frac{1}{\gamma} \approx \frac{1}{2}\left(\alpha + \frac{1}{3}\right)\left[1 + \sqrt{\frac{\alpha}{\alpha + 2}}\right] + \frac{1}{6}\sqrt{\frac{\alpha}{\alpha + 2}}.$$

<u>Cubic temperature profile</u>

We now seek to approximate the solution of (7.10) by an expression of the form

$$\overline{T}(\bar{x},\bar{t}) = A_1(\bar{t})(\bar{x} - \overline{X}) + A_2(\bar{t})(\bar{x} - \overline{X})^2 + A_3(\bar{t})(\bar{x} - \overline{X})^3. \tag{7.57}$$

From the surface condition $\overline{T}(0,\bar{t}) = 1$ and (7.55) we obtain two constraints on the functions $A_1(\bar{t})$, $A_2(\bar{t})$ and $A_3(\bar{t})$

$$-A_1(\bar{t})\overline{X}(\bar{t}) + A_2(\bar{t})\overline{X}(\bar{t})^2 - A_3(\bar{t})\overline{X}(\bar{t})^3 = 1, \quad A_2(\bar{t}) = \frac{A_1(\bar{t})^2}{2\alpha}. \tag{7.58}$$

A third equation may be obtained by requiring the heat equation to be satisfied at $\bar{x} = 0$. Differentiating the surface condition gives

$$\frac{\partial \overline{T}}{\partial \bar{t}}(0,\bar{t}) = 0,$$

and therefore

$$\frac{\partial^2 \overline{T}}{\partial \bar{x}^2}(0,\bar{t}) = 0.$$

From this equation and (7.57) we readily obtain

$$A_3(\bar{t}) = \frac{A_2(\bar{t})}{3\overline{X}(\bar{t})},$$

so that using (7.58) we have

$$A_1(\bar{t})^2\overline{X}(\bar{t})^2 - 3\alpha A_1(\bar{t})\overline{X}(\bar{t}) - 3\alpha = 0.$$

Altogether we find

$$A_1(\bar{t}) = \frac{3\alpha}{2\overline{X}(\bar{t})}\left[1 - \sqrt{1 + \frac{4}{3\alpha}}\right], \quad A_2(\bar{t}) = \frac{A_1(\bar{t})^2}{2\alpha}, \quad A_3(\bar{t}) = \frac{A_1(\bar{t})^2}{6\alpha\overline{X}(\bar{t})}. \tag{7.59}$$

Now from (7.57) and the heat-balance condition (7.53) we have

$$\frac{\mathrm{d}}{\mathrm{d}\bar{t}}\left\{\alpha\overline{X}(\bar{t}) - \frac{A_1(\bar{t})\overline{X}(\bar{t})^2}{2} + \frac{A_2(\bar{t})\overline{X}(\bar{t})^3}{3} - \frac{A_3(\bar{t})\overline{X}(\bar{t})^4}{4}\right\}$$

$$+ A_1(\bar{t}) - 2A_2(\bar{t})\overline{X}(\bar{t}) + A_3(\bar{t})\overline{X}(\bar{t})^2 = 0,$$

which together with (7.59) again yields the correct functional form of the boundary motion, this time with the constant $\frac{1}{\gamma}$ being approximated by

$$\frac{1}{\gamma} \approx \frac{\left\{\left(4\alpha^2 + 9\alpha + 3\right) + \left(4\alpha^2 + 3\alpha\right)\sqrt{1 + \frac{4}{3\alpha}}\right\}}{4(3 + 2\alpha)}.$$

The solution for β non-zero and the quadratic temperature profile is summarized in the following example.

Example 7.6 For β non-zero and the quadratic profile (7.54) show that the appropriate solution for $A_1(\bar{t})$, ensuring $\overline{T} \geq 0$ is given by

$$A_1(\bar{t}) = \frac{\left\{\alpha[\overline{X}(\bar{t}) + \beta] - \sqrt{\alpha^2[\overline{X}(\bar{t}) + \beta]^2 + 2\alpha\overline{X}(\bar{t})[\overline{X}(\bar{t}) + 2\beta]}\right\}}{\overline{X}(\bar{t})[\overline{X}(\bar{t}) + 2\beta]},$$

and hence deduce from the heat-balance condition (7.53) the following boundary motion

$$\bar{t} = \frac{1}{12(\alpha + 2)}\left\{\left[3\alpha(\overline{X} + \beta) + 2(\overline{X} + 2\beta)\right]\sqrt{\alpha^2(\overline{X} + \beta)^2 + 2\alpha\overline{X}(\overline{X} + 2\beta)}\right.$$

$$+ (\alpha + 2)(3\alpha + 1)\overline{X}^2 + 2(3\alpha^2 + 12\alpha + 4)\beta\overline{X} - \alpha(3\alpha + 4)\beta^2$$

$$- 8(\alpha + 2)\beta^2 \log\left[\frac{\alpha(\overline{X} + \beta) + 2(\overline{X} + 2\beta) + \sqrt{\alpha^2(\overline{X} + \beta)^2 + 2\alpha\overline{X}(\overline{X} + 2\beta)}}{2(\alpha + 2)\beta}\right]$$

$$\left. - \frac{4\alpha^{3/2}\beta^2}{\sqrt{\alpha + 2}} \log\left[\frac{(\overline{X} + \beta)\sqrt{\alpha^2 + 2\alpha} + \sqrt{\alpha^2(\overline{X} + \beta)^2 + 2\alpha\overline{X}(\overline{X} + 2\beta)}}{\alpha\beta + \beta\sqrt{\alpha^2 + 2\alpha}}\right]\right\}.$$

For β non-zero equation (7.55) still holds, so that we still have

$$A_2(\bar{t}) = \frac{A_1(\bar{t})^2}{2\alpha}.$$

From this equation, the surface condition and the assumed quadratic temperature profile we can deduce the quadratic equation

$$A_1(\bar{t})^2\overline{X}(\bar{t})[\overline{X}(\bar{t}) + 2\beta] - 2\alpha A_1(\bar{t})[\overline{X}(\bar{t}) + \beta] - 2\alpha = 0,$$

and the appropriate root ensuring $\overline{T} \geq 0$ is as given.

If we introduce $E(\bar{t})$ such that

$$E(\bar{t}) = A_1(\bar{t})\overline{X}(\bar{t}),$$

then it is not difficult to show that (7.53), which still holds for β non-zero, becomes

$$\frac{d}{d\bar{t}}\left[\frac{\overline{X}}{2}\left(\frac{E^2}{3\alpha} - E\right) + \alpha\overline{X}\right] = \frac{1}{\overline{X}}\left(\frac{E^2}{\alpha} - E\right), \qquad (7.60)$$

and

$$E(\bar{t}) = \frac{\left\{\alpha(\overline{X} + \beta) - \sqrt{\alpha^2(\overline{X} + \beta)^2 + 2\alpha\overline{X}(\overline{X} + 2\beta)}\right\}}{(\overline{X} + 2\beta)}.$$

Further, if we introduce $F(\bar{t})$ defined by

$$F(\bar{t}) = E(\bar{t}) - \alpha,$$

then we have

$$\overline{X}(\bar{t}) = \frac{-2\beta F(F + \alpha)}{(F^2 - b^2)},$$

where $b = \sqrt{\alpha^2 + 2\alpha}$ and (7.60) becomes

$$\frac{d}{d\bar{t}}\left[\frac{F(F + \alpha)(F^2 - \alpha F + 4\alpha^2)}{3(F^2 - b^2)}\right] = \frac{(F^2 - b^2)}{2\beta^2}.$$

On differentiating out and rearranging this equation the problem of integrating the heat-balance condition reduces to the problem of integrating the following

$$\frac{3}{4\beta^2}\,d\bar{t} = \left\{\frac{F}{(F^2 - b^2)} - \frac{2\alpha^3}{(F^2 - b^2)^2} - \frac{2\alpha^2(\alpha + 2)(2\alpha + 1)F}{(F^2 - b^2)^3} - \frac{4\alpha^4(\alpha + 2)}{(F^2 - b^2)^4}\right\}dF.$$

On integrating this equation we obtain

$$\frac{3\bar{t}}{2\beta^2} = \left\{ \log|F^2 - b^2| - \frac{\alpha^2 F}{(\alpha + 2)(F^2 - b^2)} + \frac{\left[\alpha^2(\alpha + 2)(2\alpha + 1) + 2\alpha^3 F\right]}{\left(F^2 - b^2\right)^2} \right.$$

$$\left. + \frac{1}{2}\left(\frac{\alpha}{\alpha + 2}\right)^{\frac{3}{2}} \log\left|\frac{F + b}{F - b}\right| \right\} + \text{integration constant.}$$

Now after considerable simplification and choosing the constant of integration such that $\overline{X}(0) = 0$, we can deduce the stated result.

7.6 Enthalpy formulation and numerical solution

The *enthalpy formulation* for heat-diffusion moving boundary problems is a weak formulation which is similar to that used for the heat equation for finite element methods and which eliminates explicit reference to the moving boundary. The absence of explicit reference to the moving boundary makes the formulation particularly attractive in numerical schemes where, otherwise, one of the main difficulties is tracking the moving boundary. The *enthalpy* 'function' $H(T)$ is essentially the heat content per unit volume of material, usually defined with reference to some arbitrary reference temperature T_{ref}, so that the enthalpy $H(T)$ represents the quantity of heat required to raise a unit volume from temperature T_{ref} to T. The arbitrary nature of T_{ref} reflects the fact that there is no absolute measure of energy. In the context of heat conduction, with no phase changes or moving boundaries, the enthalpy is simply

$$H(T) = \rho \int_{T_{\text{ref}}}^{T} c(\theta)\,\mathrm{d}\theta,$$

where ρ is the constant density and $c(T)$ the specific heat. For constant specific heat c, we generally choose T_{ref} to be zero so that

$$H(T) = \rho c T.$$

In either event, by an argument similar to that given in Section 1.2 we can write the one dimensional heat equation as

$$\frac{\partial H}{\partial t} = \frac{\partial}{\partial x}\left(k(T)\frac{\partial T}{\partial x}\right)$$

and, as in Section 1.6, the three dimensional heat equation becomes

$$\frac{\partial H}{\partial t} = \underline{\nabla} \cdot (k(T)\underline{\nabla} T).$$

For phase change problems the situation is slightly more complicated because of the latent heat. For purposes of illustration we consider the two phase Stefan problem (7.2)-(7.4). Taking our reference temperature as zero, we find that for temperatures T_1 less than the fusion temperature T_f (that is, in the solid phase), the enthalpy $H(T_1)$ is given by

$$H(T_1) = \rho c_1 T_1, \quad T_1 < T_f, \tag{7.61}$$

where c_1 is the specific heat of the solid. For temperatures T_2 greater than the fusion temperature T_f (that is, in the liquid phase), the enthalpy is given by

$$H(T_2) = \rho c_1 T_f + \rho L + \rho c_2(T_2 - T_f), \quad T_2 > T_f, \tag{7.62}$$

where L is the latent heat of fusion and c_2 is the specific heat of the liquid. This expression arises as follows. To heat a unit volume from temperature zero to temperature $T_2 > T_f$ we must first raise the solid to its melting point, requiring $\rho c_1 T_f$ units of heat, then melt the solid completely, requiring a further ρL units of heat, and finally raise the liquid from temperature T_f to T_2, requiring another $\rho c_2(T_2 - T_f)$ units of heat. At the fusion temperature T_f we have the inequality

$$\rho c_1 T_f \leq H(T_f) \leq \rho(c_1 T_f + L), \tag{7.63}$$

since at the fusion temperature a unit volume of material can exist as a pure solid with enthalpy $\rho c_1 T_f$, pure liquid with enthalpy $\rho(c_1 T_f + L)$ or as some mixture of both with an overall enthalpy in between these two extremes. Thus, for the two phase problem (7.2)-(7.4) the enthalpy $H(T)$ is defined by

$$H(T) = \begin{cases} \rho c_1 T & T < T_f, \\ [\rho c_1 T_f, \rho(c_1 T_f + L)] & T = T_f, \\ \rho(c_1 T_f + L + c_2(T - T_f)) & T > T_f, \end{cases} \tag{7.64}$$

where T denotes the temperature in either phase and where $[a, b]$ is meant to imply that $H(T)$ has any value in the interval $a \leq H(T) \leq b$. We emphasize that only the magnitude ρL of the jump in enthalpy at the phase change temperature is important. We could easily choose a non-zero reference temperature to define enthalpy, thereby adding some constant factor to each of the terms in (7.64) which would not invalidate the following. Note also that in a strict sense the enthalpy is not a function of temperature since it is not uniquely determined for $T = T_f$, but is rather a relation on T (see Figure 7.6). However for any given value of H there corresponds one and only one temperature T, so that temperature is a function of enthalpy.

One of the major advantages of the enthalpy formulation of a moving boundary problem is that explicit reference to the moving boundary is eliminated and the problem is reduced to a fixed boundary problem, which is achieved at the expense of slightly complicating the governing partial differential equation. To motivate this formulation we return to the physics underlying the heat equation as discussed in Section 1.2. There we equate the time rate of change of heat in a small control slice with the instantaneous net heat flux into the control slice. For the problem (7.2)-(7.4) thermal energy is still conserved so that this principle remains valid and in terms of the enthalpy we have

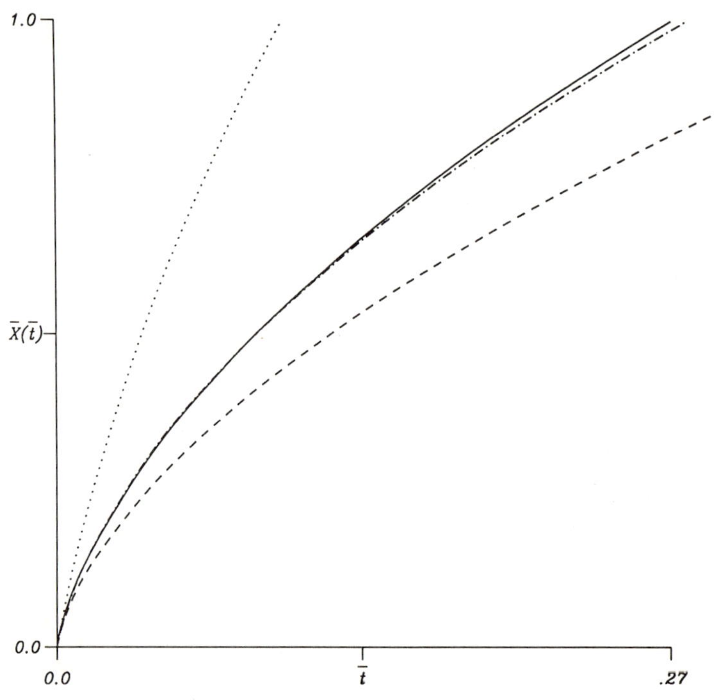

Figure 7.6. Graph of enthalpy-temperature relation.

$$\frac{\mathrm{d}}{\mathrm{d}t} \int_{x}^{x + \Delta x} H(T(\xi,t))\,\mathrm{d}\xi = k(T(x + \Delta x,t))\frac{\partial T}{\partial x}(x + \Delta x,t) - k(T(x,t))\frac{\partial T}{\partial x}(x,t),$$

where here the thermal conductivity $k(T)$ is a function of temperature given by

$$k(T) = \begin{cases} k_1 & T < T_f, \text{ (solid)}, \\ k_2 & T > T_f, \text{ (liquid)}, \end{cases} \tag{7.65}$$

corresponding to the conductivities in the solid and liquid phases. At $T = T_f$ the conductivity is not well defined, for it depends on what phase the material is in. Assuming that we can apply the mean value theorem to the previous integral we find that

$$\frac{\partial}{\partial t} H(T(\zeta,t)) = \frac{1}{\Delta x}\left[k(T(x + \Delta x,t))\frac{\partial T}{\partial x}(x + \Delta x,t) - k(T(x,t))\frac{\partial T}{\partial x}(x,t)\right],$$

where $x \le \zeta \le x + \Delta x$, and further *assuming* that we can take the limit $\Delta x \to 0$, this becomes

$$\frac{\partial H}{\partial t} = \frac{\partial}{\partial x}\left(k(T)\frac{\partial T}{\partial x}\right). \tag{7.66}$$

These assumptions are valid provided $T \neq T_f$, that is provided we are not on the moving boundary $x = X(t)$, since away from the moving boundary (7.66) is simply the heat equation for the solid and liquid regions as appropriate. On the moving boundary $x = X(t)$, there is a jump in $H(T)$ of magnitude ρL (this is how, implicitly, we find the moving boundary from the enthalpy), so that $\frac{\partial H}{\partial t} = \frac{\mathrm{d}H}{\mathrm{d}T}\frac{\partial T}{\partial t}$, is not well defined in the usual sense and neither the limiting process nor the mean value theorem are valid at $x = X(t)$ because of the jump in enthalpy. Nevertheless (7.66) with $H(T)$ defined by (7.64) and $k(T)$ defined by (7.65), together with appropriate boundary conditions on the temperature T and specification of the initial enthalpy H at time zero, summarizes the heat flow in both phases without explicitly involving the moving boundary. Accordingly there is considerable motivation to find some alternative interpretation of (7.66) which is mathematically tenable.

Before proceeding to do this it is convenient to non-dimensionalize the full two phase problem. We introduce non-dimensional variables

$$\bar{x} = \frac{x}{a}, \quad \overline{X}(\bar{t}) = \frac{X(t)}{a}, \quad \bar{t} = \frac{\kappa_1 t}{a^2}, \quad U_0 = \frac{k_2}{k_1}\left(\frac{T_f - T_\ell}{T_f - T_0}\right),$$

$$\overline{T}_1(\bar{x},\bar{t}) = \frac{T_f - T_1(x,t)}{T_f - T_0}, \quad \overline{T}_2(\bar{x},\bar{t}) = \frac{k_2}{k_1}\left(\frac{T_f - T_2(x,t)}{T_f - T_0}\right),$$

$$\alpha = \frac{L}{c_1(T_f - T_0)}, \quad \bar{c} = \frac{k_1 c_2}{k_2 c_1}, \tag{7.67}$$

in terms of which problem (7.2)-(7.4) becomes (Problem 11)

$$\left.\begin{array}{ll} \dfrac{\partial \overline{T}_1}{\partial \bar{t}} = \dfrac{\partial^2 \overline{T}_1}{\partial \bar{x}^2}, & 0 < \bar{x} < \overline{X}(\bar{t}), \\[3mm] \bar{c}\,\dfrac{\partial \overline{T}_2}{\partial \bar{t}} = \dfrac{\partial^2 \overline{T}_2}{\partial \bar{x}^2}, & \overline{X}(\bar{t}) < \bar{x} < \infty, \end{array}\right\} \tag{7.68}$$

subject to the boundary and initial conditions

$$\overline{T}_1(0,\bar{t}) = 1, \quad \overline{T}_2(\infty,\bar{t}) = U_0, \qquad \overline{T}_2(\bar{x},0) = U_0, \quad \overline{X}(0) = 0, \tag{7.69}$$

and the conditions on the moving boundary

$$\overline{T}_1(\overline{X}(\bar{t}),\bar{t}) = \overline{T}_2(\overline{X}(\bar{t}),\bar{t}) = 0, \quad \dfrac{\partial \overline{T}_1}{\partial \bar{x}}(\overline{X}(\bar{t}),\bar{t}) - \dfrac{\partial \overline{T}_2}{\partial \bar{x}}(\overline{X}(\bar{t}),\bar{t}) = -\alpha\,\dfrac{\mathrm{d}\overline{X}}{\mathrm{d}\bar{t}}. \tag{7.70}$$

The non-dimensional enthalpy $\overline{H}(\overline{T})$ for problem (7.68)-(7.70) can be taken to be

$$\overline{H}(\overline{T}) = \left\{\begin{array}{ll} \overline{T} + \alpha & \overline{T} > 0, \\ [0,\alpha] & \overline{T} = 0, \\ \bar{c}\,\overline{T} & \overline{T} < 0, \end{array}\right. \tag{7.71}$$

where the Stefan number α is interpreted as a non-dimensional latent heat. The non-dimensional equivalent of (7.66) describing the heat flow becomes

$$\dfrac{\partial \overline{H}}{\partial \bar{t}} = \dfrac{\partial^2 \overline{T}}{\partial \bar{x}^2}, \quad 0 < \bar{x} < \infty, \tag{7.72}$$

which, from (7.69) and (7.71) has boundary and initial conditions

$$\overline{T}(0,\bar{t}) = 1, \quad \overline{T}(\infty,\bar{t}) = U_0, \tag{7.73}$$

$$\overline{H}(\bar{x},0) = \bar{c}\,U_0. \tag{7.74}$$

We note that for the enthalpy equation (7.72) requires two temperature boundary conditions and one initial enthalpy condition, because the temperature appears in (7.72) as a second spatial derivative while the enthalpy appears as a first time derivative. In view of (7.71) and provided the initial temperature is not equal to the fusion temperature, specifying the initial enthalpy is equivalent to specifying the initial temperature. However, if the initial temperature is equal to the fusion temperature then the liquid-solid constitution must also be prescribed to uniquely determine the enthalpy. As for the dimensional enthalpy equation (7.66), (7.72) is meaningful in the usual sense only when $\bar{x} \neq \bar{X}(\bar{t})$, since the enthalpy $\overline{H}(\overline{T})$ and $\frac{\partial \overline{T}}{\partial \bar{x}}$ are discontinuous at $\bar{x} = \bar{X}(\bar{t})$. Although we make no essential use of the following weak formulation, as for generalized functions (7.72) becomes meaningful for all positions and times if we employ a space of test functions. This weak formulation also forms the basis of theoretical proofs of the existence and uniqueness of solutions which is beyond our scope.

Let $\bar{t}_0 > 0$ be an arbitrary fixed, non-zero non-dimensional time. A function $\phi(\bar{x}, \bar{t})$ is called a test function if $\phi(\bar{x}, \bar{t}_0) = 0$ and ϕ, $\frac{\partial \phi}{\partial \bar{t}}$, $\frac{\partial \phi}{\partial \bar{x}}$ and $\frac{\partial^2 \phi}{\partial \bar{x}^2}$ are all integrable in the sense that the integral

$$\int_0^{\bar{t}_0} \int_0^\infty \left(|\phi(\bar{x}, \bar{t})| + \left|\frac{\partial \phi}{\partial \bar{t}}(\bar{x}, \bar{t})\right| + \left|\frac{\partial \phi}{\partial \bar{x}}(\bar{x}, \bar{t})\right| + \left|\frac{\partial^2 \phi}{\partial \bar{x}^2}(\bar{x}, \bar{t})\right| \right) d\bar{x}\, d\bar{t},$$

exists and is finite, which in particular implies that $\phi(\bar{x}, \bar{t})$ and all its derivatives vanish at $\bar{x} = \infty$. We now suppose that (7.72) is replaced by the integral condition

$$\int_0^{\bar{t}_0} \int_0^\infty \phi(\bar{x}, \bar{t}) \left(\frac{\partial \overline{H}}{\partial \bar{t}} - \frac{\partial^2 \overline{T}}{\partial \bar{x}^2} \right) d\bar{x}\, d\bar{t} = 0,$$

for every test function $\phi(\bar{x}, \bar{t})$. Then, integrating by parts once in $\bar{t}$ and twice in $\bar{x}$ we have

$$\int_0^{\bar{t}_0} \int_0^\infty \left(\overline{H} \frac{\partial \phi}{\partial \bar{t}} + \overline{T} \frac{\partial^2 \phi}{\partial \bar{x}^2} \right) d\bar{x}\, d\bar{t} = - \int_0^\infty \phi(\bar{x}, 0) \overline{H}(\bar{x}, 0)\, d\bar{x}$$

$$+ \int_0^{\bar{t}_0} \left(\phi(0, \bar{t}) \frac{\partial \overline{T}}{\partial \bar{x}}(0, \bar{t}) - \overline{T}(0, \bar{t}) \frac{\partial \phi}{\partial \bar{x}}(0, \bar{t}) \right) d\bar{t},$$

$$(7.75)$$

using the facts that $\phi(\bar{x}, \bar{t}_0) = \phi(\infty, \bar{t}) = \frac{\partial \phi}{\partial \bar{x}}(\infty, \bar{t}) = 0$ for any test function. Finally since $\frac{\partial \overline{T}}{\partial \bar{x}}(0, \bar{t})$ is unknown, we also require our test functions to satisfy the condition $\phi(0, \bar{t}) = 0$, thereby eliminating this unknown in (7.75). We now define a *weak solution* of problem (7.72)-(7.74) to be a pair of functions $\overline{H}(\bar{x}, \bar{t})$ and $\overline{T}(\bar{x}, \bar{t})$ which are related by (7.71) and are such that

$$\int_0^{\bar{t}_0} \int_0^\infty \left(\overline{H} \frac{\partial \phi}{\partial \bar{t}} + \overline{T} \frac{\partial^2 \phi}{\partial \bar{x}^2} \right) d\bar{x}\, d\bar{t} = -\bar{c} U_0 \int_0^\infty \phi(\bar{x}, 0)\, d\bar{x} - \int_0^{\bar{t}_0} \frac{\partial \phi}{\partial \bar{x}}(0, \bar{t})\, d\bar{t}, \quad (7.76)$$

for any $\bar{t}_0 > 0$ and for every test function $\phi(\bar{x}, \bar{t})$ such that $\phi(0, \bar{t}) = \phi(\bar{x}, \bar{t}_0) = 0$. Note that in (7.76) we have applied the initial condition (7.74) and the boundary condition $(7.73)_1$, and that (7.76) involves only $\overline{H}$ and $\overline{T}$ and not their (possibly undefined) derivatives.

Example 7.6 Show that the classical solution of (7.68)-(7.70) is also a weak solution of (7.72)-(7.74). That is, show that it satisfies (7.76) for any $\bar{t}_0 > 0$ and all test functions $\phi(\bar{x}, \bar{t})$ such that $\phi(0, \bar{t}) = \phi(\bar{x}, \bar{t}_0) = 0$.

If $\overline{T}_1$, $\overline{T}_2$ and $\overline{X}(\bar{t})$ satisfy (7.72)-(7.74), then the overall temperature $\overline{T}$ is given by

$$\overline{T}(\bar{x}, \bar{t}) = \begin{cases} \overline{T}_1(\bar{x}, \bar{t}) > 0, & \bar{x} < \overline{X}(\bar{t}), \\ 0, & \bar{x} = \overline{X}(\bar{t}), \\ \overline{T}_2(\bar{x}, \bar{t}) < 0, & \bar{x} > \overline{X}(\bar{t}), \end{cases} \tag{7.77}$$

and the enthalpy $\overline{H}$ is determined by

$$\overline{H}(\bar{x}, \bar{t}) = \begin{cases} \overline{T}_1(\bar{x}, \bar{t}) + \alpha, & \bar{x} < \overline{X}(\bar{t}), \\ [0, \alpha], & \bar{x} = \overline{X}(\bar{t}), \\ \bar{c}\,\overline{T}_2(\bar{x}, \bar{t}), & \bar{x} > \overline{X}(\bar{t}). \end{cases} \tag{7.78}$$

With $\overline{T}$ and $\overline{H}$ so defined, we have for any test function $\phi(\bar{x}, \bar{t})$

$$\int_0^{\bar{t}_0} \int_0^\infty \left(\overline{H} \frac{\partial \phi}{\partial \bar{t}} + \overline{T} \frac{\partial^2 \phi}{\partial \bar{x}^2} \right) d\bar{x}\, d\bar{t} = \int_0^{\bar{t}_0} \int_0^{\overline{X}(\bar{t})} \left((\overline{T}_1 + \alpha) \frac{\partial \phi}{\partial \bar{t}} + \overline{T}_1 \frac{\partial^2 \phi}{\partial \bar{x}^2} \right) d\bar{x}\, d\bar{t}$$

$$+ \int_0^{\bar{t}_0} \int_{\overline{X}(\bar{t})}^\infty \left(\bar{c}\,\overline{T}_2 \frac{\partial \phi}{\partial \bar{t}} + \overline{T}_2 \frac{\partial^2 \phi}{\partial \bar{x}^2} \right) d\bar{x}\, d\bar{t}. \tag{7.79}$$

Using the identities

$$\int_0^{\bar{t}_0} \int_0^{\overline{X}} \overline{T}_1 \frac{\partial^2 \phi}{\partial \bar{x}^2} \, d\bar{x}\, d\bar{t} = \int_0^{\bar{t}_0} \int_0^{\overline{X}} \phi \frac{\partial^2 \overline{T}_1}{\partial \bar{x}^2} \, d\bar{x}\, d\bar{t} - \int_0^{\bar{t}_0} \phi(\overline{X}, t) \frac{\partial \overline{T}_1}{\partial \bar{x}} (\overline{X}, \bar{t}) \, d\bar{t}$$

$$- \int_0^{\bar{t}_0} \frac{\partial \phi}{\partial \bar{x}} (0, \bar{t}) \, d\bar{t},$$

$$\int_0^{\bar{t}_0} \int_{\overline{X}}^\infty \overline{T}_2 \frac{\partial^2 \phi}{\partial \bar{x}^2} \, d\bar{x}\, d\bar{t} = \int_0^{\bar{t}_0} \int_{\overline{X}}^\infty \phi \frac{\partial^2 \overline{T}_2}{\partial \bar{x}^2} \, d\bar{x}\, d\bar{t} + \int_0^{\bar{t}_0} \phi(\overline{X}, \bar{t}) \frac{\partial \overline{T}_2}{\partial \bar{x}} (\overline{X}, \bar{t}) \, d\bar{t},$$

which arise from integration by parts and the boundary conditions which apply to $\overline{T}_1$, $\overline{T}_2$ and ϕ, and

$$\int_0^{\bar{t}_0}\int_0^{\overline{X}}(\overline{T}_1 + \alpha)\frac{\partial\phi}{\partial\bar{t}}\,\mathrm{d}\bar{x}\,\mathrm{d}\bar{t} = -\int_0^{\bar{t}_0}\int_0^{\overline{X}}\phi\frac{\partial\overline{T}_1}{\partial\bar{t}}\,\mathrm{d}\bar{x}\,\mathrm{d}\bar{t} - \alpha\int_0^{\bar{t}_0}\phi(\overline{X},\bar{t})\frac{\mathrm{d}\overline{X}}{\mathrm{d}\bar{t}}\,\mathrm{d}\bar{t},$$

$$(7.80)$$

$$\int_0^{\bar{t}_0}\int_{\overline{X}}^{\infty}\bar{c}\overline{T}_2\frac{\partial\phi}{\partial\bar{t}}\,\mathrm{d}\bar{x}\,\mathrm{d}\bar{t} = -\bar{c}\int_0^{\bar{t}_0}\int_{\overline{X}}^{\infty}\phi\frac{\partial\overline{T}_2}{\partial\bar{t}}\,\mathrm{d}\bar{x}\,\mathrm{d}\bar{t} - \bar{c}U_0\int_0^{\infty}\phi(\bar{x},0)\,\mathrm{d}\bar{x},$$

(see Problem 14), we see that (7.79) reduces to

$$\int_0^{\bar{t}_0}\int_0^{\infty}\left(\overline{H}\frac{\partial\phi}{\partial\bar{t}} + \overline{T}\frac{\partial^2\phi}{\partial\bar{x}^2}\right)\mathrm{d}\bar{x}\,\mathrm{d}\bar{t} = -\bar{c}U_0\int_0^{\infty}\phi(\bar{x},0)\,\mathrm{d}\bar{x} - \int_0^{\bar{t}_0}\frac{\partial\phi}{\partial\bar{x}}(0,\bar{t})\,\mathrm{d}\bar{t}$$

$$- \alpha\int_0^{\bar{t}_0}\phi(\overline{X},\bar{t})\frac{\mathrm{d}\overline{X}}{\mathrm{d}\bar{t}}\,\mathrm{d}\bar{t} - \int_0^{\bar{t}_0}\phi(\overline{X},\bar{t})\left[\frac{\partial\overline{T}_1}{\partial\bar{x}}(\overline{X},\bar{t}) - \frac{\partial\overline{T}_2}{\partial\bar{x}}(\overline{X},\bar{t})\right]\mathrm{d}\bar{t},$$

$$(7.81)$$

where we have used the fact that $\overline{T}_1$ and $\overline{T}_2$ satisfy the heat equations $(7.68)_1$ and $(7.68)_2$ respectively. If we now invoke the Stefan condition $(7.70)_2$, we see that (7.81) reduces to (7.76).

If the boundary condition $(7.73)_1$ is replaced by the Newton cooling condition

$$\overline{T}(0,\bar{t}) - \beta\frac{\partial\overline{T}}{\partial\bar{x}}(0,\bar{t}) = 1,$$

$$(7.82)$$

then the interpretation applied to the weak enthalpy problem (7.72), (7.82), $(7.73)_2$ and (7.74) is to find functions $\overline{H}(\bar{x},\bar{t})$ and $\overline{T}(\bar{x},\bar{t})$ related by (7.71) and such that

$$\int_0^{\bar{t}_0}\int_0^{\infty}\left(\overline{H}\frac{\partial\phi}{\partial\bar{t}} + \overline{T}\frac{\partial^2\phi}{\partial\bar{x}^2}\right)\mathrm{d}\bar{x}\,\mathrm{d}\bar{t} = -\bar{c}U_0\int_0^{\infty}\phi(\bar{x},0)\,\mathrm{d}\bar{x} - \int_0^{\bar{t}_0}\frac{\partial\phi}{\partial\bar{x}}(0,\bar{t})\,\mathrm{d}\bar{t},$$

for any time $\bar{t}_0 > 0$ and all test functions $\phi(\bar{x},\bar{t})$ such that $\phi(\bar{x},\bar{t}_0) = 0$ and

$$\phi(0,\bar{t}) - \beta\frac{\partial\phi}{\partial\bar{x}}(0,\bar{t}) = 0.$$

We impose this condition on the test functions to eliminate the unknown quantities in (7.75). We note that the weak formulation of this problem is valid whether or not freezing has commenced, for if there is no phase change present, then the enthalpy equation simply reduces to the heat equation for the liquid. Therefore unlike the classical statement of the problem we do not have to treat the problem in two parts (see

Section 7.1). As we have noted, the enthalpy formulation leads to particularly effective numerical techniques for treating many classical moving boundary problems. We will only consider the finite difference enthalpy method, however. Since finite difference methods are not well suited to semi-infinite spatial domains and since in any case equations (7.68)-(7.70) admit a well known exact solution (see Problem 12), we consider instead the following non-dimensional problem posed on the region $[0, 1]$,

$$
\left.
\begin{aligned}
&\frac{\partial \overline{T}_1}{\partial \overline{t}} = \frac{\partial^2 \overline{T}_1}{\partial \overline{x}^2}, \quad 0 < \overline{x} < \overline{X}(\overline{t}), \\[2mm]
&\overline{c}\,\frac{\partial \overline{T}_2}{\partial \overline{t}} = \frac{\partial^2 \overline{T}_2}{\partial \overline{x}^2}, \quad \overline{X}(\overline{t}) < \overline{x} < 1, \\[2mm]
&\overline{T}_1(0,\overline{t}) - \beta \frac{\partial \overline{T}_1}{\partial \overline{x}}(0,\overline{t}) = 1, \quad \frac{\partial \overline{T}_2}{\partial \overline{x}}(1,\overline{t}) = 0, \\[2mm]
&\overline{T}_1(\overline{X}(\overline{t}),\overline{t}) = \overline{T}_2(\overline{X}(\overline{t}),\overline{t}) = 0, \\[2mm]
&-\alpha \frac{d\overline{X}}{d\overline{t}} = \frac{\partial \overline{T}_1}{\partial \overline{x}}(\overline{X}(\overline{t}),\overline{t}) - \frac{\partial \overline{T}_2}{\partial \overline{x}}(\overline{X}(\overline{t}),\overline{t}), \\[2mm]
&\overline{T}_2(\overline{x},0) = U_0, \quad \overline{X}(0) = 0,
\end{aligned}
\right\}
\tag{7.83}
$$

which has the corresponding enthalpy formulation

$$
\left.
\begin{aligned}
&\frac{\partial \overline{H}}{\partial \overline{t}} = \frac{\partial^2 \overline{T}}{\partial \overline{x}^2}, \quad 0 < \overline{x} < 1, \\[2mm]
&\overline{T}(0,\overline{t}) - \beta \frac{\partial \overline{T}}{\partial \overline{x}}(0,\overline{t}) = 1, \quad \frac{\partial \overline{T}}{\partial \overline{x}}(1,\overline{t}) = 0, \\[2mm]
&\overline{H}(\overline{x},0) = \overline{c}\,U_0,
\end{aligned}
\right\}
\tag{7.84}
$$

where $\overline{H}(\overline{x},\overline{t})$ and $\overline{T}(\overline{x},\overline{t})$ are related by (7.71). A weak solution of problem (7.84) is a pair of functions $\overline{H}(\overline{x},\overline{t})$ and $\overline{T}(\overline{x},\overline{t})$ related by (7.71) and such that

$$
\int_0^{\overline{t}_0} \int_0^1 \left(\overline{H}\frac{\partial \phi}{\partial \overline{t}} + \overline{T}\frac{\partial^2 \phi}{\partial \overline{x}^2} \right) d\overline{x}\, d\overline{t} = -\overline{c}\,U_0 \int_0^1 \phi(\overline{x},0)\, d\overline{x} - \int_0^{\overline{t}_0} \frac{\partial \phi}{\partial \overline{x}}(0,\overline{t})\, d\overline{t}, \tag{7.85}
$$

for any time $\overline{t}_0 > 0$ and all test functions $\phi(\overline{x},\overline{t})$ which satisfy $\phi(\overline{x},\overline{t}_0) = \phi(0,\overline{t}) - \beta\frac{\partial \phi}{\partial \overline{x}}(0,\overline{t}) = \frac{\partial \phi}{\partial \overline{x}}(1,\overline{t}) = 0$, and a test function for the finite interval $[0, 1]$ is any function $\phi(\overline{x},\overline{t})$ which has integrable first time and second space partial derivatives on $[0, 1]$. The following finite difference scheme which formally arises by discretizing

(7.84) is well defined since it involves only the values of $\overline{H}$ and $\overline{T}$ and not their derivatives and can actually be shown to converge to the weak solution of (7.85). We obtain the explicit finite difference representation of $(7.84)_1$ using forward difference approximations in time and central difference approximations in space, namely

$$\frac{\partial \overline{H}}{\partial \bar{t}} = \frac{1}{\Delta \bar{t}}\left(\overline{H}_{ij+1} - \overline{H}_{ij}\right) + O\left(\Delta \bar{t}\right),$$

$$\frac{\partial \overline{T}}{\partial \bar{x}} = \frac{1}{2\Delta \bar{x}}\left(\overline{T}_{i+1j} - \overline{T}_{i-1j}\right) + O\left((\Delta \bar{x})^2\right),$$

$$\frac{\partial^2 \overline{T}}{\partial \bar{x}^2} = \frac{1}{(\Delta \bar{x})^2}\left(\overline{T}_{i+1j} + \overline{T}_{i-1j} - 2\overline{T}_{ij}\right) + O\left((\Delta \bar{x})^2\right),$$

where

$$\overline{H}_{ij} = \overline{H}(i\Delta \bar{x}, j\Delta \bar{t}), \quad \overline{T}_{ij} = \overline{T}(i\Delta \bar{x}, j\Delta \bar{t}),$$

and we freely use the conventions and notation of Sections 6.2 and 6.3. Thus to order $O\left((\Delta \bar{x})^2 + \Delta \bar{t}\right)$ the enthalpy equation $(7.84)_1$ is approximated by the finite difference system

$$\tilde{H}_{ij+1} = \tilde{H}_{ij} + \delta\left(\tilde{T}_{i+1j} + \tilde{T}_{i-1j} - 2\tilde{T}_{ij}\right), \quad 1 \le i \le N - 1, j \ge 0, \qquad (7.86)$$

where $\delta = \Delta \bar{t}/(\Delta \bar{x})^2$ and we use the notation $\tilde{H}_{ij}$ and $\tilde{T}_{ij}$ to distinguish the finite difference approximations from the exact values $\overline{H}_{ij}$ and $\overline{T}_{ij}$. At the endpoints we use central difference approximations for the partial derivatives in the boundary conditions and assume that (7.86) is valid for $i = 0$ and $i = N$ as in Chapter 6 to obtain

$$\tilde{H}_{0j+1} = \tilde{H}_{0j} + 2\delta\left[\tilde{T}_{1j} - \left(1 + \frac{\Delta \bar{x}}{\beta}\tilde{T}_{0j}\right) + \frac{\Delta \bar{x}}{\beta}\right],$$
$$j \ge 0. \qquad (7.87)$$
$$\tilde{H}_{Nj+1} = \tilde{H}_{Nj} + 2\delta\left[\tilde{T}_{N-1j} - \tilde{T}_{Nj}\right],$$

Finally we find from (7.71) that $\tilde{H}_{ij}$ and $\tilde{T}_{ij}$ are related by

$$\tilde{T}_{ij} = \begin{cases} \tilde{H}_{ij} - \alpha, & \tilde{H}_{ij} > \alpha, \\ 0, & \tilde{H}_{ij} \in [0, \alpha], \\ \dfrac{\tilde{H}_{ij}}{\bar{c}}, & \tilde{H}_{ij} < 0. \end{cases} \qquad (7.88)$$

Thus starting with the initial enthalpy $\tilde{H}_{i0} = \bar{c}U_0$ we find the initial temperature $\tilde{T}_{i0}$ using (7.88), then applying (7.86) and (7.87) we find $\tilde{H}_{i1}$, and so on repeatedly applying

(7.88) to find $\tilde{T}_{ij}$ from $\tilde{H}_{ij}$ and then using (7.86) and (7.87) to find $\tilde{H}_{ij+1}$ from $\tilde{T}_{ij}$. As for the explicit finite difference method for the ordinary heat equation this procedure is stable only for $\delta < 0.5$, or even smaller for small β. As previously noted it can be shown that as $\Delta\bar{x}$ and $\Delta\bar{t}$ tend to zero in such a way that δ remains constant and small enough for the scheme to remain stable, the finite difference approximations $\tilde{H}_{ij}$ and $\tilde{T}_{ij}$ tend to the exact solution of the weak problem (7.85).

As for the explicit finite difference method for the heat equation (see Section 6.2), the restriction that $\delta \leq 0.5$ is a severe constraint, seriously limiting the size of the time step $\Delta\bar{t}$. Therefore many implicit finite difference schemes have been developed for the enthalpy equation. Due to the relationship between $\tilde{H}_{ij}$ and $\tilde{T}_{ij}$, however the resulting difference equations are non-linear and a strategy somewhat more sophisticated than the linear methods employed in Section 6.3 for the implicit finite difference heat equation are necessary. For this reason we do not pursue these methods here, but refer the interested reader to one of the specialist texts listed in the bibliography.

Finally there arises the matter of tracking the moving boundary. For a procedure based on the enthalpy method this is a task to be carried out after the solution is found and is largely independent of the particular numerical method used to solve the enthalpy equation. Although one is tempted, at first sight, to track the moving boundary by tracking the zeros of the temperature, experience has shown that this leads to poor and erratic approximations to the moving boundary. A particularly simple and effective means of tracking the moving boundary is based on the idea of locating the jump in enthalpy which occurs at the moving boundary. However, since a finite difference procedure only returns the nodal values of the temperature and enthalpy, it is not possible to explicitly search for the jump itself so we proceed as follows. Provided the time step $\Delta\bar{t}$ is not too large we may assume that the moving boundary will next pass through the point $i\Delta\bar{x}$, since it is not unreasonable to suppose that we know the node through which it has just passed. We consider the interval $[(i - \frac{1}{2})\Delta\bar{x}, (i + \frac{1}{2})\Delta\bar{x}]$ containing this node (see Figure 7.7). The enthalpy of this node is approximately $\tilde{H}_{ij}\Delta\bar{x}$ at time step j. Suppose that the freezing front has entered the interval and that a fraction ϵ of the total material in the interval has frozen. Then, assuming that the temperature at this node is the constant fusion temperature zero, we can calculate that the total enthalpy of this interval is approximately $\epsilon\alpha\Delta\bar{x}$, since the fraction $(1 - \epsilon)$ of liquid in the interval contributes zero total enthalpy. Equating the two terms we have

$$\tilde{H}_{ij}\Delta\overline{x} = \epsilon\alpha\Delta\overline{x},$$

and noting that the moving boundary passes through the node $i\Delta\overline{x}$ when $\epsilon = \frac{1}{2}$ (that is when half the material in the interval has frozen), we find that

$$\tilde{H}_{ij} = \frac{\alpha}{2}, \tag{7.89}$$

which is one criterion used for determining when the moving boundary is at position $i\Delta\overline{x}$.

Normally we are not fortunate enough to have (7.89) occur at all, because it is highly unlikely that the moving boundary will move through the space nodes at exact integer multiples of the time step. Therefore we adopt the strategy of watching the node that the moving boundary is next going to pass through until such time as we have

$$\tilde{H}_{ij} < \frac{\alpha}{2}, \quad \tilde{H}_{ij+1} > \frac{\alpha}{2},$$

that is until the moving boundary passes through the node during the time interval $[j\Delta\overline{t}, (j + 1)\Delta\overline{t}]$. We then assume that the enthalpy varies linearly with time between time steps, that is that for $0 \le \theta \le 1$ we have

$$\overline{H}(i\Delta\overline{x}, (j + \theta)\Delta\overline{t}) \approx (1 - \theta)\tilde{H}_{ij} + \theta\tilde{H}_{ij+1},$$

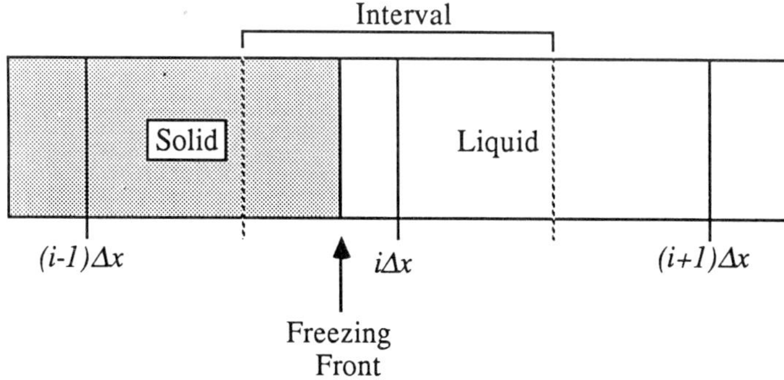

Figure 7.7. Determination of the numerical boundary location criterion (7.89).

so that the time $(j + \theta)\Delta\bar{t}$ at which the enthalpy at node $i\Delta\bar{x}$ is equal to $\frac{\alpha}{2}$ is determined by solving

$$\tilde{H}_{ij} + \theta\left(\tilde{H}_{ij+1} - \tilde{H}_{ij}\right) = \frac{\alpha}{2},$$

from which we find that θ is given by

$$\theta = \frac{\frac{\alpha}{2} - \tilde{H}_{ij}}{\tilde{H}_{ij+1} - \tilde{H}_{ij}}.$$

Using this strategy to track the moving boundary from the enthalpy vector results in reasonably accurate and sensible approximations for the boundary motion (see Figures 7.3, 7.4 and 7.5), but naturally the accuracy depends on the accuracy of the enthalpy values.

PROBLEMS

1. Verify that

$$\overline{T}(\overline{x},\overline{t}) = \alpha\left[e^{-V_0(\overline{x} - V_0\overline{t})} - 1\right],$$

is a solution of

$$\frac{\partial \overline{T}}{\partial \overline{t}} = \frac{\partial^2 \overline{T}}{\partial \overline{x}^2},$$

such that

$$\overline{T}(\overline{X}(\overline{t}),\overline{t}) = 0, \quad \frac{\partial \overline{T}}{\partial \overline{x}}(\overline{X}(\overline{t}),\overline{t}) = -\alpha\frac{d\overline{X}}{d\overline{t}},$$

on the moving boundary $\overline{X}(\overline{t}) = V_0\overline{t}$, where V_0 is the constant speed.

2. Verify that

$$\overline{T}(\overline{r},\overline{t}) = \frac{\alpha}{V_0\overline{r}}\left\{(2 + 2V_0 - V_0\overline{r} - 2V_0^2\overline{t})e^{V_0(\overline{r} + V_0\overline{t} - 1)} - (2 + V_0\overline{r})\right\},$$

is a solution of

$$\frac{\partial \overline{T}}{\partial \overline{t}} = \frac{\partial^2 \overline{T}}{\partial \overline{r}^2} + \frac{2}{\overline{r}}\frac{\partial \overline{T}}{\partial \overline{r}},$$

such that

$$\overline{T}(\overline{R}(\overline{t}),\overline{t}) = 0, \quad \frac{\partial \overline{T}}{\partial \overline{r}}(\overline{R}(\overline{t}),\overline{t}) = -\alpha\frac{d\overline{R}}{d\overline{t}},$$

on the moving boundary $\overline{R}(\overline{t}) = 1 - V_0\overline{t}$ where V_0 is the constant speed.

3. Consider a spherical container of radius a, or an infinite circular cylindrical container of the same radius. Assuming that these contain a liquid uniformly at its fusion temperature T_f and that at time $t = 0$ the temperature at the container surface is lowered to T_0 show that, assuming a Newton heat loss at the surface, the moving boundary problem for the solidified region is

$$\frac{\partial T_1}{\partial t} = \kappa_1\left(\frac{\partial^2 T_1}{\partial r^2} + \frac{m}{r}\frac{\partial T_1}{\partial r}\right), \quad R(t) < r < a,$$

$$-k_1\frac{\partial T_1}{\partial r}(a,t) = h[T_1(a,t) - T_0], \quad T_1(R(t),t) = T_f,$$

$$k_1\frac{\partial T_1}{\partial r}(R(t),t) = \rho L\frac{dR}{dt}, \quad R(0) = a,$$

where $T_1(r,t)$ denotes the temperature in the solidified region, $R(t)$ is the radius of the solidification front, κ_1, k_1 h, ρ and L have their usual meaning and $m = 2$ for the spherical container and $m = 1$ for the cylindrical container.

4.   Continuation.   In the previous problem introduce non-dimensional variables defined by

$$\bar{r} = \frac{r}{a}, \quad \bar{R}(\bar{t}) = \frac{R(t)}{a}, \quad \bar{t} = \frac{\kappa_1 t}{a}, \quad \bar{T}(\bar{r},\bar{t}) = \frac{T_f - T_1(r,t)}{T_f - T_0},$$

and show that the moving boundary problem becomes

$$\left.\begin{array}{c} \dfrac{\partial \bar{T}}{\partial \bar{t}} = \dfrac{\partial^2 \bar{T}}{\partial \bar{r}^2} + \dfrac{m}{\bar{r}} \dfrac{\partial \bar{T}}{\partial \bar{r}}, \quad \bar{R}(\bar{t}) < \bar{r} < 1, \\[2ex] \bar{T}(1,\bar{t}) + \beta \dfrac{\partial \bar{T}}{\partial \bar{r}}(1,\bar{t}) = 1, \quad \bar{T}(\bar{R}(\bar{t}),\bar{t}) = 0, \\[2ex] \dfrac{\partial \bar{T}}{\partial \bar{r}}(\bar{R}(\bar{t}),\bar{t}) = -\alpha \dfrac{\mathrm{d}\bar{R}}{\mathrm{d}\bar{t}}, \quad \bar{R}(0) = 1, \end{array}\right\}$$

where the constants α and β are defined by

$$\alpha = \frac{L}{c_1(T_f - T_0)}, \quad \beta = \frac{k_1}{ha}.$$

Show also that in the non-dimensional variables the temperature is normalized such that

$$0 \le \bar{T}(\bar{r},\bar{t}) \le 1.$$

5.   Continuation.   For the previous problem, show that the pseudo steady state approximations to the temperature and boundary motion for the cylinder and sphere are given respectively by

$$\left.\begin{array}{c} \bar{T}_{\text{pss}}(\bar{r},\bar{t}) = \dfrac{[\log \bar{r} - \log \bar{R}(\bar{t})]}{[\beta - \log \bar{R}(\bar{t})]}, \\[3ex] \bar{t}_{\text{pss}}(\bar{R}) = \dfrac{\alpha}{4}[2\bar{R}^2 \log \bar{R} + (1 + 2\beta)(1 - \bar{R}^2)], \end{array}\right\} \quad \text{(cylinder)}$$

and

$$\left.\begin{array}{c} \bar{T}_{\text{pss}}(\bar{r},\bar{t}) = \dfrac{\left\{1 - \dfrac{\bar{R}(\bar{t})}{\bar{r}}\right\}}{[1 + (\beta - 1)\bar{R}(\bar{t})]}, \\[3ex] \bar{t}_{\text{pss}}(\bar{R}) = \dfrac{\alpha}{6}(1 - \bar{R})\left[(1 + 2\beta)(1 + \bar{R}) + 2(\beta - 1)\bar{R}^2\right]. \end{array}\right\} \quad \text{(sphere)}$$

6. **Continuation.** By two integrations of the heat equation in the form

$$\bar{r}^{m}\frac{\partial \overline{T}}{\partial \bar{t}} = \frac{\partial}{\partial \bar{r}}\left(\bar{r}^{m}\frac{\partial \overline{T}}{\partial \bar{r}}\right), \quad \overline{R}(\bar{t}) < \bar{r} < 1,$$

from $\overline{R}(\bar{t})$ to an arbitrary position $\bar{r} > \overline{R}(\bar{t})$, deduce the integro-partial differential relations

$$\left.\begin{array}{l}
\overline{T}(\bar{r},\bar{t}) = \dfrac{\partial}{\partial \bar{t}}\displaystyle\int_{\overline{R}(\bar{t})}^{\bar{r}} \xi\,(\log \bar{r} - \log \xi)[\alpha + \overline{T}(\xi,\bar{t})]\,d\xi, \\[2em]
\bar{t} = \displaystyle\int_{\overline{R}(\bar{t})}^{1} \xi\,(\beta - \log \xi)[\alpha + \overline{T}(\xi,\bar{t})]\,d\xi,
\end{array}\right\} \text{(cylinder)},$$

and

$$\left.\begin{array}{l}
\overline{T}(\bar{r},\bar{t}) = \dfrac{\partial}{\partial \bar{t}}\displaystyle\int_{\overline{R}(\bar{t})}^{\bar{r}} \xi^{2}\left(\dfrac{1}{\xi} - \dfrac{1}{\bar{r}}\right)[\alpha + \overline{T}(\xi,\bar{t})]\,d\xi, \\[2em]
\bar{t} = \displaystyle\int_{\overline{R}(\bar{t})}^{1} \xi[1 + (\beta - 1)\xi][\alpha + \overline{T}(\xi,\bar{t})]\,d\xi.
\end{array}\right\} \text{(sphere)}.$$

Following the method described in Example 7.5, show that in both cases

$$\overline{T}(\bar{r},\bar{t}) \le \overline{T}_{pss}(\bar{r},\bar{t}),$$

where the pseudo steady state temperature is given in the previous question.

7. **Continuation.** Use the inequalities

$$0 \le \overline{T}(\bar{r},\bar{t}) \le \overline{T}_{pss}(\bar{r},\bar{t}),$$

and the formal integrals for the boundary motion given in the previous problem to deduce that

$$0 \le \bar{t}(\overline{R}) - \bar{t}_{pss}(\overline{R}) \le \frac{1}{4}\left\{(1 + 2\beta + \overline{R}^{2}) - \frac{(1 + 2\beta + 2\beta^{2} - \overline{R}^{2})}{(\beta - \log \overline{R})}\right\}, \quad \text{(cylinder)}$$

$$0 \le \bar{t}(\overline{R}) - \bar{t}_{pss}(\overline{R}) \le \frac{(1 - \overline{R})^{2}[1 + 2\beta + (\beta - 1)\overline{R}]}{6[1 + (\beta - 1)\overline{R}]}, \quad \text{(sphere)}$$

where $\bar{t}_{pss}(\overline{R})$ for the cylinder and sphere are given in Problem 5.

8. For problem (7.11), show that the integral iteration scheme (7.44) and (7.45) with

$$\overline{T}_0^{\dagger}(\overline{x}, \overline{X}) = 1 - \frac{\overline{x}}{\overline{X}},$$

yields

$$\overline{T}_n^{\dagger}(\overline{x}, \overline{X}) = \sum_{k=1}^{2n+1} a_k^n \left(1 - \frac{\overline{x}}{\overline{X}}\right)^k, \quad \overline{t}_{n+1}(\overline{X}) = b_n \frac{\overline{X}^2}{2},$$

where the constants a_k^n and b_n are given by

$$a_1^0 = 1, \quad a_1^{n+1} = \frac{\alpha}{b_n}, \quad a_2^{n+1} = \frac{a_1^n}{2b_n},$$

$$a_k^{n+1} = \frac{1}{k\,b_n}\left\{a_{k-1}^n - \left(\frac{k-2}{k-1}\right)a_{k-2}^n\right\}, \quad k = 3, 4, \ldots, 2n+2,$$

$$a_{2n+3}^{n+1} = \frac{-(2n+1)a_{2n+1}^n}{(2n+2)(2n+3)b_n}, \quad b_n = \alpha + 2\sum_{k=1}^{2n+1} \frac{a_k^n}{(k+1)(k+2)}.$$

Further, show *numerically* that:

(i)   for each k, the sequence $\left(a_k^n\right)$ converges as n tends to infinity,

(ii) $\overline{T}_n^{\dagger}(\overline{x}, \overline{X})$ and b_n tend respectively to the exact solution $\overline{T}(\overline{x}, \overline{t})$ and $\frac{1}{\gamma}$ as n tends to infinity (see (7.16) and (7.17)).

9.   For the problem of planar solidification (7.10), with β zero, show that if the heat-balance condition (7.53) is replaced by

(i)   the Stefan condition $(7.10)_4$,

(ii)   the integral formula (7.38),

then assuming a quadratic temperature profile, show that the correct boundary motion $\overline{X}(\overline{t}) = \sqrt{2\gamma\overline{t}}$ is still obtained, but with $\frac{1}{\gamma}$ being approximated respectively by

(i) $\dfrac{1}{\gamma} \approx \dfrac{\alpha}{2}\left[1 + \sqrt{1 + \dfrac{2}{\alpha}}\right],$

(ii) $\dfrac{1}{\gamma} \approx \dfrac{\alpha}{6}\left[5 + \sqrt{1 + \dfrac{2}{\alpha}}\right] + \dfrac{1}{6}.$

10.   Continuation. For the problem of planar solidification (7.10) with β non-zero show that if the heat-balance condition (7.53) is replaced by

(i)   the Stefan condition $(7.10)_4$,

(ii) the integral formula (7.38),

then assuming a quadratic temperature profile deduce respectively the following approximate boundary motions

(i) $\bar{t} \approx \frac{1}{4}\Big\{ \alpha \overline{X}(\overline{X} + 2\beta) + (\overline{X} + \beta)\sqrt{\alpha^2(\overline{X} + \beta)^2 + 2\alpha\overline{X}(\overline{X} + 2\beta)} - \alpha\beta^2$

$$-2\beta^2\sqrt{\frac{\alpha}{\alpha + 2}}\,\log\left[\frac{(\overline{X} + \beta)\sqrt{\alpha^2 + 2\alpha} + \sqrt{\alpha^2(\overline{X} + \beta)^2 + 2\alpha\overline{X}(\overline{X} + 2\beta)}}{\alpha\beta + \beta\sqrt{\alpha^2 + 2\alpha}}\right]\Big\},$$

(ii) $\bar{t} \approx \dfrac{\overline{X}}{12(\overline{X} + 2\beta)^2}\Big\{ \alpha\Big(5\overline{X}^3 + 30\beta\overline{X}^2 + 59\beta^2\overline{X} + 40\beta^3\Big) + \overline{X}(\overline{X} + 2\beta)(\overline{X} + 4\beta)$

$$+\Big(\overline{X}^2 + 5\beta\overline{X} + 8\beta^2\Big)\sqrt{\alpha^2(\overline{X} + \beta)^2 + +2\alpha\overline{X}(\overline{X} + 2\beta)}\Big\}.$$

11. Consider the semi-infinite half-space $0 \le x < \infty$ initially in a liquid phase and uniformly at constant temperature T_ℓ which is above the fusion temperature T_f. Suppose that at time $t = 0$ and subsequently the surface $x = 0$ is maintained at constant temperature T_0 which is below T_f and assume that there is perfect thermal contact at $x = 0$. Solidification occurs immediately and the temperatures $T_1(x, t)$ and $T_2(x, t)$ in the solid and liquid regions, respectively, satisfy equations (7.1)–(7.5). Show that in terms of the non-dimensional variables and parameters defined by (7.67) the two phase moving boundary problem (7.2)-(7.4) becomes the non-dimensional problem (7.68)-(7.70).

12. Continuation. Verify that the non-dimensional problem (7.68)-(7.70) admits Neumann's solution

$$\overline{T}_1(\bar{x}, \bar{t}) = 1 - \text{erf}\left(\frac{\bar{x}}{2\sqrt{\bar{t}}}\right)\Big/ \text{erf}\left(\sqrt{\frac{\gamma}{2}}\right),$$

$$\overline{T}_2(\bar{x}, \bar{t}) = U_0\left\{1 - \text{erfc}\left(\frac{\sqrt{\bar{c}}\,\bar{x}}{2\sqrt{\bar{t}}}\right)\Big/ \text{erfc}\left(\sqrt{\frac{\bar{c}\gamma}{2}}\right)\right\},$$

with the moving boundary $\overline{X}(\bar{t}) = \sqrt{2\gamma\bar{t}}$ where γ is the positive root of the transcendental equation

$$\alpha\sqrt{\frac{\pi\gamma}{2}} = \left\{ \frac{e^{-\gamma/2}}{\text{erf}\left(\sqrt{\gamma/2}\right)} + \frac{U_0\sqrt{\bar{c}}\,e^{-\bar{c}\gamma/2}}{\text{erfc}\left(\sqrt{\bar{c}\gamma/2}\right)} \right\}.$$

13. Show that the weak interpretation of the non-dimensional enthalpy problem

$$
\begin{aligned}
\frac{\partial\overline{H}}{\partial\bar{t}} &= \frac{\partial^2\overline{T}}{\partial\bar{x}^2}, \quad 0 < \bar{x} < \infty, \\[2mm]
\overline{T}(0,\bar{t}) - \beta\frac{\partial\overline{T}}{\partial\bar{x}}(0,\bar{t}) &= \overline{T}_0(\bar{t}), \quad \frac{\partial\overline{T}}{\partial\bar{x}}(\infty,\bar{t}) = 0, \\[2mm]
\overline{H}(\bar{x},0) &= f(\bar{x}),
\end{aligned}
$$

where $\overline{T}_0(\bar{t})$ and $f(\bar{x})$ are arbitrary functions of non-dimensional time and position respectively, is to find functions $\overline{H}(\bar{x},\bar{t})$ and $\overline{T}(\bar{x},\bar{t})$ formally related by (7.71) and such that

$$\int_0^{\bar{t}_0}\int_0^{\infty}\left(\overline{H}\frac{\partial\phi}{\partial\bar{t}} + \overline{T}\frac{\partial^2\phi}{\partial\bar{x}^2}\right)d\bar{x}\,d\bar{t} = -\int_0^{\infty}\phi(\bar{x},0)f(\bar{x})\,d\bar{x} - \int_0^{\bar{t}_0}\phi(0,\bar{t})\overline{T}_0(\bar{t})\,d\bar{t},$$

for any fixed time $\bar{t}_0 > 0$ and all test functions $\phi(\bar{x},\bar{t})$ such that $\phi(\bar{x},\bar{t}_0) = \phi(0,\bar{t}) - \beta\frac{\partial\phi}{\partial\bar{x}}(0,\bar{t}) = 0$.

14. Let $\overline{T}_1$, $\overline{T}_2$ and $\overline{X}(\bar{t})$ satisfy the non-dimensional freezing problem (7.68)-(7.70). Derive (7.80) by evaluating each of the integrals in two ways,

(i)
$$\int_0^{\bar{t}_0}\left(\frac{d}{d\bar{t}}\int_0^{\overline{X}(\bar{t})}(\overline{T}_1 + \alpha)\phi\,d\bar{x}\right)d\bar{t},$$

(ii)
$$\int_0^{\bar{t}_0}\left(\frac{d}{d\bar{t}}\int_{\overline{X}(\bar{t})}^{\infty}\bar{c}\overline{T}_2\phi\,d\bar{x}\right)d\bar{t},$$

where $\bar{t}_0 > 0$ is a fixed time and $\phi(\bar{x},\bar{t})$ is a test function such that $\phi(\bar{x},\bar{t}_0) = \phi(0,\bar{t}) = 0$.

15. Show that the non-dimensional single phase Stefan problem (7.10) admits the enthalpy formulation

$$
\frac{\partial \overline{H}}{\partial \bar{t}} = \frac{\partial^2 \overline{T}}{\partial \bar{x}^2}, \quad 0 < \bar{x} < \infty,
$$

$$
\overline{T}(0,\bar{t}) - \beta \frac{\partial \overline{T}}{\partial \bar{x}}(0,\bar{t}) = 1, \quad \frac{\partial \overline{T}}{\partial \bar{x}}(\infty,\bar{t}) = 0,
$$

$$
\overline{H}(\bar{x},0) = 0,
$$

where the temperature $\overline{T}$ and enthalpy $\overline{H}$ are related by

$$
\overline{H}(\overline{T}) = \begin{cases} \overline{T} + \alpha & \overline{T} > 0, \\ [0,\alpha] & \overline{T} = 0. \end{cases}
$$

16. Continuation. By integrating the enthalpy equation given in the previous question, deduce that

$$
\overline{T}(\bar{x},\bar{t}) = \frac{\partial}{\partial \bar{t}} \int_{\bar{x}}^{\infty} (\xi - \bar{x})\overline{H}(\xi,\bar{t})\,d\xi,
$$

$$
\bar{t} = \int_{0}^{\infty} (\xi + \beta)\overline{H}(\xi,\bar{t})\,d\xi.
$$

Show that these integrals are equivalent to (7.36) and (7.38) respectively. $\Big[$ Hint, recall that for $x > \overline{X}(\bar{t})$ the material is in the liquid phase at its fusion temperature and hence has non-dimensional enthalpy $\overline{H} = 0.\Big]$

17. Consider a polymorphous material (that is a material capable of existing in several distinct phases) which can exist in one of two solid phases and in one liquid phase. Assume that there are two constant phase change temperatures, T_{f_1} and T_{f_2} such that $T_{f_1} < T_{f_2}$. Now suppose that the semi-infinite half-space $0 \le x < \infty$ is filled with this material, initially in liquid form but uniformly at the constant fusion temperature T_{f_2}. At time $t = 0$ and thereafter the surface $x = 0$ is maintained at the constant temperature T_0 which is less than T_{f_1}. With the usual notation and assuming perfect thermal contact, solidification begins immediately with two moving phase change fronts $X_1(t)$ and $X_2(t)$, such that for $0 \le x < X_1(t)$ we have solid in phase 1, for $X_1(t) < x < X_2(t)$ there is solid in phase 2, and for $x > X_2(t)$ there is liquid at the fusion temperature. If $T_1(x,t)$ and $T_2(x,t)$ denote the temperatures in solid phases 1 and 2 respectively then

$$
T_0 \le T_1 \le T_{f_1} \le T_2 \le T_{f_2}.
$$

Assuming all three phases have the same density, show that in terms of the non-dimensional quantities defined by

$$\bar{x} = \frac{x}{a}, \quad \overline{X}_1(\bar{t}) = \frac{X_1(t)}{a}, \quad \overline{X}_2(\bar{t}) = \frac{X_2(t)}{a}, \quad \bar{t} = \frac{\kappa_1 t}{a^2},$$

$$\overline{T}_1(\bar{x},\bar{t}) = \frac{T_{f_1} - T_1(x,t)}{T_{f_1} - T_0}, \quad \overline{T}_2(\bar{x},\bar{t}) = \frac{k_2}{k_1}\left(\frac{T_{f_1} - T_2(x,t)}{T_{f_1} - T_0}\right),$$

$$\bar{c} = \frac{k_1 c_2}{k_2 c_1}, \quad U_0 = \frac{k_2}{k_1}\left(\frac{T_{f_1} - T_{f_2}}{T_{f_1} - T_0}\right),$$

$$\alpha_1 = \frac{L_1}{c_1\left(T_0 - T_{f_1}\right)}, \quad \alpha_2 = \frac{L_2}{c_1\left(T_0 - T_{f_1}\right)},$$

the moving boundary problem for $\overline{T}_1(\bar{x},\bar{t})$ and $\overline{T}_2(\bar{x},\bar{t})$ becomes

$$\frac{\partial \overline{T}_1}{\partial \bar{t}} = \frac{\partial^2 \overline{T}_1}{\partial \bar{x}^2}, \quad 0 < \bar{x} < \overline{X}_1(\bar{t}),$$

$$\bar{c}\frac{\partial \overline{T}_2}{\partial \bar{t}} = \frac{\partial^2 \overline{T}_2}{\partial \bar{x}^2}, \quad \overline{X}_1(\bar{t}) < \bar{x} < \overline{X}_2(\bar{t}),$$

with boundary conditions

$$\overline{T}_1(0,\bar{t}) = 1, \quad \overline{T}_1(\overline{X}_1(\bar{t}),\bar{t}) = 0,$$

$$\overline{T}_2(\overline{X}_1(\bar{t}),\bar{t}) = 0, \quad \overline{T}_2(\overline{X}_2(\bar{t}),\bar{t}) = U_0,$$

and Stefan conditions

$$-\alpha_1\frac{d\overline{X}_1}{d\bar{t}} = \frac{\partial \overline{T}_1}{\partial \bar{x}}(\overline{X}_1(\bar{t}),\bar{t}) - \frac{\partial \overline{T}_2}{\partial \bar{x}}(\overline{X}_1(\bar{t}),\bar{t}),$$

$$-\alpha_2\frac{d\overline{X}_2}{d\bar{t}} = \frac{\partial \overline{T}_2}{\partial \bar{x}}(\overline{X}_2(\bar{t}),\bar{t}),$$

where initially $\overline{X}_1(0) = \overline{X}_2(0) = 0$ and observe that the non-dimensional temperatures satisfy the inequalities

$$U_0 \leq \overline{T}_2 \leq 0 \leq \overline{T}_1 \leq 1.$$

18. Continuation. Verify that the previous problem admits the following similarity solution

$$\overline{T}_1(\overline{x},\overline{t}) = 1 - \operatorname{erf}\left(\overline{x}/2\sqrt{\overline{t}}\right)\Big/ \operatorname{erf}\left(\sqrt{\gamma_1/2}\right),$$

$$\overline{T}_2(\overline{x},\overline{t}) = U_0 \frac{\left[\operatorname{erfc}\left(\sqrt{\overline{c}}\,\overline{x}/2\sqrt{\overline{t}}\right) - \operatorname{erf}\left(\sqrt{\overline{c}\gamma_1/2}\right)\right]}{\left[\operatorname{erf}\left(\sqrt{\overline{c}\gamma_2/2}\right) - \operatorname{erf}\left(\sqrt{\overline{c}\gamma_1/2}\right)\right]},$$

$$\overline{X}_1(\overline{t}) = \sqrt{2\gamma_1\overline{t}}, \quad \overline{X}_2(\overline{t}) = \sqrt{2\gamma_2\overline{t}},$$

where γ_1 and γ_2 denote positive constants which satisfy the coupled transcendental equations

$$\alpha_1\sqrt{\frac{\pi\gamma_1}{2}} = \left\{ \frac{e^{-\gamma_1/2}}{\operatorname{erf}\left(\sqrt{\gamma_1/2}\right)} + \frac{U_0\sqrt{\overline{c}}\,e^{-\overline{c}\gamma_1/2}}{\left[\operatorname{erf}\left(\sqrt{\overline{c}\gamma_2/2}\right) - \operatorname{erf}\left(\sqrt{\overline{c}\gamma_1/2}\right)\right]} \right\},$$

$$\alpha_2\sqrt{\frac{\pi\gamma_2}{2}} = \frac{-U_0\sqrt{\overline{c}}\,e^{-\overline{c}\gamma_2/2}}{\left[\operatorname{erf}\left(\sqrt{\overline{c}\gamma_2/2}\right) - \operatorname{erf}\left(\sqrt{\overline{c}\gamma_1/2}\right)\right]}.$$

19. Continuation. Verify that the solution given in the previous problem satisfies the integral

$$(1 - U_0)\overline{t} = \int_0^{\overline{X}_1(\overline{t})} \xi\left[\overline{T}_1(\xi,\overline{t}) + \alpha_1 + \alpha_2 - \overline{c}U_0\right]\mathrm{d}\xi$$

$$+ \int_{\overline{X}_1(\overline{t})}^{\overline{X}_2(\overline{t})} \xi\left[\overline{T}_2(\xi,\overline{t}) + \alpha_2 - \overline{c}U_0\right]\mathrm{d}\xi.$$

Can you interpret this integral in terms of the enthalpy function for this problem?

TABLE OF LAPLACE TRANSFORMS

The Laplace transform of a function $\psi(t)$ is denoted by $\hat\psi(p)$ and defined to be

$$\hat\psi(p) = \int_0^\infty e^{-pt}\psi(t)\,dt.$$

In the following table we use x, κ, μ and α to denote positive real numbers and β to denote an arbitrary real number. The Laplace transform variable p is taken to be complex and it is understood that p lies to the right of the largest pole of $\hat\psi(p)$ in the complex plane. We use the conventional notation q for the quantity $\sqrt{p/\kappa}$.

$\psi(t)$	$\hat\psi(p)$
$C_1\psi_1(t) + C_2\psi_2(t)$	$C_1\hat\psi_1(p) + C_2\hat\psi_2(p)$
$\dfrac{d\psi}{dt}(t)$	$p\hat\psi(p) - \psi(0)$
$\int_0^t \psi(\tau)\,d\tau$	$\hat\psi(p)/p$
$t\psi(t)$	$-\dfrac{d\hat\psi}{dp}(p)$
$e^{-\beta t}\psi(t)$	$\hat\psi(p + \beta)$
$H(t - \alpha)\psi(t - \alpha)$	$e^{-\alpha p}\hat\psi(p)$
$(\psi_1 * \psi_2)(t)$	$\hat\psi_1(p)\hat\psi_2(p)$
$\dfrac{1}{\sqrt{\pi\kappa t}}\int_0^\infty \psi(\xi)e^{-\xi^2/4\kappa t}\,d\xi$	$\hat\psi(q)/\kappa q$
1	$1/p$
t	$1/p^2$
$t^\beta,\ \beta > -1$	$\Gamma(\beta + 1)/p^{\beta + 1}$
$e^{\beta t}$	$1/(p - \beta)$
$\sin\beta t$	$\beta/(p^2 + \beta^2)$
$\cos\beta t$	$p/(p^2 + \beta^2)$
$H(t - \alpha)$	$e^{-\alpha p}/p$
$\delta(t - \alpha)$	$e^{-\alpha p}$

$\psi(t)$	$\hat{\psi}(p)$
$\dfrac{x}{2\sqrt{\pi\kappa t^3}}\,e^{-x^2/4\kappa t}$	e^{-qx}
$\sqrt{\dfrac{\kappa}{\pi t}}\,e^{-x^2/4\kappa t}$	e^{-qx}/q
$\operatorname{erfc}\left(\dfrac{x}{2\sqrt{\kappa t}}\right)$	e^{-qx}/p
$2\sqrt{\dfrac{\kappa t}{\pi}}\,e^{-x^2/4\kappa t}-x\operatorname{erfc}\left(\dfrac{x}{2\sqrt{\kappa t}}\right)$	e^{-qx}/pq
$\left(t+\dfrac{x^2}{2\kappa}\right)\operatorname{erfc}\left(\dfrac{x}{2\sqrt{\kappa t}}\right)-x\sqrt{\dfrac{t}{\pi\kappa}}\,e^{-x^2/4\kappa t}$	e^{-qx}/p^2
$\sqrt{\dfrac{\kappa}{\pi t}}\,e^{-x^2/4\kappa t}-\mu\kappa e^{\mu x+\kappa\mu^2 t}\operatorname{erfc}\left(\dfrac{x}{2\sqrt{\kappa t}}+\mu\sqrt{\kappa t}\right)$	$e^{-qx}/(q+\mu)$
$\kappa e^{\mu x+\kappa\mu^2 t}\operatorname{erfc}\left(\dfrac{x}{2\sqrt{\kappa t}}+\mu\sqrt{\kappa t}\right)$	$e^{-qx}/q(q+\mu)$
$\dfrac{1}{\mu}\operatorname{erfc}\left(\dfrac{x}{2\sqrt{\kappa t}}\right)-\dfrac{1}{\mu}e^{\mu x+\kappa\mu^2 t}\operatorname{erfc}\left(\dfrac{x}{2\sqrt{\kappa t}}+\mu\sqrt{\kappa t}\right)$	$e^{-qx}/p(q+\mu)$
$\kappa(1+\mu x+2\mu^2\kappa t)e^{\mu x+\kappa\mu^2 t}\operatorname{erfc}\left(\dfrac{x}{2\sqrt{\kappa t}}+\mu\sqrt{\kappa t}\right)$ $-2\mu\sqrt{\dfrac{\kappa^3 t}{\pi}}\,e^{-x^2/4\kappa t}$	$e^{-qx}/(q+\mu)^2$
$\dfrac{1}{2}e^{\alpha t}\left\{e^{-x\sqrt{\alpha/\kappa}}\operatorname{erfc}\left(\dfrac{x}{2\sqrt{\kappa t}}-\sqrt{\alpha t}\right)\right.$ $\left.+e^{x\sqrt{\alpha/\kappa}}\operatorname{erfc}\left(\dfrac{x}{2\sqrt{\kappa t}}+\sqrt{\alpha t}\right)\right\}$	$e^{-qx}/(p-\alpha)$
$\dfrac{1}{2}\sqrt{\dfrac{\kappa}{\alpha}}e^{\alpha t}\left\{e^{-x\sqrt{\alpha/\kappa}}\operatorname{erfc}\left(\dfrac{x}{2\sqrt{\kappa t}}-\sqrt{\alpha t}\right)\right.$ $\left.-e^{x\sqrt{\alpha/\kappa}}\operatorname{erfc}\left(\dfrac{x}{2\sqrt{\kappa t}}+\sqrt{\alpha t}\right)\right\}$	$e^{-qx}/q(p-\alpha)$
$\dfrac{1}{2}e^{\alpha t}\left\{\left(t-\dfrac{x}{2\sqrt{\kappa\alpha}}\right)e^{-x\sqrt{\alpha/\kappa}}\operatorname{erfc}\left(\dfrac{x}{2\sqrt{\kappa t}}-\sqrt{\alpha t}\right)\right.$ $\left.+\left(t+\dfrac{x}{2\sqrt{\kappa\alpha}}\right)e^{x\sqrt{\alpha/\kappa}}\operatorname{erfc}\left(\dfrac{x}{2\sqrt{\kappa t}}+\sqrt{\alpha t}\right)\right\}$	$e^{-qx}/(p-\alpha)^2$

BIBLIOGRAPHY

Abramowitz M. and Stegun I.A., Handbook of Mathematical Functions, Dover, New York, 1972.

Carslaw, H.S., An Introduction to the Theory of Fourier's Series and Integrals, 3rd edition (revised), Dover, New York, 1950.

Carslaw, H.S. and Jaeger, J.C., Conduction of Heat in Solids, 2nd edition, Clarendon Press, Oxford, 1980.

Crank, J., The Mathematics of Diffusion, 2nd edition, Clarendon Press, Oxford, 1975.

Crank, J., Free and Moving Boundary Problems, Clarendon Press, Oxford, 1984.

Elliott, C.M. and Ockendon, J.R., Weak and Variational Methods for Moving Boundary Problems, Pitman, London, 1982.

Gel'fand, I.M. and Shilov, G.E., Generalized Functions, Academic Press, New York, 1964.

Herivel, J., Joseph Fourier the man and the physicist, Clarendon Press, Oxford, 1975.

Hill, J.M., One Dimensional Stefan Problems: An Introduction, Pitman Monographs and Surveys in Pure and Applied Mathematics, 31, Longman, London, 1987.

John, F., Partial Differential Equations, 4th edition, Springer-Verlag, London, 1982.

Johnson, L.W. and Riess, R.D., Numerical Analysis, 2nd edition, Addison-Wesley, Massachusetts, 1982.

Oberhettinger, F. and Badii, L., Tables of Laplace Transforms, Springer-Verlag, New York, 1970.

Richtmyer R.D., and Morton, K.W., Difference Methods for Initial-Value Problems, 2nd edition, Interscience, New York, 1967.

Strang, G. and Fix, G.J., An Analysis of the Finite Element Method, Prentice-Hall, New Jersey, 1973.

Whittaker, E.T. and Watson, G.N., A Course of Modern Analysis, 4th edition, Cambridge University Press, 1963.

Widder, D.V., The Laplace Transform, 2nd edition, Princeton University Press, Princeton, 1946.

Widder, D.V., The Heat Equation, Academic Press, New York, 1975.

Index